18th Nordic Laser Materials Processing Conference (18th NOLAMP)

IOP Conference Series: Materials Science and Engineering
Volume 1135

Lulea, Sweden
18-20 January 2022

Editors:

J. Frostevarg
A. Da Silva
M.-L. Palmblad
L. Andersson

ISBN: 978-1-7138-4614-7
ISSN: 1757-8981

Printed from e-media with permission by:

Curran Associates, Inc.
57 Morehouse Lane
Red Hook, NY 12571

Some format issues inherent in the e-media version may also appear in this print version.

Institute of Physics
Dirac House, Temple Back
Bristol BS1 6BE UK

Phone: 44 1 17 929 7481
Fax: 44 1 17 920 0979

techtracking@iop.org

Additional copies of this publication are available from:

Curran Associates, Inc.
57 Morehouse Lane
Red Hook, NY 12571 USA
Phone: 845-758-0400
Fax: 845-758-2633
Email: curran@proceedings.com
Web: www.proceedings.com

TABLE OF CONTENTS

IOP Conf. Series: Materials Science and Engineering 1135 (2021) 011001 doi:10.1088/1757-899X/1135/1/011001

Preface of the 18th NOLAMP conference

About the conference:

The first conference on **No**rdic **La**ser **M**aterial **P**rocessing, NOLAMP was held in Oslo in 1987. Then the laser material processing was a rather new and revolutionary way to manufacture products with high quality and high efficiency. The laser research in the Nordic countries was in its pioneering age and the enthusiasm among researchers for the new possibilities was unrivalled. The intention for creating this conference, which was originally suggested by Dr. Bernt Thorstensen at SI in Oslo, Norway, was to establish a fruitful cooperation among Nordic laser researchers. Also NOLAMP intended to provide a forum for young researchers to present on-going works and to establish contacts in the scientific community as well as in industry.

Since then, these biannual conferences have fulfilled its intentions and has provided a useful meeting ground for research and industry involved in laser technology in the Nordic countries. The NOLAMP conference has also expanded significantly and now attracts people from all over Europe. NOLAMP is now for the fifth time arranged in Sweden and this time it will be held in Luleå in the northern part of Sweden. With this new location we hope to attract a strong participation from industry presenting industrial R&D work and new applications, making this conference especially successful.

Due to the development of covid-19, the related vaccination programs and the likelihood of having the conference physically with all participants present, the consortium has decided to postpone the conference to 18-20th January 2022 in Luleå, Sweden. This will be the first winter NOLAMP, which will also be an experience for many to be in the "true north" in the middle of the winter. This involves high probabilities of much snow, cold temperatures, ice roads, aurora borealis and even some brief sunlight during mid-day.

Conference date and location:

18-20 January 2022 in Luleå, Sweden

Conference topics:

Laser materials processing on:

- ☐ Additive manufacturing
- ☐ Welding
- ☐ Cutting
- ☐ Surface treatment
- ☐ Laser systems
- ☐ Other processing

Keynote speakers:

- ☐ **Frank Brückner**, Head of Additive Manufacturing (AM) at the Fraunhofer Institute for Material and Beam Technology IWS Dresden, Fraunhofer IWS, Dresden, Germany: "***Trends in AM by Laser Direct Energy Deposition***"
- ☐ **Rickard Brånemark**, MD, PhD, presently Research Scientist at the Biomechatronics Group at MIT Media Lab, Cambridge, MA, Associate Professor at the Department of Orthopaedics, Gothenburg University, Gothenburg, Sweden, Professor Emeritus at the Department of Orthopaedics, University of California, San Francisco, CA, Chairman and Founder of Integrum

IOP Conf. Series: Materials Science and Engineering 1135 (2021) 011001 doi:10.1088/1757-899X/1135/1/011001

AB
"***Does laser processing of titanium implant surfaces improve clinical results?***"

- ☐ **Andreas Graichen**, group chief of digitalization and industrialization in additive manufacturing, Siemens Energy, Finspång, Sweden
  "***A journey of implementing additive manufacturing in large scale production, challenges and mitigation strategies***"
- ☐ **Johnny K Larsson,** (now retired) Volvo Cars and Autokropolis engineering:
  "***From the Early Days of Automotive Laser Processing to Recent Innovations and beyond.....***"

Sponsors:

-

Committee members:

- ☐ Jan Frostevarg, Luleå University of Technology (LTU), Sweden, jan.frostevarg@ltu.se
- ☐ Odd M Akselsen, Sintef, Norway, Odd.M.Akselsen@sintef.no
- ☐ Heidi Piili, Lappeenranta-Lahti University of Technology (LUT), Finland, Heidi.Piili@lut.fi
- ☐ Morten Kristiansen, Aalborg University (AAU), Denmark, morten@mp.aau.dk

Editors:

- ☐ Main editor: Jan Frostevarg, Luleå University of Technology (LTU), Sweden, jan.frostevarg@ltu.se, +46705204750
- ☐ Adrien Da Silva, Luleå University of Technology (LTU), Sweden, adrien.da.silva@ltu.se
- ☐ Marie-Louise Palmblad, Luleå University of Technology (LTU), Sweden, Marie-Louise.Palmblad@ltu.se
- ☐ Lars Andersson, Luleå University of Technology (LTU), Sweden, lars.andersson@ltu.se

Reviewers:

- ☐ Heidi Piili, Lappeenranta-Lahti University of Technology (LUT), Finland, Heidi.Piili@lut.fi
- ☐ Morten Kristiansen, Aalborg University (AAU), Denmark, morten@mp.aau.dk,
- ☐ Alexander Kaplan, Luleå University of Technology (LTU), Sweden, alexander.kaplan@ltu.se
- ☐ Odd M Akselsen, Sintef, Norway, Odd.M.Akselsen@sintef.no
- ☐ Ivan Bunaziv, Sintef, Norway, Ivan.Bunaziv@sintef.no
- ☐ Antti Salminen, University of Turku (UT), Finland, antti.salminen@utu.fi
- ☐ Morgan Nilsen, High School West (HV), Sweden, morgan.nilsen@hv.se
- ☐ Fredrik Sikström, High School West (HV), Sweden, fredrik.sikstrom@hv.se
- ☐ Jesper Sundqvist, Luleå University of Technology (LTU), Sweden, jesper.sundqvist@ltu.se
- ☐ Jörg Volpp, Luleå University of Technology (LTU), Sweden, jorg.volpp@ltu.se
- ☐ Himani Näsström, Luleå University of Technology (LTU), Sweden, himani.naesstroem@ltu.se

IOP Conf. Series: Materials Science and Engineering 1135 (2021) 011002 doi:10.1088/1757-899X/1135/1/011002

Peer review declaration

All papers published in this volume of IOP Conference Series: Materials Science and Engineering have been peer reviewed through processes administered by the Editors. Reviews were conducted by expert referees to the professional and scientific standards expected of a proceedings journal published by IOP Publishing.

- **Type of peer review:**

Double-blind

- **Conference submission management system:**

The submissions was sent mainly by e-mail to: nolamp18@ltu.se, a shared inbox for the conference main editor and administrators. Decisions and reviewing comments was also sent back from the same inbox.

- **Number of submissions received:**

40 abstracts

31 full manuscript submissions

- **Number of submissions sent for review:**

40 abstracts sent for review

31 full manuscript submissions

- **Number of submissions accepted:**

28 full manuscripts

- **Acceptance Rate (Number of Submissions Accepted / Number of Submissions Received X 100):**

70%

- **Average number of reviews per paper:**

3

IOP Conf. Series: Materials Science and Engineering 1135 (2021) 011002 doi:10.1088/1757-899X/1135/1/011002

- **Total number of reviewers involved:**

10

- **Any additional info on review process:**

The reviewing process consisted of a double-blind review with up to three rounds of reviews on the manuscript depending on the reviewers comments on the papers. In addition a first review was made on the submitted abstracts. The main editor distributed the manuscripts randomly for reviewing to the reviewers, all professional scientists in laser-materials processing (scope of the conference).

Aims and scope:

The NOLAMP conference is aimed at sharing research and scientific findings about laser materials processing, where emphasis is on a fundamental level to implementation level, theoretical and practical, materials and processing, life-cycle analysis and systems engineering and concepts for new methods.

Manuscripts should offer clear evidence of novelty and significance. Papers focusing on only parameter optimization or computational issues are not suitable. Similarly, papers focussed on an application rather than the laser processing method fall outside the conference's scope. The scope of the journal is defined to include the following:

- development in all types of lasers and optical systems directed towards materials processing
- Laser methods in manufacturing
- applications of lasers to materials processing, especially on
    - Additive manufacturing
    - Welding
    - Cutting
    - Surface treatment
    - Laser systems
- developments in optical computing and optical information processing
- research and development in the field of laser safety including studies of hazards resulting from the applications of lasers (laser safety, hazards of laser fume)

Reviewing criteria:

Contributions that falls outside the conference scope, showing results of using a system without deep analysis or only presents parameter studies will be rejected. This is first at the abstract submission level, whereby accepted abstracts contributions are allowed to prepare and submit full manuscripts for peer-review, where three reviewers read and comment on each paper, up to three review rounds before being finally accepted. Before final acceptance, the manuscripts also needs to have proper formatting according to the templates and guidelines given by the journal and presented on the conference homepage (10 page limit). Manuscripts that fail to meet the requirements on a high scientific level or

IOP Conf. Series: Materials Science and Engineering 1135 (2021) 011002 doi:10.1088/1757-899X/1135/1/011002

meet the reviewer comments will be rejected. After authors receive the reviewer comments the manuscripts needs to be adjusted accordingly and resubmitted before re-submission deadline along with a separate file describing how the reviewing comments where met.

After final acceptance, final submission with ORCID-iD and all author information is added in a final submission file, to be sent from the conference editorial team to the journal for publishing.

- **Contact person for queries:**
  Name : Jan Frostevarg
  Affiliation: Luleå University of Technology (LTU)
  Email : jan.frostevarg@ltu.se

Laser Metal Wire drop-by-drop Deposition: a material and dilution investigation

Adrien Da Silva[1], Keivan Amiri Kasvayee[2], Jan Frostevarg[1], Jan Zachrisson[2], Alexander F.H. Kaplan[1]

[1] Department of Engineering Sciences and Mathematics, Luleå University of Technology, 971 87 Luleå, Sweden
[2] R&D filler metals, ESAB AB, Lindholmsallén 9, 402 77 Göteborg, Sweden

adrien.da.silva@ltu.se

Abstract. Additive Manufacturing has become a field of high interest in the industry, mostly due to its strong freedom of design and its flexibility. Numerous Additive Manufacturing techniques exist and present different advantages and disadvantages. The technique investigated in this research is a drop-by-drop deposition alternative to Laser Metal Wire Deposition. This technique is expected to induce a better control over the power input in the material, resulting in a better power efficiency and tailorable material properties. The aim of this research is to investigate selected material properties of the structures produced with the drop-by-drop deposition technique. Multi-drops structures were deposited from 316L, Inconel 625 (NW6625) and AlSi5 (AW4043) wires. Two drop deposition methods were investigated: (i) a contactless recoil pressure driven detachment for 316L and Inconel 625, (ii) a contact-based surface tension driven detachment for AlSi5. A material characterization including optical microscopy, EDS and hardness measurements was performed in transverse and longitudinal cross-sections. The microstructure of the deposited material, the dilution with the substrate and the heat affected zone were analysed. The contactless detachment showed a higher dilution than the contact-based technique due to the laser irradiating the substrate between two drop detachments, which melts the substrate that then mixes with the deposited drops.

1. Introduction

Additive Manufacturing (AM) regroups numerous different techniques that offer different manufacturing possibilities and material properties. Certain AM techniques such as Laser Powder Bed Fusion or Electron Beam Powder Bed Fusion offer a high accuracy with a fine microstructure but also have a low building rate, while other techniques such as Directed Energy Deposition, Wire-Arc Additive Manufacturing (WAAM) and Laser Metal Wire Deposition (LMWD) offer a high building rate with a coarser microstructure and lower precision [1,2].

In regular LMWD, the wire is usually fed from the front and a defocused laser beam is used to melt both the wire and the substrate in a common melt pool that solidifies into a track [3]. One limitation of this process is that it is direction dependent. However, this direction dependence can be eliminated by using a coaxial wire feeding system where the laser beam is split coaxially to a vertically fed wire and focused at the intersection between the wire and the plate [4]. Another limitation of LMWD is that a consequent part of the laser beam energy is used to melt the substrate instead of the wire. In order to improve the energy efficiency of the process, it has been showed that feeding an aluminium wire vertically and melting it with a laser beam few millimeters above the substrate could improve the energy efficiency [5]. In this case, the laser beam was used only to melt the wire, while the heat condition through the molten deposited material was enough to cause a sufficient remelting of the previous layer and guarantee a correct attachment.

The arc processes, such as arc welding and WAAM, have been further optimized by modulating the arc and the wire feeding, resulting in three different material deposition modes. The continuous arc mode consists of continuously feeding the wire with a continuous arc that melts both the wire and the substrate. The pulsed arc mode consists of continuously feeding the wire and pulsing the arc in order to detach one droplet from the wire for each arc pulse. The Cold Metal Transfer mode consists of pulsing the arc when the wire is at a high position to generate a droplet and detaching the droplet by feeding the wire, where the drop detaches by contact with the melt pool, then the wire is retracted and a new arc pulse generates a new drop. These different improvements involving drop generation enabled to increase the energy efficiency and reduce the heat input in the substrate or previous layer, which is essential for building structures with WAAM [6,7].

While the wire-arc processes have been extensively improved, very little development has yet been achieved for wire-laser processes. The present study focuses on a drop-by-drop alternative to LMWD under development in a related research. This technique consists of depositing material drop by drop from the wire instead of continuously melting the wire with the substrate, where incremental wire feeding is applied to detach a drop and the laser power is kept constant. The drop generation from a metal wire with the help of the laser beam was first theorized and demonstrated by Kokalj et al. (2006) where it was called Laser Droplet Formation Process [8]. In this process, metal drops in a size range of 1 mm to 2 mm are generated by melting a wire with a low laser power, and they are detached by a high power laser pulse. In subsequent studies this technique was then called Laser Droplet Generation (LDG) and it has been used for multi-material joining of wires, foils and plates with nickel and silver droplets [9].

This research is dedicated to the material properties obtained with different drop-by-drop alternatives to LMWD, especially the dilution between the added material and the substrate. The chemical dilution has a limited importance in additively manufactured structures as it affects mostly the first layer [10]. However, it is a good indicator of the excess energy input that contributes to melting the substrate or the previous layer instead of melting the added material [11]. The dilution was found to be 10% to 30% with WAAM and about 8% with LWMD at optimized parameters [12,13]. The aim of this study is to analyze the dilution of the drops with the substrate and the material properties obtained in order to know if the present process with only incremental wire feeding is sufficiently energy efficient, or if laser power modulation should also be applied to increase the energy efficiency.

2. Methods

2.1. Process

Drop-by-drop LMWD experiments were carried out with two different techniques: a contactless technique where the drops are detached from the wire, fall and land in the track (Fig. 1a), and a contact-based technique where the drops detach by contact with the track (Fig. 1b). A laser beam was used to melt the tip of the wire into a droplet, of which the size was increased by feeding the wire through the laser beam, thus melting more material into the drop. The laser used was a 5 kW Ytterbium fiber laser with an optic fiber of 200 µm diameter. The optics consisted in a collimator lens of 150 mm focal length, a quattroXX beam splitter that was used to split the laser beam into four separated beams (Fig. 1c) and a focusing lens of 250 mm focal length. All the experiments were carried out with argon shielding gas through a 20 mm diameter tube with a gas flow of 20 L/min.

For the contactless deposition technique, the wire was molten at 6 mm above the substrate (Fig. 1a) and it was fed for a feeding length of 8 mm at 5 m/min through the laser beams of which the total power was 2 600 W (*phase 1* in Fig. 2a). When the wire feeding stopped, the drop (which was about 2.6 mm

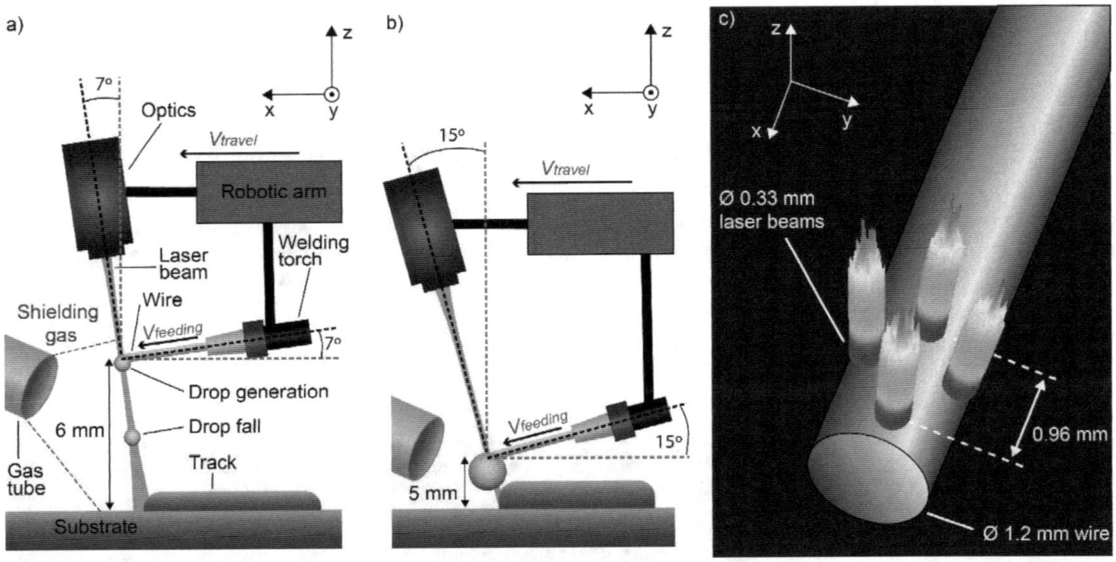

Figure 1. a) Set up for the contactless deposition, b) setup for the contact-based deposition, c) laser beams profile on the wire.

diameter) detached in the laser beam and landed on the substrate. After a short delay (*phase 2* in Fig. 2a), the robotic arm was moved 2.5 mm forward to the next drop detachment position at the speed of 0.6 m/min (*phase 3* in Fig. 2a). This technique was used to deposit 30-drop tracks from a 316L stainless steel wire and a NW6625 nickel alloy wire.

For the contact-based deposition technique, the wire was molten at 5 mm above the substrate (Fig. 1b) and it was fed for a feeding length of 20 mm at 15 m/min though the laser beams of which the total power was 5 000 W (*phase 1* in Fig. 2b). The wire feeding stopped approximately when the drop

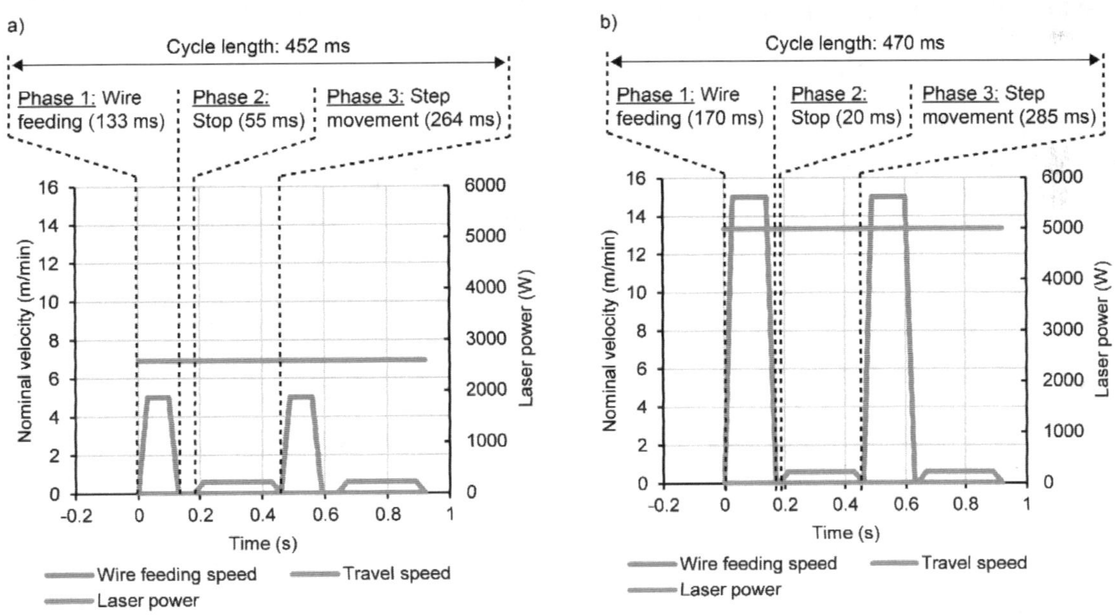

Figure 2. Representation of a drop deposition cycle for a) the contactless deposition technique, b) the contact-based deposition technique.

was large enough (about 3.5 mm diameter) to touch the substrate and detach by contact. After a short delay (*phase 2* in Fig. 2b), the robotic arm was moved 3 mm forward to the next drop detachment position at the speed of 0.6 m/min (*phase 3* in Fig. 2b). This technique was used to deposit tracks with up to 4 drops from an AlSi5 aluminium alloy wire. The contact-based deposition technique was developed specifically for aluminium alloys, since their thermodynamics properties and surface tensional properties under oxidation did not allow drop detachment with the contactless deposition technique.

Fig. 3 shows a high-speed imaging frame sequence for each of these two drop deposition methods when the drop detaches from the wire and attaches to the substrate. A specificity of the contactless method is that the laser beams irradiate the substrate between two drop generations (*phases 2* and *3*) which creates a melt pool in which the drops land (Fig. 3a). While with the contact-based method with aluminium, the laser beams irradiate the wire after each drop detachment, which does not directly create a melt pool on the substrate.

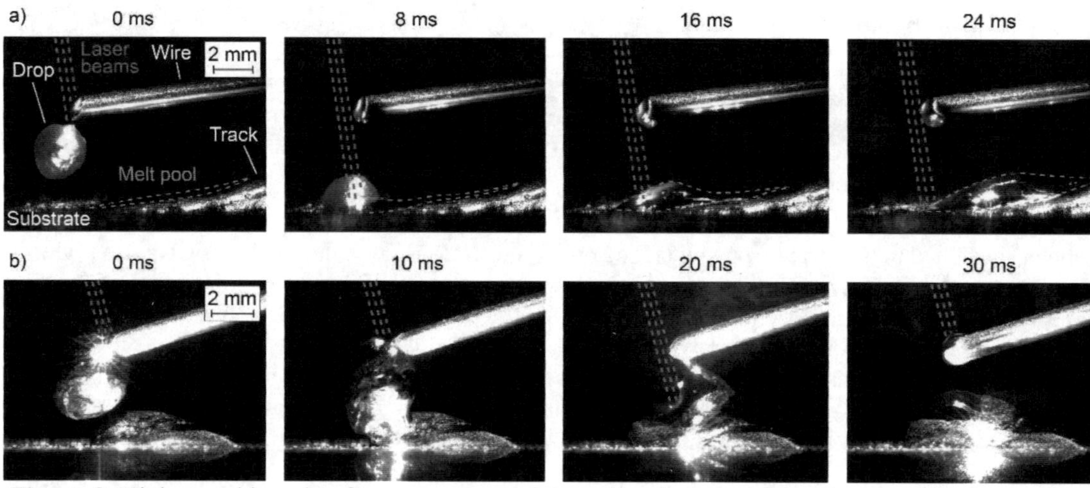

Figure 3. High-speed imaging frame sequence of drop detachment and attachment with a) contactless deposition from a 316L wire, b) contact-based deposition from an AlSi5 wire.

2.2. Material analysis

Three tracks were produced with three different materials commonly used in AM, of which the compositions are shown in Table 1. Two 30-drop tracks were deposited with the contactless technique from a 316L wire and a NW6625 wire on a DOMEX 350LA substrate. Two tracks (4-drop and 2-drop long) were deposited with the contact-based technique from an AlSi5 wire on a 1000 series aluminium substrate.

Longitudinal and transversal cross-sections were cut out of the tracks produced in order to analyze the three materials deposited with two different techniques. Optical microscopy was used to capture macro- and micrographs of the samples with different etchants. For the 316L and NW6625 samples, Klorpikrin etchant was used for the macrograph and Villela reagent was used for the micrograph. On the NW6625 sample, the substrate was masked in order to etch only the track. For the AlSi5 samples, Na-OH 20% was used for both the macro- and micrographs.

Energy-Dispersive X-ray Spectroscopy (EDS) measurements were taken on a vertical line crossing both the track and the substrate in order to measure the chemical composition throughout the tracks. Vickers hardness was measured on vertical lines crossing the track and the substrate, including the Heat Affected Zone (HAZ), where each measurement point was achieved with a 1 Kg load.

Two methods were investigated to calculate the dilution of the drops with the substrate. The first method consisted of using EDS to measure inside the tracks the average content of chemical elements that were originally present only in the wire and not in the substrate. For the 316L track, these targeted elements were Cr, Ni and Mo, for the NW6625 track these elements were Cr, Ni, Nb and Mo and for the AlSi5 track the only targeted element was Si. Comparing the chemical contents of these elements measured in the track to the ones of the original wire gives the dilution ratio. The second method consisted of measuring on the transversal macrographs the track area above and below the substrate upper surface, where the area above the substrate represents the wire material added by drop transfer and the area below the substrate represents the substrate material molten during the process. The ratio of the area below the substrate to the total track area gives the dilution ratio.

Table 1. Chemical compositions (wt%) of the wires and substrates used.

	C	Si	Mn	P	S	Cr	Ni	Mo	Nb	Zn	Al	Fe
316L wire	0.02	0.8	0.7	-	-	18.0	12.0	2.5	-	-	-	Bal.
NW6625 wire	0.025	0.2	0.15	≤0.015	≤0.015	21.5	61	8.7	3.5	-	-	4
AlSi5 wire	-	5.0	≤0.05	-	-	-	-	-	-	≤0.1	Bal.	≤0.4
DOMEX 350LA substrate	0.10	0.040	≤0.90	≤0.030	≤0.025	-	-	-	≤0.1	-	≥0.10	Bal.
1000 series aluminium substrate	-	-	-	-	-	-	-	-	-	-	Bal.	0.4

3. Results

3.1. Stainless steel 316L deposited with the contactless technique

Figure 4 shows the track produced with 316L and its material analysis. The different drops deposited are visible in the macrographs and a HAZ of 1 mm to 1.5 mm is visible in the substrate under the track (Fig 4b and c). The microstructure was bainitic in the two drops on both sides of the drop-to-drop interface (Fig. 4.d).

In Figure 4.c, it was measured that 46% of the track is above the substrate and 54% is below the substrate (white dashed line), which gives a dilution ratio of 54% in this cross-section. The line EDS measurements (yellow line in Fig. 4b) exhibited in Figure 4.e show an average chemical composition in the track of 9.5% Cr, 5.6% Ni, 1.4% Mo and 82% Fe. The Cr, Ni and Mo contents in the track are respectively 53%, 47% and 50% of the contents in the original wire (Table 1), which gives a dilution ratio of 47% to 53% in this measurement line.

However, the chemical composition is not exactly uniform in the track, the EDS measurement line crosses the upper part of drop n°20 and the lower part of drop n°19, which have slight variations of chemical composition (Fig. 4e). In this case, the upper part of drop n°20 contains 10.9% Cr, 6.4% Ni and 1.6% Mo, which is the result of a 39% to 47% dilution ratio with the substrate, while the lower part of drop n°19 contains 8.1% Cr, 4.8% Ni and 1.1% Mo which represents a 55% to 61% dilution ratio.

Figure 4f shows the line hardness measurements (blue line in Fig. 4b) crossing two different drops and the HAZ in the substrate. The hardness is relatively uniform in the track regardless of the drop change, with an average hardness value of 389 HV1. The hardness of the substrate outside the HAZ is about 138 HV1 while the hardness in the HAZ is about 152 HV1, so there might be a slight hardening in the HAZ.

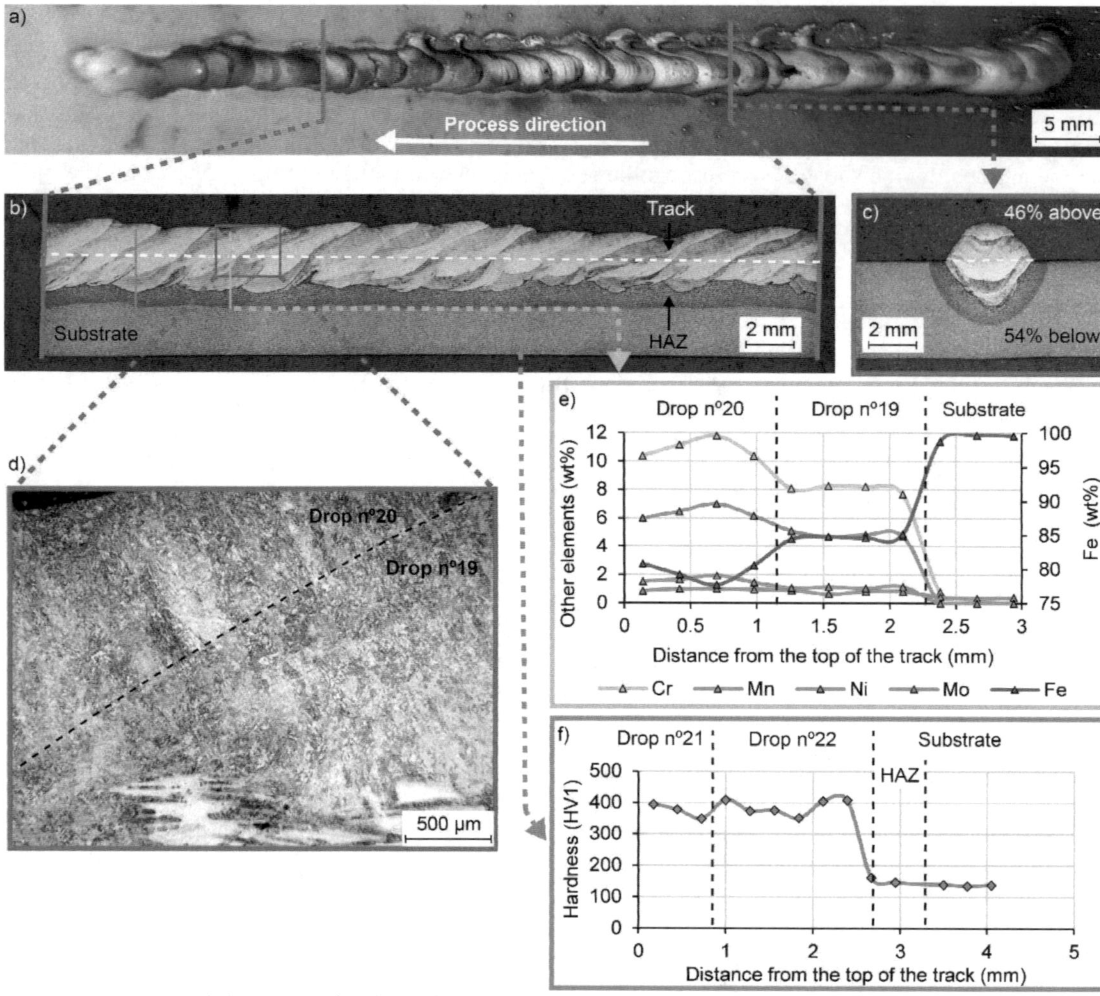

Figure 4. Material characterization of the 316L track deposited with the contactless method, with: a) photography of the track as-built, b) macrograph of the longitudinal cross section, c) macrograph of the transversal cross-section, d) micrograph of the microstructure, e) line EDS measurement, f) line Vickers hardness measurement.

3.2. Nickel alloy NW6625 deposited with the contactless technique

Figure 5 shows the track produced with NW6625 and its material characterisation. The different drops are hardly visible in the macrographs, probably because of a different etching reaction to this chemical composition (Fig. 5b and c). The microstructure was austenitic (Fig. 5d).

In Figure 5c, 56% of the track is above the substrate and 44% below the substrate, indicating a dilution ratio of 44% in this cross-section. In the line EDS measurement shown in Figure 5e, the average chemical composition in the track is 9.4% Cr, 28% Ni, 1.7% Nb, 4.2% Mo and 56% Fe. The Cr, Ni, Nb and Mo contents are respectively 44%, 46%, 49% and 48% of the contents in the original wire (Table 1), which indicates a dilution ratio of 51% to 56% in this measurement line.

With NW6625, the chemical composition is also non-uniform in the track, which can be observe where the EDS measurements change from the upper part of drop n°24 to the lower part of drop n°23 (Fig. 5e). The upper part of drop n°24 contains 8.1% Cr, 25% Ni, 1.4% Nb and 3.7% Mo, which represents a 57% to 62% dilution with the substrate, while the lower part of drop n°23 contains 10% Cr, 29% Ni, 1.8% Nb and 4.4% Mo, which indicates a 49% to 53% dilution ratio.

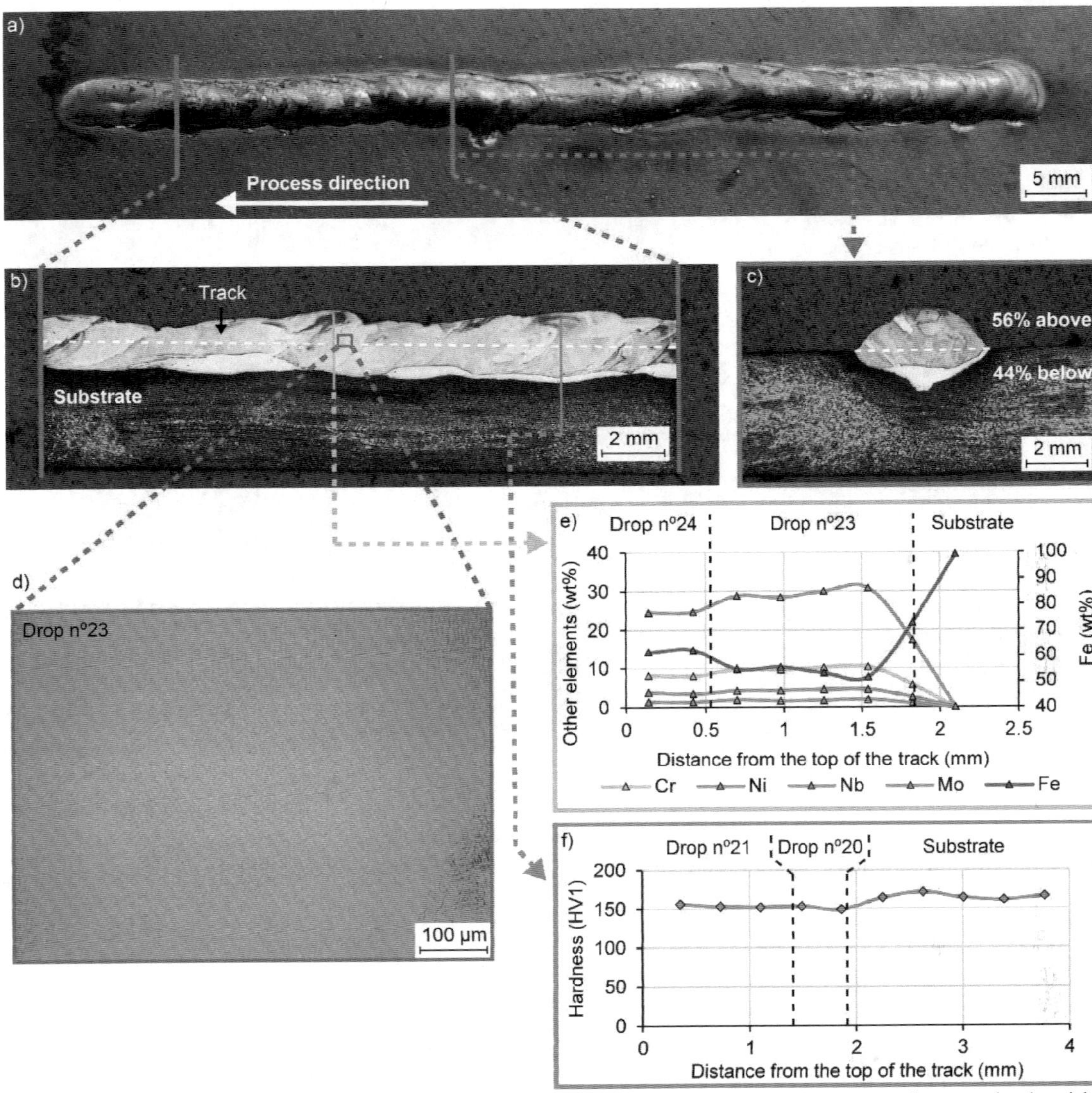

Figure 5. Material characterization of the NW6625 track deposited with the contactless method, with: a) photography of the track as-built, b) macrograph of the longitudinal cross section, c) macrograph of the transversal cross-section, d) micrograph of the microstructure, e) line EDS measurement, f) line Vickers hardness measurement.

The line hardness measurements exhibited in Figure 5f show a very uniform hardness over the track that does not seem to be affected by the change of drop. The average hardness in the track is 158 HV1 while the average hardness in the substrate is 173 HV1.

3.3. Aluminium alloy AlSi5 deposited with the contact-based technique

Figure 6 shows the tracks produced with an AlSi5 wire and their material analysis. The different drops deposited are distinguishable in the macrographs with a Partially Melted Zone (PMZ) at the interface between two drops (Fig. 6c and d). The PMZ occurs when an aluminium alloy is heated between its liquidus and solidus temperature, where the eutectic phase is liquid and the primary α aluminium phase is still solid, it is commonly observable at interlayer interfaces in additively manufactured structures [5].

The microstructure of the track in Figure 6.e shows tightly packed primary α aluminium dendrites (bright regions) separated by a thin eutectic phase (dark mesh). It can be noted that numerous pores are present inside the material (black dots in Figure 6c and d), their spherical shape indicate that they are most likely gas pores.

In figure 6d, 72% of the track is above the substrate and 28% is below the substrate, thus the dilution with the substrate is about 28% in this cross-section. In Figure 6f, the line EDS measurements show that the average chemical composition in the track is 4.3% Si and 95.7% Al. The Si content in the track is 86% of the content in the original wire (Table 1), thus the dilution with the substrate is about 14% in this cross-section.

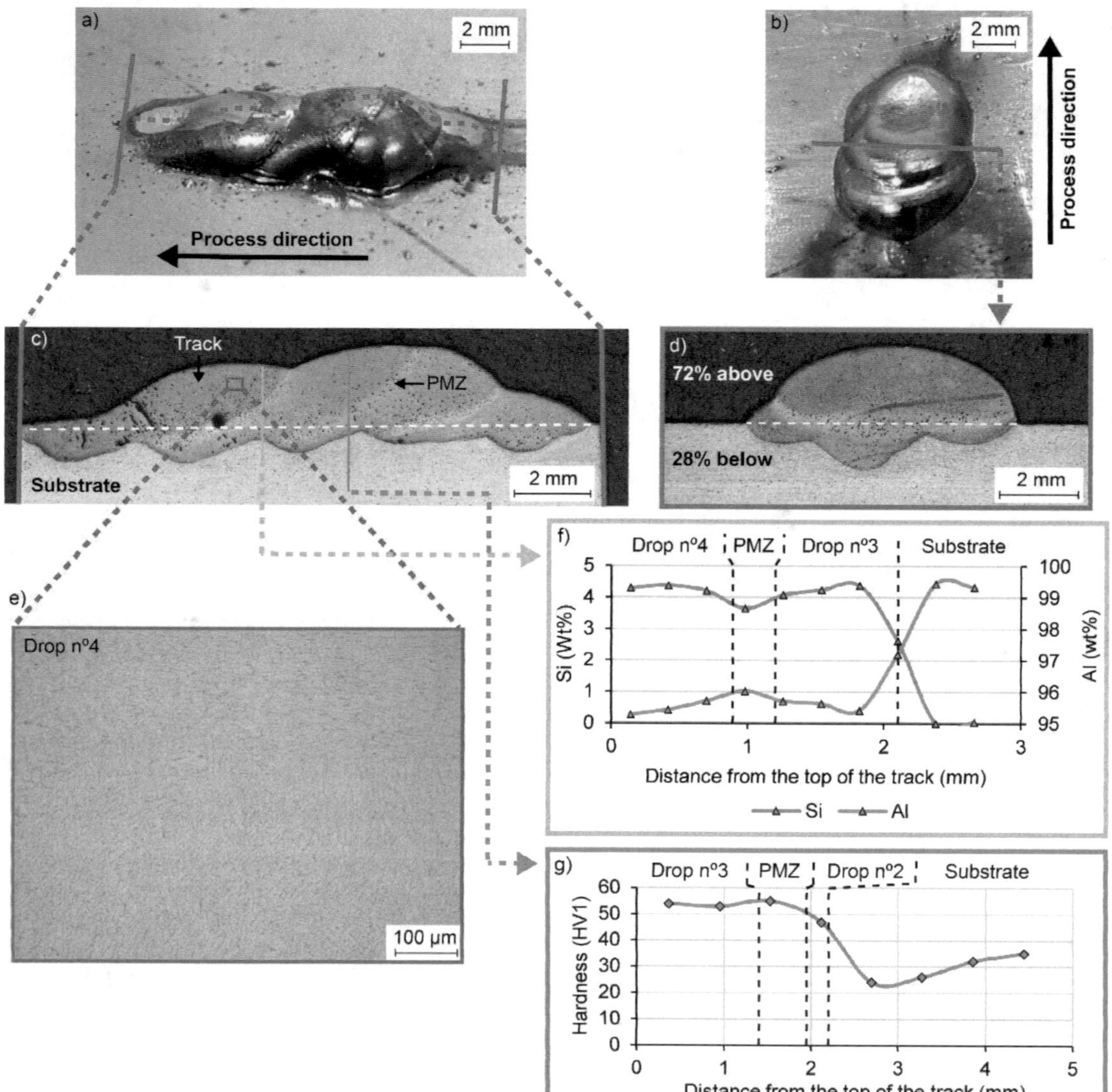

Figure 6. Material characterization of the AlSi5 tracks deposited with the contact-based method, with: a) photography of the 4-drop track, b) photography of the 2-drop track, c) macrograph of the longitudinal cross section, d) macrograph of the transversal cross-section where the black dots are gas porosities, e) micrograph of the microstructure, f) line EDS measurement, g) line Vickers hardness measurement.

In this track, no difference of chemical composition was observed from one drop to another. However, there seem to be a small decrease of Si content in the PMZ, where it is 3.8% compared to 4.4% outside of the PMZ. This decrease of Si content in the PMZ is most likely due to the partial dilution of the liquid eutectic phase present in the PMZ (containing most of the Si) with the newly added liquid drop.

The hardness measured in the track is relatively uniform with an average value of 52 HV1, while the average hardness measured in the substrate is 30 HV1, where there seem to be a softening effect close to the deposited drops (Fig. 6g).

4. Discussion

The two methods investigated to measure the dilution showed some variations of results that could be attributed to the different locations where the measurements were taken. Relatively small variations of chemical composition (thus dilution) were observed from one drop to another with the contactless deposition technique. 316L and NW6625 showed inverse variations of composition from the upper part of a drop to the lower part of another drop. It means that these are most likely not variations within the lower part and upper part of a same drop but rather variations from one drop to another. These variations of dilution might depend on the landing conditions of the drops, especially their contact area with the pre-existing melt pool. This effect could be reduced by improving the accuracy of drop positioning. However, the hardness properties were not affected by these local changes of chemical compositions, thus it might not have a considerable effect on the final mechanical properties.

Overall, it is noticeable that the contactless deposition technique investigated with 316L and NW6625 induces considerably more dilution than the contact-based deposition technique investigated with AlSi5. This is very likely due to melt pool generated by the laser beam on the substrate between two drop detachments with the contactless technique. Such a large melt pool might not be required for the drop attachment, it was shown in this study that it induces an extensive dilution between the drop and the substrate, and it might also induce too much remelting of the previous layer when building AM structures. Thus, a next development step will be apply laser power modulation to this process, for instance by reducing the laser power during *phases 2* and *3* (Fig. 2), which will also reduce the energy consumed by the process.

5. Conclusions

In the present study where two drop-by-drop deposition techniques were investigated with three different materials, the following conclusions can be drawn:

- The contactless drop deposition technique with 316L and NW6625 involved a melt pool formed with approximately the same quantity of material molten from the substrate as in the added drops, with a dilution of 44% to 56%. This is due to the laser beam irradiation on the substrate between two drop depositions, which intensively melts the substrate and contributes to its dilution with the added drops.
- The contact-based deposition technique with AlSi5 involved a lower dilution with the substrate, of about 14% to 28%, because the laser beam does not act on the substrate and no melt pool is actively formed for the drops dilute with. The small quantity of substrate molten is due to the heat conduction through the aluminium drop after landing. Many gas pores are present in the material, which could be due to the technique used or to the material.
- The contactless drop deposition with 316L and NW6625 showed a difference of dilution from one drop to another of 16% to 35%. It is most likely due to slightly different landing positions of the drops, resulting in different contact areas with the melt pool, which induces a difference of mixing between the drop and the melt pool. While the contact-based drop deposition with AlSi5 showed a homogeneous chemical composition throughout a drop with a localised

decrease of silicon content in the partially melted zone, probably due to the mixing of the liquid eutectic phase with the newly deposited drop.

- While the dilution induced by the contact-based deposition technique can be acceptable, the contactless technique in its current development stage results in a too high melting of the substrate which cause a high dilution. Incremental wire feeding with constant laser power might result in a too high power input in the deposited material. As a next step, laser power modulation should be applied in order to reduce the heat input between two drop depositions and increase the process efficiency.

Acknowledgments
This study was performed within the projects LAM-4D (no. 2018-04324) and LAM-4D stage 2 (no. 2019-04872) funded by Vinnova, Sweden's innovation agency, as well as the European project SAMOA (no. 18079) funded by EIT Raw Materials.

ORCID iDs
A Da Silva https://orcid.org/0000-0002-0649-0130
K A Kasvayee https://orcid.org/0000-0002-5635-8023
J Frotevarg https://orcid.org/0000-0003-4265-1541
A F H Kaplan https://orcid.org/0000-0002-3569-6795

References

[1] Frazier W E 2014 Metal additive manufacturing: A review *J. Mater. Eng. Perform.* **23** 1917–28
[2] Lewandowski J J and Seifi M 2016 Metal Additive Manufacturing: A Review of Mechanical Properties *Annu. Rev. Mater. Res.* **46** 151–86
[3] Heralić A 2012 *Monitoring and Control of Robotized Laser Metal-Wire Deposition* (Göteborg: Chalmers University of Technology)
[4] Heralić A, Christiansson A K, Ottosson M and Lennartson B 2010 Increased stability in laser metal wire deposition through feedback from optical measurements *Opt. Lasers Eng.* **48** 478–85
[5] Da Silva A, Wang S, Volpp J and Kaplan A F H 2020 Vertical laser metal wire deposition of Al-Si alloys *Procedia CIRP* **94** 341–5
[6] Derekar K S, Addison A, Joshi S S, Zhang X, Lawrence J, Xu L, Melton G and Griffiths D 2020 Effect of pulsed metal inert gas (pulsed-MIG) and cold metal transfer (CMT) techniques on hydrogen dissolution in wire arc additive manufacturing (WAAM) of aluminium *Int. J. Adv. Manuf. Technol.* **107** 311–31
[7] Frostevarg J, Kaplan A F H and Lamas J 2014 Comparison of CMT with other arc modes for laser-arc hybrid welding of steel *Weld. World* **58** 649–60
[8] Kokalj T, Klemenčič J, Mužič P, Grabec I and Govekar E 2006 Analysis of the laser droplet formation process *J. Manuf. Sci. Eng. Trans. ASME* **128** 307–14
[9] Govekar E, Jerič A, Weigl M and Schmidt M 2009 Laser droplet generation: Application to droplet joining *CIRP Ann. - Manuf. Technol.* **58** 205–8
[10] Li C, Gu H, Wang W, Wang S, Ren L and Wang Z 2020 Effect of Heat Input on Formability, Microstructure, and Properties of Al–7Si–0.6Mg Alloys Deposited by CMT-WAAM Process *Appl. Sci.* **10**
[11] Demir A G and Biffi C A 2019 Micro laser metal wire deposition of thin-walled Al alloy components: Process and material characterization *J. Manuf. Process.* **37** 362–9
[12] Lin Z 2019 *Wire and Arc Additive Manufacturing of Thin Structures Using Metal-Cored Wire Consumables* (Delft: Delft University of Technology)
[13] Abioye T E, Folkes J and Clare A T 2013 A parametric study of Inconel 625 wire laser deposition *J. Mater. Process. Tech.* **213** 2145–51

A novel method for approximating local changes in the surface absorption for laser marking using 3D laser scanning

Anders F. Mikkelstrup[1], Anders N. Thomsen[1], Morten Kristiansen[1]

[1]Department of Materials and Production, Aalborg University, Fibigerstraede 16, 9220 Aalborg East, Denmark

afm@mp.aau.dk

Abstract. Laser marking is a non-contact technique, which achieves colouring by using a laser beam to increase surface oxidation. Controlling the amount of heat induced into the part is essential in ensuring the desired degree of oxidisation. However, the induced heat is not only dependent on the process parameters, but also on the surface absorption, which in turn is dependent on the material, laser wavelength, and surface quality, i.e., current degree of oxidation and contaminants as well as surface roughness. This paper proposes a method for correlating backscatter from a 3D laser scanner with the surface absorption of sheet metal parts. The purpose is to determine local changes in the surface absorption caused by surface oxidation and contamination. The method utilises a 3D laser scanner, which projects a laser line at the surface and measures the resulting backscatter at an angle. The proposed solution applies a bi-directional reflectance model to reduce the influence of varying scanning angles. The method's sensitivity to variations in surface treatments is investigated and validated against backscatter spectroscopy measurements. The results show that the proposed method can identify changes in the absorption. However, these were, in some cases, more than 70% higher compared to spectroscopy measurements.

1. Introduction

Laser marking is a method for increasing the thickness of the oxide layer by increasing the temperature. By changing the thickness, light is refracted differently, which gives colour changes. This can be used for aesthetic and communicative purposes in parts of metal. Controlling the colouring is a matter of process settings [1] and surface properties. The surface absorption describes the energy coupling efficiency of the surface. It thus is defined as the ratio between the incident energy from the laser and the absorbed energy by the surface [2]. As laser marking is dependent on controlling the induced heat into the material, knowledge of the absorption plays an essential role in the process stability and quality. It is well known that surface absorption is influenced by contamination, oxidation and roughness. Typically, surface contamination of sheet metal components originates from fingerprints and oil residue from manufacturing. Oxidation arises from oxygen reacting with the surface and will for stainless steel present itself as a thin layer of chromium oxide. The oxidation process will naturally occur over time but can be expedited by inducing heat, which is the principle behind laser marking. Abedi and

Hoseinpour [3] found that by coating the surface of an AISI 304L sample with an oxide layer, the surface absorption of wavelengths exceeding 1000 nm could be more than doubled.

Roughening the surface of the sample showed comparable results. Niu et al. [4] and Bergström et al. [5] discovered that the surface morphology and surface structure drastically influences the absorption, especially at incident angles below 60° in which a rougher surface generally leads to an increase in the absorption. Niu et al. [4] further concluded that the effect of the incident angle is dependent on the structure of the surface. However, the impact of the incident angle generally becomes less prominent as the surface roughness is increased. Moreover, the polarisation of the laser and the type of laser is known to affect the correlation between the incident angle and the absorption [6]. In addition, Niu et al. [4] investigated the influence of the laser wavelength in the spectrum from 0.4 μm to 4 μm and concluded that the absorption generally decreases with increasing wavelength.

The surface absorption in the solid state of a material can be estimated through a number of theoretical methods and indirect and direct measurement techniques [7]. Theoretical approaches like the classic Drude model [8] and Hagen and Rubens [9] rely on knowledge of parameters such as surface roughness, oxidation, contamination, and material properties.

Indirect measurement techniques are based on computational models to estimate the surface absorption. These methods take into account parameters such as the surface roughness [10], beam intensity [11], angle of incidence [12] and surface temperature [13]. As a result, they generally offer an improved estimation of the surface absorption compared to the purely theoretical methods.

The direct measurement techniques are both accurate and straightforward experimental methods, which can be split into two groups: Methods that directly measure the absorption (calorimetry methods [14]) and methods that indirectly measure the absorption by the radiative properties of the surface, e.g., reflectance and emission (radiometric methods) [7,15]. However, calorimetry methods cannot provide a local estimation of the surface absorption compared to methods that indirectly measure the absorption by the radiative properties of the surface, such as reflectometry and emittance spectroscopy. Kügler and Vollertsen [16] used an integrated sphere to perform reflectometry spectroscopy to determine the absorption of stainless steel. The experimental set-up consisted of a 300 mW laser diode with a wavelength of 1030 nm, which was utilised to radiate the sample, while a Si-photodiode captured the reflected radiation. The integrated sphere spectrophotometry provides simple and highly accurate measurements. Though, as each sample must be placed inside the integrated sphere, it is not sustainable in a flexible production set-up and is further limited in the physical size of the sample.

A range of theoretical models, indirect and direct measurement techniques have been developed for estimating absorption. However, none can provide a local estimation of the absorption for use in a flexible laser marking set-up. As a result, it becomes challenging to locally control the induced heat into the material, which, as mentioned earlier, is a prerequisite for laser marking to obtain a stable process and quality.

The literature review indicates a need for a solution that can determine local changes in the absorption in a flexible manufacturing step-up. The main contribution of this paper is to investigate the feasibility of using a 3D laser scanner to provide an approximation of the absorption as a preceding step for laser marking of stainless steel. By correlating the backscatter from a 3D laser scanner, it is possible to approximate local changes in the surface reflectivity. As the surface reflection is highly dependent on incident angle and perspective, the received backscatter intensity is corrected based on the Cook-Torrance bi-directional surface reflectance (BRDF) model. A series of laser marking experiments are performed to oxidate the surface and create varying levels of reflectivity across the surface of the part to validate the developed method. The results are further compared with the measured values from spectroscopy. The remainder of the paper is organised as follows: In Section 2, the experimental set-up for laser marking and geometrical and intensity measurements is presented. In Section 3, the developed method for correcting intensity values and identifying areas with increased or decreased surface reflectivity is shown. In Section 4, the results are presented and discussed. Section 5 includes the concluding remarks of the paper.

2. Experimental set-up

The experimental set-up is presented in Figure 1 and Figure 2, which, respectively, illustrates laser processing and geometrical and reflectivity measurements of a part. The laser processing set-up consists of a 3 kW IPG YLS-3000 SM laser connected to a modified HighYag processing head. The focal length is 470 mm with a collimated beam diameter of 11.05 mm. The beam quality is 1.2 M^2 with a wavelength of 1076 nm. The measurement set-up consists of a Wenglor MLWL153 3D line scanner, which is applied to acquire a 3D surface representation of the processed parts. The scanner uses a class 3R laser beam with a wavelength of 407 nm and a CMOS sensor to determine the position of the projected laser. The CMOS sensor can further measure the backscatter intensity from the projected laser in a 10-bit greyscale resolution.

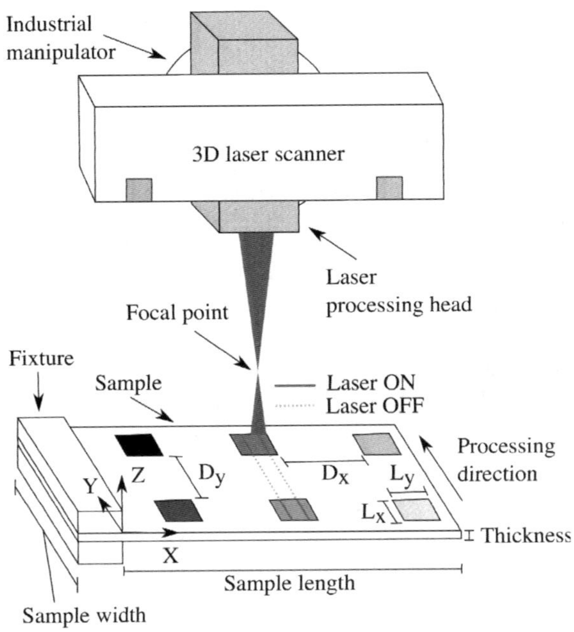

Figure 1. Laser processing set-up for laser marking.

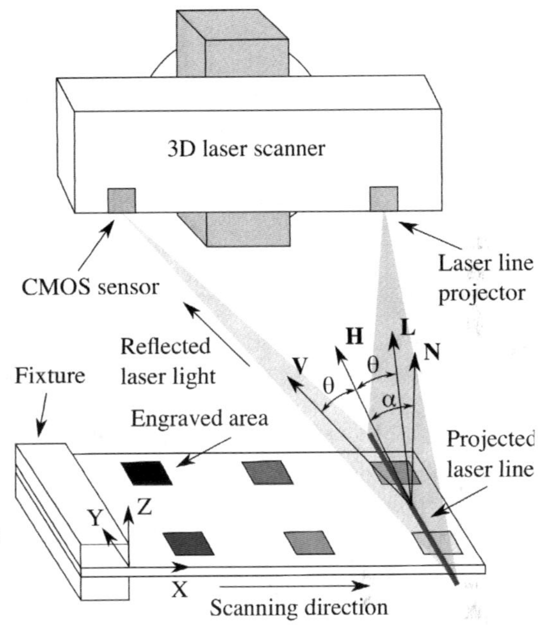

Figure 2. Measurement set-up for geometrical and intensity measurements. The projected laser line is observed by the CMOS sensor at a varying angle depending on the distance to the measured surface.

It should be noted that the specifications of both the laser and the optical components in the scanner are not disclosed by the manufacturer and, therefore, unknown to the authors. The processing head and the 3D line scanner are mounted onto and manipulated by a KUKA KR120 R2700 industrial robot for accurate and flexible positional control. Additionally, spectroscopy measurements have been performed using a LAMBDA 1050+ UV/Vis/NIR spectrophotometer with an integrated sphere to capture diffuse and specular reflectance at normal incidence. Each measured reflectivity is taken as an average of three measurements with a spacing of approximately 5 mm apart. The measured spot size was 3 x 3 mm.

The experiments consisted of three main phases: laser marking, laser scanning and spectroscopy. Spectroscopy and 3D scanning were done preceding laser marking to establish a set of reference values. Subsequently, laser marking was performed on the samples at varying speeds and power levels to generate different oxidation levels and different surface reflections. Finally, spectroscopy was performed again, along with 3D scanning of the samples. Table 1 states the parameters used for laser marking and 3D scanning, while Table 2 includes an overview of the experimental samples. The area # of Table 1 refers to the processed/engraved area, as illustrated in figures 2 and 3.

IOP Conf. Series: Materials Science and Engineering 1135 (2021) 012002 doi:10.1088/1757-899X/1135/1/012002

Table 1. Overview of samples for laser marking and reflectance measurements.

Area #	Scanning speed [mm/min]	Power [W]	Pulse frequency [Hz]	Duty cycle [%]	Spot size [mm]	Overlap [%]	3D scanning speed [mm/min]	3D scanning exposure time [ns]
1	7000	2300	5000	11	1	80	15000	400
2	7300	2300	5000	11	1	80	15000	400
3	7000	2100	5000	11	1	80	15000	400
4	7300	2100	5000	11	1	80	15000	400

Table 2. Overview of samples for laser marking and reflectance measurements. All samples are of rolled AISI 304L stainless steel plates.

Sample #	Treatment	Size [mm x mm x mm]	Reflection [%] $\lambda = 407$ nm	Reflection [%] $\lambda = 1076$ nm
1	Untreated	100 x 100 x 2	48.60	58.84
2	Untreated	100 x 100 x 2	48.65	62.83
3	Treated with WD-40	100 x 100 x 2	35.77	46.65
4	Treated with WD-40	100 x 100 x 2	37.90	49.65
5	Ground P80 – linear pattern	100 x 100 x 2	53.28	62.68
6	Ground P80 – random pattern	100 x 100 x 2	51.44	60.52

3. Method

The proposed method is based on measuring backscatter from a 3D scanner to locally approximate changes in the reflectivity of the laser light from the 3D scanner. The 3D scanner uses laser triangulation principles, which rely on geometric relations between the projected beam and the CMOS in the scanner. As the beam is projected in an opposite V-shape from the scanner onto the surface, the incident angle of the beam changes across the scanning width (Y-direction of Figure 2). For a perfect diffuse or Lambertian surface, the reflected light is scattered homogenously into the hemisphere, and the surface will hence appear equally bright regardless of the observed perspective. Lambert's cosine law defines a directly proportional relationship between the cosine of the incident angle and surface normal and the amount of reflected light in the given direction. Thus, it is apparent from Lambert's law that the measured backscatter intensity will not be constant across a flat surface. However, perfectly diffuse surfaces are rare in practice, as the majority of surfaces compose of an additional specular component that represents light being concentrated around the mirror direction. This is especially problematic when scanning reflective surfaces, i.e., rolled steel and other metals, as light can be reflected directedly into the sensor, resulting in high local backscatter intensities.

The ratio between the diffuse and specular components of reflected light depends on the optical properties of the surface, i.e. surface roughness and material composition [17]. The remainder of light that is not reflected will either be transmitted or absorbed by the material, hence equation (1):

$$I_i = I_d + I_s + I_a + I_t \tag{1}$$

I_i is the radiance of incident light, I_d is the diffused radiance, I_s is the specular radiance, I_a is the absorbed radiance, and I_t is the transmitted radiance, which for opaque materials like steel is typically zero. Hence, the absorbed radiance I_a can be approximated by determining I_i, I_d and I_s.

The 3D scanner is utilised to acquire 3D geometrical data of the surface and the corresponding backscatter intensities at each point in the shape of a point cloud. The reflection characteristics of the surface can be determined by modelling the specular and diffusion reflection using a bi-directional reflectance model (BRDF). The result is a rendered surface, which will act as a reference for determining changes in reflectivity across the actual surface. This approach aims to correct the received backscatter intensities and minimise the geometrical effects resulting from changes in incident angles. The Cook-

Torrance model, proposed by Cook and Torrance [18], is a BRDF based on optics theory and is well suited for rendering metallic surfaces. Furthermore, it can be generalised for numerous surfaces and represent both diffuse and specular reflections. Other mathematically simpler BRDF models are only able to represent diffuse surfaces (Oren-Nayer [19]) or have been shown not to be realistic enough for applicability (Blinn-Phong [20]). The Cook-Torrance model is further physically plausible, meaning that it obeys the laws of energy conservation and the Helmholtz Reciprocity. It should, however, be noted that the Cook-Torrance model assumes that the light is unpolarised and that the surface is isotropic, which can be a problematic assumption when dealing with ground surfaces. Reflection from other objects is also not considered. As stated by Cook and Torrance [18], the bi-directional reflectance can be expressed as the ratio between the reflected intensity from one direction and the incident radiance from another direction. Hence, the reflected intensity is given by equation (2):

$$R = \frac{I_r}{I_i(\mathbf{N} \cdot \mathbf{L})\mathrm{d}\omega_i} \tag{2}$$

Where I_r is the reflected intensity, I_i is the incident radiance, while $\mathrm{d}\omega_i$ is the differential solid angle of the light source i. **N** is the surface normal, and **L** is the unit vector in the direction of the incident light, as illustrated in Figure 2. The Cook-Torrance BRDF represents the reflection as the sum of the diffuse R_d and specular R_s reflection, controlled by the material-dependent specular s and diffuse d fractions as given by equation (3).

$$R = sR_s + dR_d, \qquad \text{where } s + d = 1 \tag{3}$$

Based on equation (2), the intensity reflected into the sensor I_r is then given by equation (4).

$$I_r = RI_i(\boldsymbol{N} \cdot \boldsymbol{L})\mathrm{d}\omega_i \tag{4}$$

By introducing the ambient light term I_a and its corresponding reflectivity R_a, the intensity of the reflected light I_r can be expressed as the sum of the ambient term and the sum of the contribution from each light source. The ambient term consists of light that is not directedly pointed at the surface. In this case, the laser from the 3D scanner is represented as a number of individual light sources with small solid angles $\mathrm{d}\omega_i$, which combined make up the entire width of the projected beam. To reduce computational complexity, it is assumed that each light source l only radiates the single point (or pixel of the rendered surface) that it hits. In this case, the reflectance intensity I_r is not computed over the entire hemisphere and thus only represents the reflection from a single point I_p in a single direction (direction of the CMOS sensor). I_p is then expressed in equation (5).

$$I_p = I_{ia}R_a + I_{il}(\boldsymbol{N} \cdot \boldsymbol{L}_l)\mathrm{d}\omega_{il}(sR_s + dR_d) \tag{5}$$

Computing I_p for all points in the acquired point cloud from the 3D scanner will result in a $1 \times \mathrm{n}$ vector that can be rearranged into a matrix form to be represented as a 2D rendered image of the surface reflection.

The specular reflection R_s can be expressed by combining the Fresnel term, the facet slope distribution D and the shadowing-masking term G, as given by equation (6).

$$R_s = \frac{F}{4} \frac{DG}{(\boldsymbol{N} \cdot \boldsymbol{L})(\boldsymbol{N} \cdot \boldsymbol{V})} \tag{6}$$

V is the unit vector in the direction of the observer, which in this case is the CMOS sensor in the scanner. The Fresnel term F introduces wavelength dependence to the model and expresses the reflectance of a perfectly smooth surface based on the index of refraction n. For most metals, the index of refraction has experimentally been determined and can be found in the literature [21]. The Fresnel term is approximated using the cheaper Schlick approximation [22], given by equation (7).

$$F = F_0 + (1 - F_0)[1 - (\mathbf{L} \cdot \mathbf{H})]^5, \qquad \text{where } F_0 = \left(\frac{1 - n}{1 + n}\right) \tag{7}$$

The diffuse reflection R_d is assumed to be of the same colour as the specular reflection R_s and is therefore expressed as $R_d = F_0/\pi$. Assuming that the ambient light is uniformly incident on the surface, the ambient reflectivity can be defined as $R_a = \pi R_d$. Assuming that the surface consists of a distribution of microfacets D that only reflect specular light, the angular spread of the specular component s can be determined. It should be noted that it is merely micro facets whose normal are in the direction of the normalised bi-sector vector **H** that contribute to the observed specular reflection. **H** is defined such that $\cos(\theta) = \mathbf{V} \cdot \mathbf{H} = \mathbf{L} \cdot \mathbf{H}$, where θ is the angle between **V** and **H** or **L** and **H**, as illustrated in Figure 2. As given by equation (8), the Beckman distribution considers the root mean square (RMS) of the facet slopes σ and is based on Gaussian roughness assumptions. The Beckman distribution is the commonly preferred distribution in the literature [23].

$$D(\mathbf{N}) = \frac{1}{\pi\sigma^2(\mathbf{N} \cdot \mathbf{H})^4} \exp\left[\frac{-\tan^2(\alpha)}{\sigma^2}\right], \qquad \text{where } \alpha = \cos^{-1}(\mathbf{H} \cdot \mathbf{N}) \tag{8}$$

Small values of σ represent a relatively smooth surface, resulting in a highly directional specular component, while higher values of σ represent steep facet slopes and, therefore, a rougher surface. The shadowing-masking term G describes the probability of a microfacet being visible in the ingoing (shadowing) and outgoing (masking) and is thus angular dependent [24]. Walter et al. [23] recommend using the Smith approximation, equation (9), as it serves as a more realistic representation compared to the term initially proposed by Cook and Torrance [18].

$$G_1(\mathbf{V}, \mathbf{H}) = \chi^+\left(\frac{\mathbf{V} \cdot \mathbf{H}}{\mathbf{V} \cdot \mathbf{N}}\right) \begin{cases} \dfrac{3.535y + 2.181y^2}{1 + 2.276y + 2.577y^2} & \text{if } y < 1.6, \\ 1 & \text{otherwise} \end{cases} \tag{9}$$

Where $y = (\sigma \tan \langle \mathbf{V}, \mathbf{N} \rangle)^{-1}$ and $\langle \mathbf{V}, \mathbf{N} \rangle$ is the angle between the unit vectors **N** and **V**. As the Smith G approximation is a product between the two monodirectional shadowing terms G_l, G is given by equation (10).

$$G(\mathbf{L}, \mathbf{V}, \mathbf{H}) \approx G_1(\mathbf{L}, \mathbf{H}) \cdot G_1(\mathbf{V}, \mathbf{H}) \tag{10}$$

The unit vectors **N**, **H**, **V** and **L** are acquired from the 3D geometrical data from the point cloud. As the laser specifications from the 3D scanner are unknown to the authors, it is impossible to directly compute the intensity of the incident light I_i. The RMS value of the facet slope σ of the surface is equally an unknown parameter. The same applies to the intensity of the ambient light I_a and the specular and diffuse fractions s and d. The diffuse fraction d can be computed based on equation (3), hence $d = s - 1$. The RMS facet slope σ, the fraction of specular light s and the intensity of the incident and ambient light, respectively I_i and I_a, are determined through a non-linear least square problem by minimising the difference between the computed reflectance intensities I_p and the acquired backscatter intensities I_b from the 3D scan. The fitted parameters (σ, s, I_i, I_a) are then applied to generate a perfect reference surface W_r free from any local changes in reflection due to surface contamination or oxidation. A rendered surface is illustrated in Figure 3 (left).

The acquired backscatter intensities I_b from the 3D scanner, Figure 3 (right), is reshaped as a matrix W_b and then compared to the generated reference surface W_r. This results in the relative change in reflectivity (or light intensity) ΔR, given by equation (11). ΔR is illustrated as an image in Figure 3 (right).

$$\Delta R = \frac{W_r}{W_b} \cdot 100 \tag{11}$$

As it is challenging to accurately determine surface absorption values, it is chosen to approximate the relative change in reflectivity ΔR. As it is clear from equation (1), the reflectivity is directly related to the absorption. By assuming that no light is transmitted through the material, a change in the surface reflection corresponds to a change in the surface absorption.

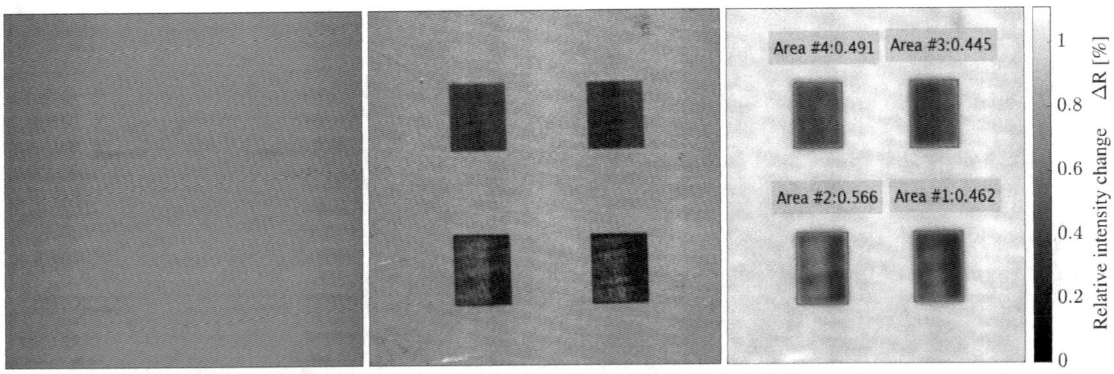

Figure 3. Left: Rendered image W_r of the surface of Sample 3 based on the Cook-Torrance BRDF model. Centre: Intensity backscatter surface W_b acquired by the 3D scanner of Sample 3. Note that the processed areas have a decreased ΔR, observed as the darker areas. Right: An image based on relative intensity change ΔR between the rendered surface W_i and the backscatter intensities W_b. Note that the values on the figure are the average ΔR within each red rectangular outline.

4. Results and discussion

The estimated relative change in reflectivity ΔR of all samples across selected, processed areas is compared with ΔR values obtained through spectroscopy. The reference values used to compute ΔR for the spectroscopy measurements is the reflectivity of the given sample before processing. Table 3 presents an overview of the measured ΔR using both the 3D scanner, spectroscopy at $\lambda = 405$ nm and $\lambda = 1076$ nm. Due to limited resources with the spectrophotometer, only samples 2 and 3 have been measured in all processed areas.

Table 3. Overview of relative change in reflectivity ΔR obtained from the 3D scanner and spectroscopy

	ΔR 3D scanning [%]				ΔR Spectroscopy [%], $\lambda = 405$ nm				ΔR Spectroscopy [%], $\lambda = 1076$ nm			
Area #	# 1	# 2	# 3	# 4	# 1	# 2	# 3	# 4	# 1	# 2	# 3	# 4
Sample 1	34	45.9	40.8	44.5		37.6		26.1		81.6		96.1
Sample 2	33.9	44.1	45	48.2	32.3	25.6	26.9	30.7	86.8	90.5	98	99.1
Sample 3	46.2	56.6	44.5	49.1	55.6	45.6	32.8	39.1	103.2	99.4	122.3	123.4
Sample 4	42.3	50.8	47.5	49.1		52.8		35.9		98.2		117.4
Sample 5	43.7	48.6	32.4	39.4		38.9		34.1		76.3		94.7
Sample 6	26.6	60.6	44.2	56.2		37.2		30.9		81.1		95.4

Table 4 presents a percentage comparison of the determined ΔR values from Table 3 based on the relation presented in equation (12).

$$C = \frac{\Delta R_{scan} - \Delta R_{spectroscopy}}{\Delta R_{spectroscopy}} \cdot 100\ \% \tag{12}$$

Table 4. 3D scanning versus spectroscopy. Percentage comparison of ΔR.

	C Spectroscopy [%], λ = 405 nm				C Spectroscopy [%], λ = 1076 nm			
Area #	# 1	# 2	# 3	# 4	# 1	# 2	# 3	# 4
Sample 2	4.95	72.27	67.29	57	-62.79	-71.71	-72.5	-51.36
Sample 3	-16.91	56.6	35.67	36.78	-55.23	-44.06	-63.61	-58.18

The results of tables 3 and 4 show that the proposed solution can identify changes in the reflection. However, there is generally a significant deviation (in some cases > 70%) in the measured ΔR values compared to spectroscopy. As isotropy is assumed in the model, a deviation is expected in the ground samples 4 and 5. The deviation is further expected to be substantial when comparing ΔR from the 3D scanner to those obtained by spectroscopy at λ = 1076 nm due to the difference in the wavelength. Therefore, the application of the proposed method should be made using a measuring laser that has a wavelength closer to that of the applied processing laser to minimise approximation error. Another source of error could be due to the presented laser marking experiments being conducted on standard industrial AISI 304L sheet metal to resemble a real industrial manufacturing scenario. This induced that the surface of the samples was not entirely homogeneous, which produces a visible variation in the absorption and thereby the marking results, as seen in Figure 4 (left). This effect was evident compared to the results presented by Antończak et al. [1] conducted on well-prepared surfaces. As the spectroscopy measurements are performed in an area of 3 x 3 mm, there is additionally a possible risk of accidentally measuring an area with considerably higher absorption.

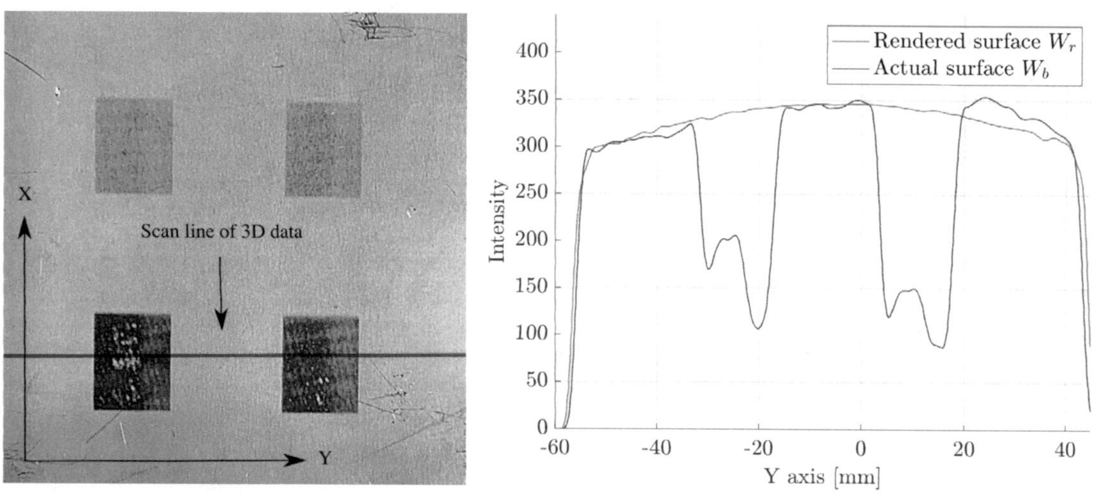

Figure 4. Left: A photograph of Sample 4 after laser marking. The different oxidation levels as a result of the induced heat are apparent. Right: A comparison of the rendered reference W_r and backscatter intensity W_b at a random scan line, indicated by the left figure. Note that the rendered surface follows the actual surface except for the oxidised areas as intended.

It can be observed from Figure 4 that the implemented model performs as intended. The rendered surface of W_r follows the actual surface W_b in the non-oxidised areas, which results in a relative change in reflectivity ΔR, which approaches one. However, in the processed/oxidised areas, ΔR drops rapidly, which is expected as the surface colour has shifted into the red area as this in the opposite spectrum compared to the scanner laser's blue colour. It can further be observed from the rendered surface of Figure 4 that the highest intensity is not measured at the centre of the field of view. The slight shift away from the centre signifies that the sample is rotated at a small angle relative to the scanner, thus underlining the necessity of correcting the intensity for geometric effects.

From additional experiments, it was concluded that the ambient light has no substantial effect on the results. The method has also been tested at incident angles above 30°. Due to the specular nature of the measured surfaces, light is reflected away from the sensor, and the reflection could not be satisfactorily approximated. The proposed model was also able to render realistic surfaces for titanium, construction steel and aluminium. Though these samples were not laser marked, and thus any change in relative reflectivity was not approximated. Highly reflective or mirror-like surfaces are expected to be problematic, as the Cook-Torrance model is not well suited for such surfaces. In addition, the model is expected to be applicable for very rough surfaces that exhibit multi-reflections, assuming an isotropy random roughness as such surfaces will likely act as diffuse reflectors. Note that performing measurements simultaneously with active laser processing should be avoided, as processing light inevitably will be reflected into the CMOS sensor of the scanner, drastically influencing the results.

5. Conclusion

The study has examined the feasibility of applying a 3D scanner to measure backscatter intensities for determining surface absorption of stainless steel samples. The proposed solution is based on the Cook-Torrance BRDF to model the reflectance across the surface to generate a reference surface free of containments and oxidation. The generated reference surface is compared to the actual surface, acquired as backscatter intensity data from the 3D scanner. The ratio between the reference surface and the actual surface defines the relative change in reflectivity or absorption. The proposed solution was proven feasible as it could identify a change in the reflectivity locally across the surface when measuring an oxidised surface from laser marking.

Comparing the approximated change in reflectivity to those acquired by spectroscopy at the same wavelength (λ = 405 nm) showed a deviation of more than 70% in some cases. At the wavelength (λ = 1076 nm) of the processing laser, the deviation was more substantial, and thus an application of the proposed solution should be made using a measurement laser with a comparable wavelength. Part of the significant deviation could be attributed to noisy spectroscopy measurement. Research into establishing a mathematical relation between the approximated reflectance and the measured reflectance from spectroscopy could potentially improve the results.

The proposed solution shows significant potential for identifying areas with a potentially large change in absorption. In a production set-up, this knowledge could be applied to reject samples before processing. Providing a clean and homogenous surface for laser marking of samples and hence a better base for model verification could aid in defining the potential for the proposed solution. The results show that further work is necessary to improve the modelled accuracy and applicability of the model. This could include fitting a complete BRDF model to backscatter data acquired from a larger variation in scanning angles, thus making it possible to model the entire reflection into the hemisphere. In addition, introducing an anisotropic term to the model would add to the flexibility of the model.

Acknowledgements

The authors would like to acknowledge Peter Kjær Kristensen from Aalborg University for guiding the spectroscopy measurements. Innovation Fund Denmark, the INTERLASE project, with number 7050-00024B is gratefully acknowledged for supporting this project. The experimental equipment used for this project was supported by the Poul Due Jensen Foundation.

ORCID iDs
A F Mikkelstrup https://orcid.org/0000-0001-8209-6787
A N Thomsen https://orcid.org/0000-0003-4865-1032
M Kristiansen https://orcid.org/0000-0001-9652-3348

References

[1] Antończak A J, Stępak B, Kozioł P E and Abramski K M 2014 The influence of process parameters on the laser-induced coloring of titanium *Appl. Phys. A Mater. Sci. Process.* **115** 1003–13
[2] Dausinger F and Shen J 1993 Energy Coupling Efficiency in Laser Surface Treatment *ISIJ Int.* **33** 925–33
[3] Abedi H R and Hoseinpour Gollo M 2019 An experimental study of the effects of surface roughness and coating of Cr2O3 layer on the laser-forming process *Opt. Laser Technol.* **109** 336–47
[4] Niu C, Zhu T and Lv Y 2019 Influence of Surface Morphology on Absorptivity of Light-Absorbing Materials *Int. J. Photoenergy* **2019** 1-9
[5] Bergström D, Powell J and Kaplan A F H 2008 The absorption of light by rough metal surfaces-A three-dimensional ray-tracing analysis *J. Appl. Phys.* **103** 1–12
[6] Davim J P 2013 Laser Beam Manufacturing *Nontraditional Machining Processes* ed J P Davim (London: Springer London) chapter 2 pp 35-96
[7] Indhu R, Vivek V, Loganathan S, Bharatish A and Soundarapandian S 2018 Overview of Laser Absorptivity Measurement Techniques for Material Processing *Lasers Manuf. Mater. Process.* **5** 458–81
[8] Drude P 1900 Zur Elektronentheorie der Metalle *Ann. Phys.* **306** 566–613
[9] Hagen E and Rubens H 1903 Über Beziehungen des Reflexions- und Emissionsvermögens der Metalle zu ihrem elektrischen Leitvermögen *Ann. Phys.* **316** 873–901
[10] Ang L K, Lau Y Y, Gilgenbach R M and Spindler H L 1997 Analysis of laser absorption on a rough metal surface *Appl. Phys. Lett.* **70** 696–8
[11] Blidegn M S K and Olsen F O 1997 Investigation into the absorptivity change in metals with increased laser power *XI Int. Symp. Gas Flow Chem. Lasers High-Power Laser Conf.* **3092** 615-8
[12] Wang H, Kawahito Y, Yoshida R, Nakashima Y and Shiokawa K 2018 A model to calculate the laser absorption property of actual surface *Int. J. Heat Mass Transf.* **118** 562–9
[13] Wang J T, Weng C I, Chang J G and Hwang C C 2000 The influence of temperature and surface conditions on surface absorptivity in laser surface treatment *J. Appl. Phys.* **87** 3245–53
[14] Rubenchik A M, Wu S S Q, Kanz V K, LeBlanc M M, Lowdermilk W H, Rotter M D and Stanley J R 2014 Temperature-dependent 780-nm laser absorption by engineering grade aluminum, titanium, and steel alloy surfaces *Opt. Eng.* **53** 122506
[15] Bergström D 2008 *The absorption of laser light by rough metal surfaces* (Luleå University of Technology)
[16] Kügler H and Vollertsen F 2018 Determining Absorptivity Variations of Multiple Laser Beam Treatments of Stainless Steel Sheets *J. Manuf. Mater. Process.* **2** 84
[17] Vukašinovíc N, Bračun D, Možina J and Duhovnik J 2010 The influence of incident angle, object colour and distance on CNC laser scanning *Int. J. Adv. Manuf. Technol.* **50** 265–74
[18] Cook R L and Torrance K E 1982 A Reflectance Model for Computer Graphics *ACM Trans. Graph.* **1** 7–24
[19] Oren M and Nayar S K 1994 Generalization of Lambert's reflectance model *Proc. 21st Annu. Conf. Comput. Graph. Interact. Tech. SIGGRAPH 1994* 239–46
[20] Blinn J F 1977 Models of light reflection for computer synthesized pictures *Proc. 4th Annu. Conf. Comput. Graph. Interact. Tech. SIGGRAPH 1977* 192–8
[21] Johnson P B and Christy R W 1974 Optical constants of transition metals: Ti, V, Cr, Mn, Fe,

Co, Ni, and Pd *Phys. Rev. B* **9** 5056–70
[22] Schlick C 1994 An Inexpensive BRDF Model for Physically-based Rendering *Comput. Graph. Forum* **13** 233–46
[23] Walter B, Marschner S, Li H and Torrance K 2007 Microfacet models for refraction through rough surfaces *Eurographics* 195–206
[24] Heitz E 2014 Understanding the Masking-Shadowing Function in Microfacet-Based BRDFs *J. Comput. Graph. Tech.* **3** 48–107

IOP Conf. Series: Materials Science and Engineering 1135 (2021) 012003 doi:10.1088/1757-899X/1135/1/012003

The bulging effect and its relevance in high power laser beam welding

Antoni Artinov[1], Xiangmeng Meng[1], Nasim Bakir[1], Ömer Üstündağ[1], Marcel Bachmann[1], Andrey Gumenyuk[1,2], Michael Rethmeier[3,1,2]

[1]Bundesanstalt für Materialforschung und -prüfung (BAM), Unter den Eichen 87, 12205 Berlin, Germany

[2]Fraunhofer Institute for Production Systems and Design Technology, Pascalstraße 8-9, 10587 Berlin, Germany

[3]Institute for Machine Tools and Factory Management, Technische Universität Berlin, Pascalstraße 8-9, 10587 Berlin, Germany

antoni.artinov@bam.de

Abstract. The present work deals with the recently confirmed widening of the weld pool interface, known as a bulging effect, and its relevance in high power laser beam welding. A combined experimental and numerical approach is utilized to study the influence of the bulge on the hot cracking formation and the transport of alloying elements in the molten pool. A technique using a quartz glass, a direct-diode laser illumination, a high-speed camera, and an infrared camera is applied to visualize the weld pool geometry in the longitudinal section. The study examines the relevance of the bulging effect on both, partial and complete penetration, as well as for different sheet thicknesses ranging from 8 mm to 25 mm. The numerical analysis shows that the formation of a bulge region is highly dependent on the penetration depth and occurs more frequently during partial penetration above 6 mm and complete penetration above 8 mm penetration depth, respectively. The location of the bulge correlates strongly with the cracking location. The obtained experimental and numerical results reveal that the bulging effect increases the hot cracking susceptibility and limits the transfer of alloying elements from the top of the weld pool to the weld root.

1. Introduction

Nowadays, the high power laser beam welding technology is an established tool with increasing capability in joining thick sheets. With available laser powers of up to 100 kW for solid-state lasers, the possible thickness of specimens welded in a single pass has increased up to 50 mm [1, 2]. The comparison of the laser beam welding process with traditional welding processes such as gas metal arc welding (GMAW) and submerged arc welding (SAW) shows the great potential offered by this welding technology. The conventional processes usually require several passes to achieve the desired welding depth, for example, in the manufacturing of thick-walled pipelines for the oil and gas industry. The application of laser beam welding may replace these processes with a single pass and offers the advantages of high efficiency, high achievable welding speed, high depth-to-width ratio, and narrow heat-affected zone [3].

Despite all advantages of the laser beam welding process, new phenomena occur with increasing welding speed and sheet thickness which are uncommon under conventional welding conditions. Recently, one such phenomena, namely the widening of the weld pool at the solid-liquid interface, known as a bulging effect, was experimentally confirmed [4, 5]. This effect represents a geometric particularity of the weld pool shape and was found to play a crucial role in the formation of defects, such as solidification cracking and porosity. The bulge results as well in a limited transport of alloying elements, e.g. filler wire transfer for wire-based laser beam welding processes, such as hybrid laser-arc welding or laser cold-wire welding. However, the optical inaccessibility of the process and the highly complicated interactions of the involved physical effects make the experimental as well as the numerical investigation of the relevant process variables, leading to the development of the bulging effect, very challenging. This applies as well to the choice of appropriate process parameters such as welding speed, laser power and focal position of the laser beam.

First studies related to the bulging effect and its influence on solidification cracking are found for the electron beam welding process. It was shown that the widened weld pool interface led to a local delay in the solidification behavior, thus increasing the cracking susceptibility in this region [6, 7]. The strong interaction between the widening of the weld pool interface and the crack formation was experimentally confirmed during hybrid laser-GMA welding of thick steel plates as well [8]. The contribution of the bulging effect to the crack formation during laser beam welding has been estimated by various means, e.g. stress distributions [9] or weld pool dynamics [10]. In a detailed study, it was shown that the bulging effect enhances the three most dominant factors, determining the cracking susceptibility of the workpiece, namely the thermal cycle, the temporal mechanical loading, and the local microstructure. Thus, it was concluded that the bulging effect results in a direct increase of the cracking susceptibility during deep penetration laser beam welding [11].

Furthermore, it has been suggested that the bulging effect may be essential for the weld pool dynamics, more specifically the transport of additional alloying elements from the top to the bottom region of the weld pool when filler material is added. As a corollary, it has been argued that the mixing of alloying elements has been limited, leading to inhomogeneous chemical composition of the weld metal and hence non-optimal weld characteristics. Several research works have shown that the widening of the weld pool shape prevents the downward transfer of additional alloying elements to the weld root and so strongly influences the weld seam quality, e.g. the Charpy impact toughness, [12, 13]. It was found that the bulging phenomenon narrows the metal transfer channel between the top and bottom regions of the weld pool, consequently contributing to the higher concentration of filler metal in the top region [14, 15].

To the best of the author's knowledge, there are few studies on the occurrence and relevance of the bulging effect during high power laser beam welding of thick sheets, although similar effects have been observed in several numerical and experimental works, e.g. [16-19]. Based on the literature review, the consequences of the bulging effect in high power laser beam welding of thick sheets are assumed to be twofold. On the one hand the widening of the weld pool interface has the consequence of increased hot cracking susceptibility and on the other hand the limited transport of alloying elements in the molten pool, resulting in an inhomogeneous chemical composition of the weld seam.

The main objective of the present work is the investigation of the two consequences of the bulging effect in deep penetration high power laser beam welding. Thus, a combined experimental and numerical approach was utilized to study the influence of the bulge on both, the hot cracking formation, and the transport of alloying elements in the molten pool. In the analysis, the sheet thickness was varied from 8 mm to 25 mm considering partial and complete penetration. Additionally, an effort has been made to obtain a critical penetration depth above which the bulging effect becomes significant for the weld seam quality.

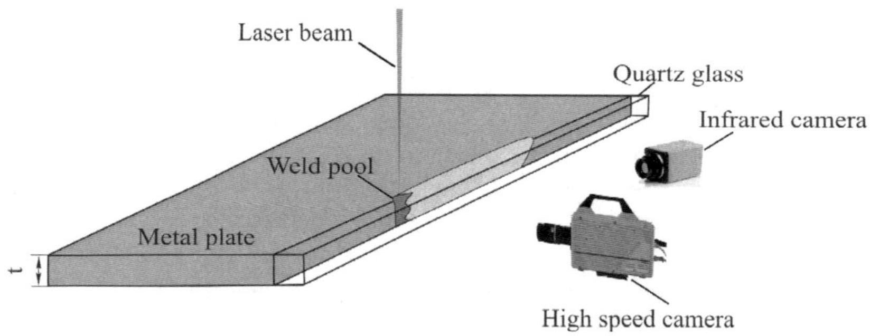

Figure 1: Schema of the experimental setup used for visualization of the weld pool interface in the longitudinal section (t was 15 mm in the complete penetration case and 25 mm in the partial penetration case) according to [5]

2. Experimental study

The experimental investigations presented in this chapter aimed for the prove of the bulge existence and its qualitative observation in the longitudinal section of the weld. The experiments were performed for two different cases, a complete penetration case of 15 mm thick S355 steel sheets and a partial penetration case of 25 mm thick S355 steel sheets. For the partial penetration case a Yb:YAG disk laser TruDisk 16002 from Trumpf has been used with a maximum output power of 16 kW at a wavelength of 1030 nm and a beam parameter product of 8 mm x mrad. In the case of complete penetration, a 20 kW fibre laser IPG YLR-20000 was used with an emission wave-length of 1070 nm and a beam parameter product of 11 mm x mrad. Both laser radiations were transmitted through an optical fiber with a core diameter of 200 µm. A specially designed setup was utilized for the experiments, which was necessary due to the optical inaccessibility of the process. To allow for the observation of the weld pool interface, a butt joint configuration with a transparent quartz glass was considered. The quartz glass used in the present study was selected according to the special requirements for the material properties. It had to provide suitable transparency for the laser wavelength of 1070 nm, meanwhile offering adequate thermal-physical properties, such as low thermal conductivity and thermal expansion coefficient, excellent thermal shock resistance, and high melting temperature. The glass-steel interface was considered as the longitudinal section of the weld. Accordingly, the laser spot was adjusted to lie on the longitudinal plane, so that half of the laser radiation was distributed to the steel plate and the other half to the quartz glass. The weld pool shape was visualized using a high-speed camera and an infrared camera. The sensor in the infrared camera allowed for a detection of a mid-wavelength infrared (MWIR) light, e.g. from 1000 nm up to 5000 nm. The high-speed camera was provided with a filter with a bandwidth of 808±10 nm to reduce the process emissions and thus make it possible to recognize the molten metal. The quartz glass was fully transparent for the complete wavelength range of the infrared camera, thus not interfering with the laser radiation and reflections. A schema of the experimental setup can be seen in Fig. 1.

The experimental setup is based on several previous works with some further improvements and modifications, for more details see [4, 5]. The corresponding process parameters and observations are described in detail in the following sections.

2.1. Complete penetration at 15 mm thick S355 steel sheets

The complete penetration laser beam welding experiment was performed at 15 mm thick S355 steel sheets with welding speed of 2 m/min, laser power of 18 kW, focal position of −5 mm, and focal diameter of 0.56 mm. The real-time observations of the longitudinal section showed that the dimensions of the weld pool interface varied with time along the middle of the penetration depth. The MWIR camera

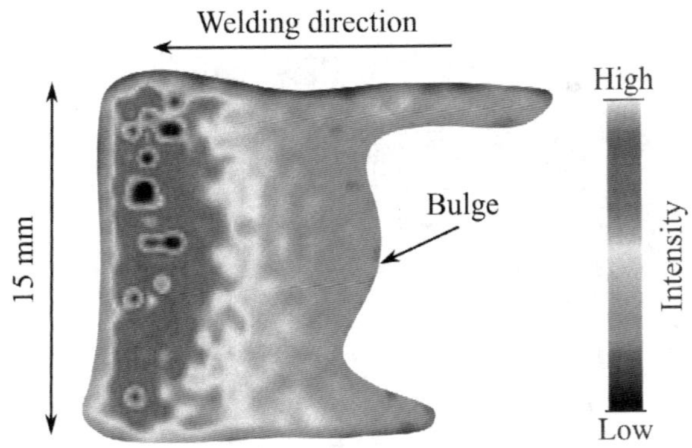

Figure 2: Longitudinal view through the quartz glass using an infrared camera according to [5]

measurement confirmed the bulge location and time variation, as exemplarily shown in Fig. 2 and discussed in more detail in [5]. The observed widening was expected to result in a solidification delay, which could subsequently lead to solidification cracking in this region. This conclusion was in good agreement with post-process results, such as radiographic films and metallographic cross-sections, since in those, the location of the bulge and the crack was nearly the same, see Section 2.2.

2.2. *Partial penetration at 25 mm thick S355 steel sheets*

The experiment for the partial penetration welding of 25 mm thick sheets was performed using 10 kW laser beam power at a welding speed of 1 m/min. The observations show that the dimensions of the weld pool are depending on the penetration depth. The areas close to the weld pool surface take a teardrop-shape. A bulge region and its temporal evolution were observed approximately in the weld root. Figure 3 a) shows exemplarily a sequence of the longitudinal profile of the weld pool. The observed flow directions are highlighted by red arrows. Two main flow circulations moved the melt from the front to the back of the keyhole ensuring mass conservation. The first circulation occurs in the upper half of the melt pool, which is mainly driven by the recoil pressure and the Marangoni stress on the melt surface. The second circulation is located in the lower half of the melt pool. The liquid metal flowed down from the front wall of the keyhole and then backwards and upwards along the liquid-solid interface, creating a circular flow at the bottom of the melt pool. The two circulations then join in the middle of the melt pool depth and form a narrow region that separates them. The fluid in the narrowed region, which comes together from the upper and lower half of the melt pool, is driven back to the vicinity of the keyhole by the two mentioned vortices. Moreover, the backflows of the melt transport the cooled material into the middle region of the melt pool, thus further reducing the temperature in this region and forming a necking. The strong melt circulation in the keyhole root leads to the elongation of the melt pool in the longitudinal direction and thus to the formation of the bulge region. This bulge promotes solidification cracking by forming a closed region filled with melt that is under tensile stress at the end of the solidification stage [20]. This as well leads to the accumulation of impurities in the final solidification stage and forms low melting phases such as (Fe-S), which additionally contributes to solidification cracking. The tensile stresses in the bulge region could be a result of the restraint condition of the welded parts or of the shrinkage restraint of the relative cold material around the weld root. Fig .3 b) shows an example for solidification crack in the weld root within the bulge region.

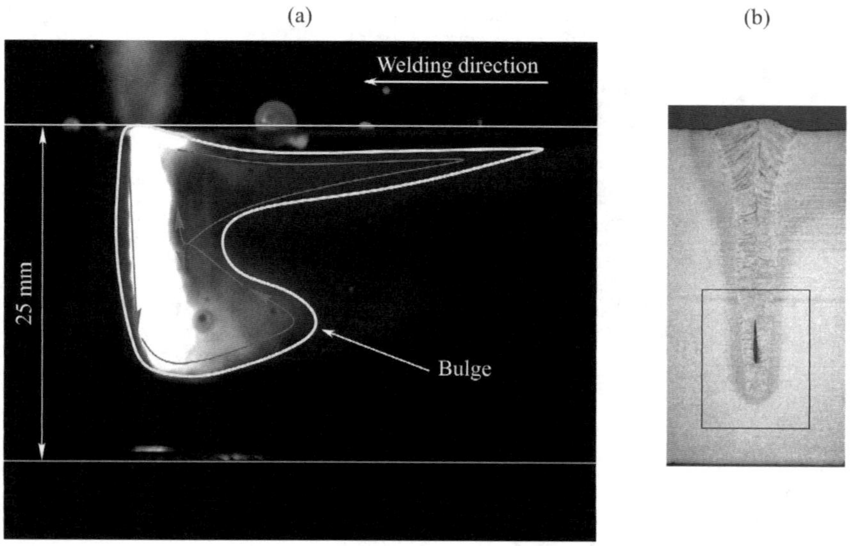

Figure 3: a) Observed melt region and flow pattern in the longitudinal section through the quartz glass using a high-speed camera b) cross-section with solidification crack in the weld root

3. Numerical modelling

This chapter deals with the mathematical description of the weld pool interface. Three numerical models are introduced, allowing for the observation of the bulging effect, more precisely its formation, and its influence on the hot-cracking formation and element transport in the weld pool. The first model was simplified by making use of a fixed keyhole geometry. This permitted for the reproduction of the bulge and the investigation of its effects on solidification cracking. This modelling approach has the advantage of a relatively fast numerical setup and short computational times; however, it is limited to the complete penetration case and does not allow for the investigation of the bulge formation. Thus, a second more sophisticated approach considering a dynamic keyhole was chosen for the investigation of the bulge formation, especially during partial penetration. Such modelling allowed to study not only the bulge formation but as well to estimate a critical penetration depth above which the bulging effect becomes significant for the weld seam quality. The third model was an extension of the dynamic keyhole model developed to simulate the wire feed laser beam welding process. Therefore, it has been applied for the study of the influence of the bulging effect on the transport of alloying elements. It is worth mentioning that the computational intensity of the second and third model exceeds this of the simplified fixed keyhole approach by an order of magnitude, thus increasing the computational time from hours to days.

3.1. Modelling of the bulging effect with a fixed keyhole geometry

The weld pool simulation proposed here aims to obtain a steady-state solution for a weld pool interface containing the bulge region, as observed experimentally in Chapter 2. It should be emphasized here that the model aims at studying the influence of the bulge on the cracking formation and not the formation of the bulge itself. Thus, the model is designed to reproduce the bulge region not quantitatively but rather qualitatively. The weld pool interface was obtained as a solution of the coupled problem, consisting of the mass and energy conservation equations, the Navier-Stokes equations, and the transport equations for the turbulent kinetic energy and turbulent dissipation rate. The model was based on previous research, whereby further improvements, and modifications, allowing for the reproduction of

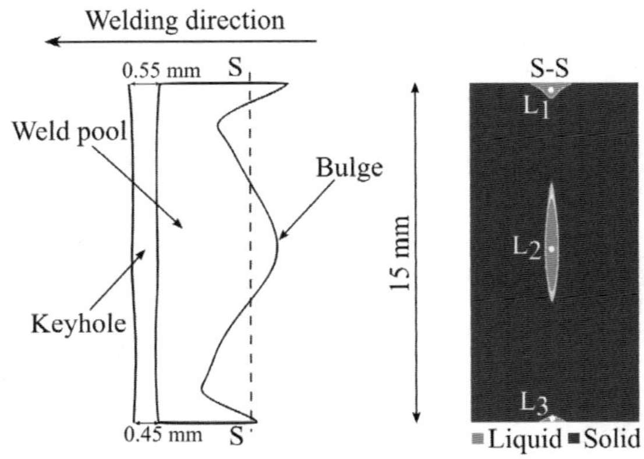

Figure 4: Results from the weld pool computation: (left) three-dimensional longitudinal section of the weld pool in the symmetry plane; (right) liquid metal in the cross-section S-S according to [28]

the bulging effect, has been made. The main assumptions are summarized here, and further details can be obtained from [21-24]. In the numerical model, the workpiece moves relative to the origin of the Cartesian coordinate system, which coincides with the origin of the laser source. A fixed keyhole geometry derived partly from the experimental results was utilized as heat source in the model. This approach was based on the equivalent heat source technique, described in more detail in [25]. The assumption made hereby is that the recoil pressure in the keyhole is perfectly balanced by the surface tension force. The surface temperature of the keyhole was set to the vaporization temperature of the material. According to results obtained with a nearly self-consistent laser beam welding models, e.g. [26, 27], this simplification can be considered as valid assumption. The model accounts for the main physical effects, such as Marangoni and natural convection, latent heat of fusion, and temperature-dependent material properties up to the vaporization temperature. A detailed analysis of the liquid zone showed how crucial the influence of the bulge on the steady-state solidification can be, and more importantly, how critical it can be for the formation of cracks. In Figure 4 the calculated longitudinal section of the weld pool geometry is shown. The weld pool interface was defined by the liquidus isotherm, highlighted by the black curve. For the analysis of the solidification behavior in the bulge region, a cross-section S-S was defined, see left hand side in Fig. 4.

According to the literature review, one of the most important factors for the crack formation is the so-called phenomenon of delayed solidification. This phenomenon, however, is found to be strongly dependent on the weld pool interface and hence on the bulging effect. As the weld pool moves along the welding direction through the cross-section S-S, the regions of liquid and solid metal were obtained from their corresponding overlap. As it can be seen on the right hand side in Fig. 4 there are three locations, highlighted by the points L_1, L_2 and L_3, containing liquid metal on the top, in the centre, and on the bottom of the cross-section. All three locations solidify at the end stage, whereby the delay on the top and bottom regions is due to the thermo-capillary driven flow. However, the solidification delay in the centre is caused by the bulging effect and solidifies at last, making it the most critical location. The presence of liquid material between the grain boundaries in the bulge region at the end stage of the steady-state solidification has a direct impact on the cracking behavior of the part, since the development of solidification cracks requires the presence of a liquid film between the grain boundaries. Thus, it can be concluded, that the bulging effect leads to lower cracking resistance in this particular area.

3.2. Modelling of the bulging effect considering a dynamic keyhole

To simulate a dynamic keyhole behavior, several physical effects have been considered in the second model. The most important of those are summarized in the following. To allow for a self-consistent energy input, a ray tracing algorithm proposed in [28] was implemented. It was used to

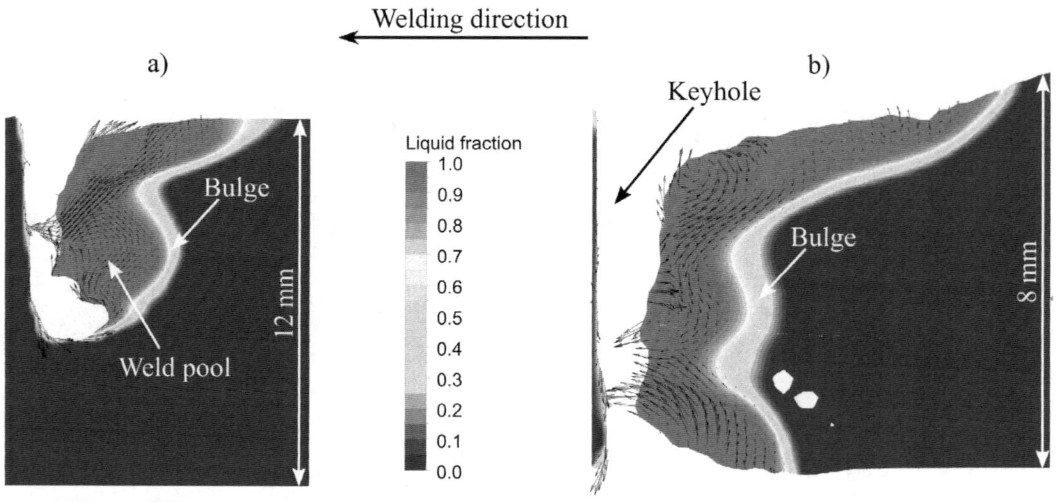

Figure 5: Exemplary results from the weld pool computation: weld pool shape with a bulge during a) partial penetration at 12 mm thick sheet (7.5 kW, 2 m/min, 0 mm focal position); b) complete penetration at 8 mm thick sheet (8 kW, 2 m/min, 0 mm focal position)

calculate the multiple reflection and the Fresnel absorption on the keyhole surface, thus directly contributing to the energy balance on it. The evaporation-induced recoil pressure, which is a dominant driving force for the formation of the keyhole was calculated according to [29]. Additionally, the vapor induced stagnation pressure and shear stresses on the keyhole wall were considered according to [30]. Further details and more complete description of the numerical model can be found in [12].

The dynamic keyhole modelling allowed to study the fluid flow and thus the formation of the bulge. For the estimation of the critical penetration depth above which the bulging effect becomes significant for the weld seam quality, various calculations with different sheet thicknesses for both partial and complete penetration has been performed. It was found that the bulge region forms more frequently during partial penetration above 6 mm penetration depth, see Fig. 5 a). In the complete penetration case, the bulge region formed already at a sheet thickness of 8 mm, see Fig. 5 b). The location of the bulge was, similar to the experimental observations, varying slightly around the middle of the plate thickness. Based on the numerical results it can be suggested that the bulging effect has a non-steady nature and forms randomly along the weld seam. The time-variation of the bulge dimensions is caused by the complex interaction of the main forces, namely the dynamic fluctuations of the keyhole due to the recoil pressure and the surface tension forces. The numerical investigation has shown that the bulging effect becomes more crucial during complete penetration laser beam welding of sheets with a thickness above 8 mm.

3.3. Modelling of the bulging effect considering a dynamic keyhole and filler wire

Commonly, accompanying with the bulging effect, the weld pool will be significantly narrowed at the middle region, which is potentially detrimental for the metal mixing. Based on the model in Section 3.2, a filler wire model was added, and the element transport equations were coupled, aiming at investigating the influence of the bulging effect on the element transport in the wire feed laser beam welding [12, 13]. The base metal used was AISI 304 stainless steel with 8.7% Ni content. The nickel-based Inconel 625 alloy with 58 % Ni content was chosen as filler wire in the current study.

A sufficient mass transfer in vertical direction is required so that the filler material can reach the weld pool bottom. Fig. 6 a) shows the flow of liquid metal in the longitudinal section during partial penetration welding. A strong circulation exists in the upper weld pool. The liquid metal is driven by the recoil

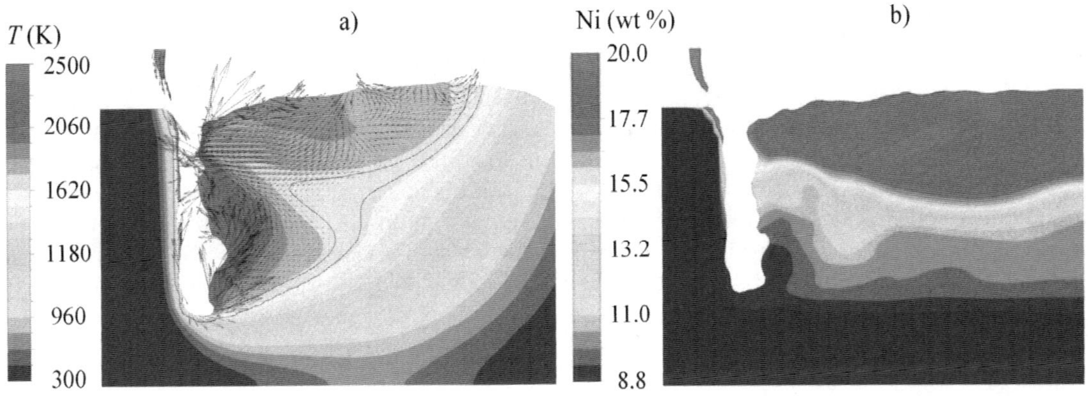

Figure 6: Simulation results of wire feed laser beam welding with bulging formation in partial penetration: a) temperature distribution and velocity field, b) Ni distribution

pressure and the Marangoni stress to flow backward along the free surface, which contributes to the elongated profile of the weld pool and the sufficient metal mixing in the upper region. It should be noted that another dominant circulation occurs in the lower weld pool. The liquid metal near the bottom of the keyhole flows backward and upward along the solid-liquid interface of the weld pool, and subsequently flows toward the keyhole rear wall. It is considered that this lower circulation may directly lead to the occurrence of the bulging effect. As the bulge fully develops, the middle region of the weld pool is significantly narrowed. The weld pool is nearly separated by the narrowed region, so that no downward flow component can be observed. Consequently, the additional Ni from the filler wire concentrates on the upper region of the weld, resulting in an inhomogeneous chemical composition along the thickness of the weld seam, see Fig.6 b).

4. Conclusions

The present work was devoted to the study of the recently confirmed widening of the weld pool interface, known as a bulging effect, and its relevance in high power laser beam welding. A combined experimental and numerical approach was utilized to investigate the influence of the bulge on the solidification cracking formation and the transport of alloying elements in the molten pool.

The following conclusions have been drawn based on the obtained results:

1) A bulge region was found to form during both partial and complete penetration laser beam welding of steel sheets.
2) The bulging effect tends to occur during deep penetration welding, thus playing a more important role in welding workpieces with higher sheet thickness. In the current study, it has been numerically estimated that the bulging will form more frequently during partial penetration above 6 mm and complete penetration above 8 mm penetration depth, respectively.
3) The bulge effect leads to an increase of the cracking susceptibility of the part due to the cause of solidification delay in the cracking region.
4) The bulge effect results in a narrowing of the weld pool interface in the middle region, which may deteriorate the homogeneity of the chemical composition in the weld seam, thus leading to non-optimal weld seam characteristics.

Acknowledgements

This work is funded by the Deutsche Forschungsgemeinschaft (DFG, German Research Foundation) with project numbers 411393804 (BA 5555/5-1) and 416014189 (BA 5555/6-1) and the research project IFG 19582N/ P1241 "Investigation of influence the restraint conditions on hot cracking in laser and laser-hybrid welding of thick structure steels" from the Research Association for steel Application (FOSTA), Düsseldorf. This was supported by the Federal Ministry of Economic Affairs and Energy through the German Federation of Industrial Research Associations (AiF) as part of the programme for promoting industrial cooperative research (IGF) on the basis of a decision by the German Bundestag. The project was carried out at Bundesanstalt für Materialforschung und –prüfung (BAM).

ORCID iDs

Antoni Artinov https://orcid.org/0000-0001-7533-5014
Xiangmeng Meng https://orcid.org/0000-0001-8996-759X
Nasim Bakir https://orcid.org/0000-0001-6847-0365
Ömer Üstündağ https://orcid.org/0000-0002-7352-429X
Marcel Bachmann https://orcid.org/0000-0003-2395-6462
Andrey Gumenyuk https://orcid.org/0000-0002-8420-5964
Michael Rethmeier https://orcid.org/0000-0001-8123-6696

References

[1] Bachmann, M., Gumenyuk, A., & Rethmeier, M. (2016). Welding with high-power lasers: trends and developments. Physics Procedia, 83, 15-25.
[2] Zhang, X., Ashida, E., Tarasawa, S., Anma, Y., Okada, M., Katayama, S., & Mitzutani, M. (2011). Welding of thick stainless steel plates up to 50 mm with high brightness lasers. Journal of Laser Applications, 23(2), 022002.
[3] Ready, J. F., & Farson, D. F. (Eds.). (2001). LIA handbook of laser materials processing.
[4] Artinov, A., Bakir, N., Bachmann, M., Gumenyuk, A., & Rethmeier, M. (2018). Weld pool shape observation in high power laser beam welding. Procedia CIRP, 74, 683-686.
[5] Artinov, A., Bakir, N., Bachmann, M., Gumenyuk, A., Na, S.J., & Rethmeier, M. (2019). On the search for the origin of the bulge effect in high power laser beam welding. Journal of Laser Applications, 31(2), 022413.
[6] Shida, T., Okumura, H., & Kawada, Y. (1979). Effects of welding parameters and prevention of defects in deep penetration electron beam welding of heavy section steel plates, Welding in the world, 17(7/8).
[7] Tsukamoto, S., & Irie, H. (1991). Mechanism of locally delayed solidification in electron beam welding, Welding international, 5(3), 177-183.
[8] Barbetta, L. D., Weingaertner, W. L., Seffer, O., Lahdo, R., & Kaierle, S. (2015). Influence of molten pool geometry and process parameters on solidification cracsk formation in hybrid laser-GMA welding of thick 5L X70 steel plates.

[9] Gebhardt, M., Gumenyuk, A., & Rethmeier, M. (2013). Numerical Analysis of Hot Cracking in Laser-Hybrid Welded Tubes, Advances in Material Science and Engineering.
[10] Wang, H., Nakanishi, M., & Kawahito, Y. (2018). Dynamic balance of heat and mass in high power density laser welding. Opt. Express 26, 6392-6399.
[11] Artinov, A., Bachmann, M., Meng, X., Karkhin, V., & Rethmeier, M. (2020). On the relationshipbetween the bulge effect and the hot cracking formation during deep penetration laser beam welding. Procedia CIRP, 94, 5-10.
[12] Meng, X., Artinov, A., Bachmann, M., & Rethmeier, M. (2020). Numerical study of additional element transport in wire feed laser beam welding. Procedia CIRP, 94, pp.722-725.
[13] Meng, X., Artinov, A., Bachmann, M., & Rethmeier, M. (2020). Theoretical study of influence of electromagnetic stirring on transport phenomena in wire feed laser beam welding. Journal of Laser Applications, 32(2), p.022026.
[14] Meng, X., Bachmann, M., Artinov, A., & Rethmeier, M. (2021). The influence of magnetic field orientation on metal mixing in electromagnetic stirring enhanced wire feed laser beam welding. Journal of Materials Processing Technology, 117135.
[15] Üstündağ, Ö., Gook, S., Gumenyuk, A., & Rethmeier, M. (2019). Mechanical properties of single-pass hybrid laser arc welded 25 mm thick-walled structures made of fine-grained structural steel. Procedia Manufacturing, 36, 112-120.
[16] Cho, W. I., Na, S. J., Thomy, C., & Vollertsen, F. (2012). Numerical simulation of molten pool dynamics in high power disk laser welding, Journal of Materials Processing Technology, 212(1), 262-275.
[17] Sohail, M., Han, S. W., Na, S. J., Gumenyuk, A., & Rethmeier, M. (2015). Numerical investigation of energy input characteristics for high-power fiber laser welding at different positions, The International Journal of Advanced Manufacturing Technology, 80(5-8), 931-946.
[18] Lu, F., Li, X., Li, Z., Tang, X., & Cui, H. (2015). Formation and influence mechanism of keyhole-induced porosity in deep-penetration laser welding based on 3D transient modelling, International Journal of Heat and Mass Transfer, 90, 1143-1152.
[19] Gao, Z., Jiang, P., Mi, G., Cao, L., & Liu, W. (2018). Investigation on the weld bead profile transformation with the keyhole and molten pool dynamic behavior simulation in high power laser welding, International Journal of Heat and Mass Transfer, 116, 1304-1313.
[20] Bakir, N., Artinov, A., Gumenyuk, A., Bachmann, & Rethmeier, M. (2018). Numerical simulation on the origin of solidification cracking in laser welded thick-walled structures. Metals. 8(6): 406.
[21] Beck, M. (1996). Modellierung des Lasertiefschweißens. Teubner.
[22] Rai, R., Kelly, S.M., Martukanitz, R.P., & DebRoy, T. (2008). A convective heat-transfer model for partial and full penetration keyhole mode laser welding of a structural steel. Metallurgical and Materials Transactions A. 39(1), 98–112.
[23] Bachmann, M., Avilov, V., Gumenyuk, A., & Rethmeier, M. (2011). Numerical simulation of full-penetration laser beam welding of thick aluminium plates with inductive support. J. Phys. D Appl. Phys. 035201.
[24] Artinov, A., Bachmann, M., & Rethmeier, M. (2018). Equivalent heat source approach in a 3D transient heat transfer simulation of full-penetration high power laser beam welding of thick metal plates, Int. J. Heat Mass Transf. 122, 1003–1013.
[25] Artinov, A., Karkhin, V., Khomich, P., Bachmann, M., & Rethmeier, M. (2019). Assessment of thermal cycles by combining thermo-fluid dynamics and heat conduction in keyhole mode welding processes. International Journal of Thermal Sciences. 145, 105981.
[26] Tan, W., Bailey, N.S., &Shin, Y.C. (2013). Investigation of keyhole plume and molten pool based on a three-dimensional dynamic model with sharp interface formulation. Journal of Physics D: Applied Physics, 46(5), 055501.
[27] Tan, W., & Shin, Y.C. (2014). Analysis of multi-phase interaction and its effects on keyhole dynamics with a multi-physics numerical model. Journal of Physics D: Applied Physics, 47(34),

345501.
[28] Cho, W. I., Na, S. J., Thomy, C., & Vollertsen, F. (2012). Numerical simulation of molten pool dynamics in high power disk laser welding. Journal of Materials Processing Technology 212, 262-275.
[29] Semak, V., & Matsunawa, A. (1997). The role of recoil pressure in energy balance during laser materials processing. Journal of Physics D-Applied Physics 30, 2541.
[30] Muhammad, S., Han, S. W., Na, S. J., Gumenyuk, A., & Rethmeier, M. (2018). Study on the role of recondensation flux in high power laser welding by computational fluid dynamics simulations. Journal of Laser Application 30, 012013.s

IOP Conf. Series: Materials Science and Engineering 1135 (2021) 012004 doi:10.1088/1757-899X/1135/1/012004

Double-sided Hybrid Laser-Arc Welding of 25 mm S690QL High Strength Steel

C Sørensen[1]*, A Nissen[2], C Brynning[2], J Nielsen[2], R Schøn[2], R Malefijt[2] and M Kristiansen[2]†

[1] Siemens Gamesa Renewable Energy, Brande, Denmark
[2] Institute of Materials and Production, Aalborg University, Aalborg, Denmark

*christian.soerensen@siemensgamesa.com, †morten@mp.aau.dk

Abstract. Hybrid Laser-Arc Welding (HLAW) technique is an enabler for the next generation high efficiency welding, but industrial adoption has been limited due to process complexity. Previously documented challenges with root cracks posed by incomplete penetration were significant; however, this work presents successful weld samples prepared from S690QL steel welded from two sides with a 16 kW disc laser. Weld travel speeds below 500 mm/min and weld line energies between 1.7 and 1.9 kJ/mm gave sound weld samples, evaluated for yield strength, elongation, hardness and Charpy-V toughness according to DS/EN ISO 10025-6:2004+A1. The results shown here indicate a significant increase in the overall efficiency of butt welds in high strength steels and further cement the HLAW process for high strength steels. It is shown that the consecutive nature of the weld procedure led to non-negligible interpass temperatures for the second weld.

Keywords: hybrid welding, toughness, interpass temperature, double-sided welding, S690QL, weldability, weld quality

1. Introduction

Demanding applications in fields such as construction and machine building utilize high strength, water-quenched steels to reduce dead mass while maintaining the structure's load-carrying capacity. Multipass welds are necessary for thicknesses beyond 5 mm, and these pose significant risks of excessive heat input, which leads to distortion and deteriorates material properties[1].

Hybrid Laser-Arc Welding (HLAW) combines a high power laser with a conventional Gas Metal Arc Welding (GMAW) torch to achieve deeper penetration, the addition of filler material and higher speed[2] compared to conventional arc welding. The laser beam's high power density results in a narrow and deep heat affected zone without the vacuum requirements necessary for electron-beam welding.

Melt containment is a critical challenge in thick section welding. Previous work on S690QL steel has achieved single pass welding on 15 mm[3] and 20 mm[4, 5]. The melt's surface tension must be sufficient to keep the liquid metal in the weld, and this prerequisite becomes harder to fulfil with increasing plate thickness[6]. Magnetic melt containment has been explored[7], which

IOP Conf. Series: Materials Science and Engineering 1135 (2021) 012004 doi:10.1088/1757-899X/1135/1/012004

could also improve filler wire dilution[8]. Just-in-time induction preheating of welds has also been investigated for S690QL[4, 9, 10], and it was found that the increased thermal input was beneficial for the mechanical properties. A similar sensitivity to thermal input has also been shown in higher grade steels with a yield strength of 960 MPa[11].

Due to the rapid cooling rates generated by the extreme power density of the laser[12, 13], there is a high likelihood for vertical cracks in the root zone of the weld. The cracks have previously been mitigated by increasing the laser energy input and preheating of the base material[14]. Double-sided welds have previously been shown in S355 steel, where the aim was an increase of the penetration capability[15] to 50 mm in steel, and up to 30 mm in aluminium[16]. It has both been shown that the mechanical properties improve from a double-sided weld[17] and that significant issues with hot cracking arise from the partial penetration nature of the weld[18] at high welding speeds.

This work addresses a gap in the previous body of work by investigating double-sided welds in S690QL high strength steel. Strenx 700E is chosen as it is the steel quality with the highest yield strength, that does not require preheating before arc welding. It is relevant to investigate whether this holds for HLAW processes as well. Interpass- or preheating is implemented directly due to the first side HLAW pass, which supports full penetration from a second pass on the other side. The principle can provide a beneficial impact for industrial applications by improving thermal input volumetric control. An experimental approach is applied to identify weld parameters that lead to successful welds, and a numerical model is used to support the hypothesis of interpass heating.

2. Experimental procedure

Weld samples were prepared by water jet cutting of 25 mm SSAB Strenx®700E plate. The steel conforms to DS/EN ISO 10025-6:2004+A1 as S690QL, alloyed according to Table 1, with a resulting carbon equivalent of 0.32 (CEV)/0.49 (CET)[19][1]. The internal weld faces were milled flat and perpendicular before lead-in and -out plates of identical material were welded on to either end of the 130 mm seam, with shims to provide the 0.3 mm gap in the I groove geometry. The samples were subsequently welded using a 1030 nm, 16 kW disc laser (Trumpf TruDisk 16002) and a Fronius TransPuls Synergic 5000 inverter welding power source. The HLAW principle is outlined in figure 1, where relevant setup parameters are annotated.

The GMAW torch and optical elements were mounted on a fixture plate driven by a Fanuc M-710 robot. ESAB OK Aristorod 69 filler wire was used in combination with 82% Ar / 18% CO2 active welding gas. Table 1 outline the fixed experimental parameters and figure 2 shows the top-view geometry of the sample.

Table 1: Chemical composition of SSAB Strenx®700 E [19] and ESAB OK Aristorod 69 [20] given as maximum percentages by mass.

	C	Si	Mn	P	S	Cr	Cu	Ni	Mo	B
Base material	0.20 %	0.60 %	1.60 %	0.020 %	0.010 %	0.8 %	0.30 %	2.0 %	0.70 %	0.005 %
ESAB OK Aristorod 69	0.089 %	0.53 %	1.54 %	-	-	0.26 %	-	1.23 %	0.24 %	-

Twenty-nine samples were welded with varying weld speeds and laser powers, where the bottom side weld was performed first. Secondly, the top side weld was performed after flipping the sample and realigning the robot path to the open groove on the second side. The second

[1] CEV $= C + Mn/6 + (Cu+Ni)/15 + (Cr+Mo+V)/5$ and CET $= C + (Mn+Mo)/10 + (Cr+Cu)/20 + Ni/40$, in mass percentages

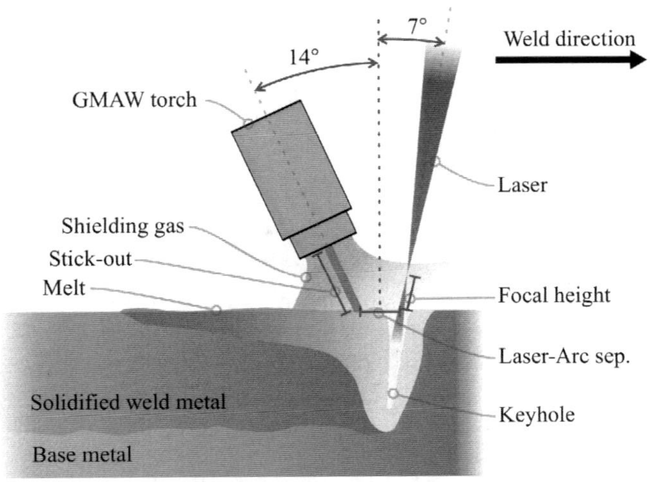

Figure 1: Principle schematic of the HLAW technology, where the primary procedure dimensions are outlined. The weld direction indicates the laser beam movement and arc source relative to a stationary weld sample.

Table 2: Fixed equipment parameters

Focal length of laser optics	600 mm
Focal height	0 mm
Laser angle (w.r.t. vert.)	7°
GMAW torch angle	14°
Wire feed	6 m min^{-1}
Wire diameter	1.6 mm
Stick-out	15 mm
Laser-Arc separation	15 mm
Gas flow	25 L min^{-1}

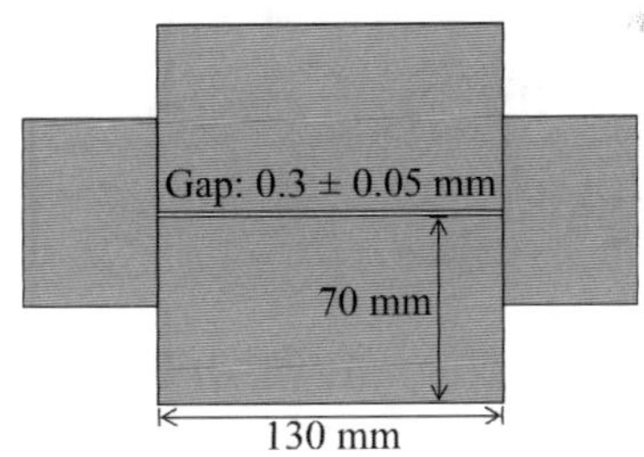

Figure 2: Weld sample geometry. Waterjet cut from 25 mm plate and milled on the internal weld face. The samples are symmetric w.r.t. the root gap. Run-in and run-out plates are shown on the sides.

weld's timing was not controlled but is estimated to be 2 to 3 minutes between the stop of the first weld and the start of the second weld.

The samples were welded with speeds between 300 mm/min and 800 mm/min, in steps of 100 mm/min. The laser power was set between 8 kW and 16 kW, in steps of 2 kW, with the remaining parameters set according to Table 2. Some samples were not finished or analyzed due to melt drooping or insufficient penetration of the weld.

The samples were subjected to tests, according to selection criteria outlined in table 3.

3. Experimental Results

Of the 29 samples, four were discarded due to insufficient penetration, and six were discarded due to drooping.

Nineteen samples were subjected to radiographic inspection, according to DS/EN ISO 10675-1 assessment grade B. As the plate thickness is ≥ 25 mm, an Image Quality Indicator (IQI) 10 is used to assess the sensitivity. IQI level 8 is identifiable on the images, corresponding to a sensitivity of 1.6%, fulfilling the requirement of 2% from DS/EN ISO 17636-2.

IOP Conf. Series: Materials Science and Engineering 1135 (2021) 012004 doi:10.1088/1757-899X/1135/1/012004

Table 3: Sample triage definition.

Step	Action	Rule-out criteria
1	Visual inspection of the first weld	Drooping rules the sample out for further processing or inspection.
2	Sectional Imaging	Incomplete penetration rules the sample out from further inspection.
3	Radiographic analysis	Identification of root cracks includes it in the following text but rules out the sample from further analysis.
	Subsequent steps are done in parallel for remaining samples	
4a	Tensile test	
4b	Hardness test	
4c	Charpy V toughness test	

The radiographic inspection discarded 14 samples due to severe pores or root cracks. Macrographs were cut and analyzed in cases of uncertain differentiation between pores and cracks. Five samples were accepted from the radiographic inspection - as indicated in figure 3. Samples were named according to the applied laser power (kW) and travel speed (mm/min): Samples *14/500*, *12/400*, *12/500*, *10/400*, *10/500* was subjected to further analysis.

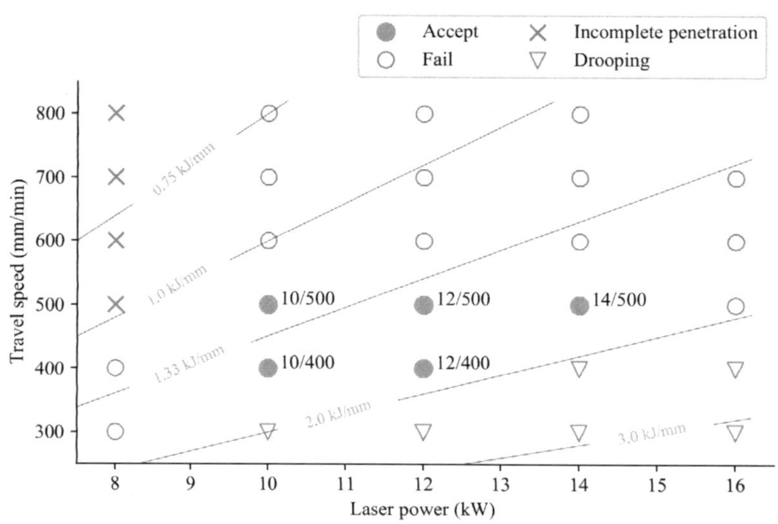

Figure 3: Investigated parameter space. Samples accepted according to DS/EN ISO 10675-1 level B are shown and labeled in green. Samples that failed, had incomplete penetration or drooping are shown in red circles, blue x's or orange triangles respectively. Labeled grey lines in the background indicate the laser line energy deposition.

Samples that were disregarded due to insufficient penetration of the first weld (less than half the plate thickness) were all welded with a laser power of 8 kW, and at speeds higher than 400 mm/min. Samples that were discarded due to drooping, was in the lower right quadrant of figure 3; above 2 kJ/mm linear energy deposition.

Five samples remained, with weld energies shown in table 4. Macrographs were made to show

Table 4: Input line energies of successful weld samples, distributed on arc, laser and in total. Note that the values reported here, are the output energies of the arc source and laser source.

Symbol		*12/400*	*14/500*	*10/400*	*12/500*	*10/500*
Q_{GMAW}	GMAW line energy (kJ/mm)	0.64	0.5	0.63	0.5	0.51
Q_{laser}	Laser line energy (kJ/mm)	1.8	1.68	1.5	1.44	1.2
Q_{tot}	Total line energy (kJ/mm)	2.44	2.18	2.13	1.94	1.71

the heat-affected zone (HAZ); these are displayed in figure 4. Due to the double-sided process, two HAZs are clearly visible. The penetration depth has a positive correlation to the laser line energy.

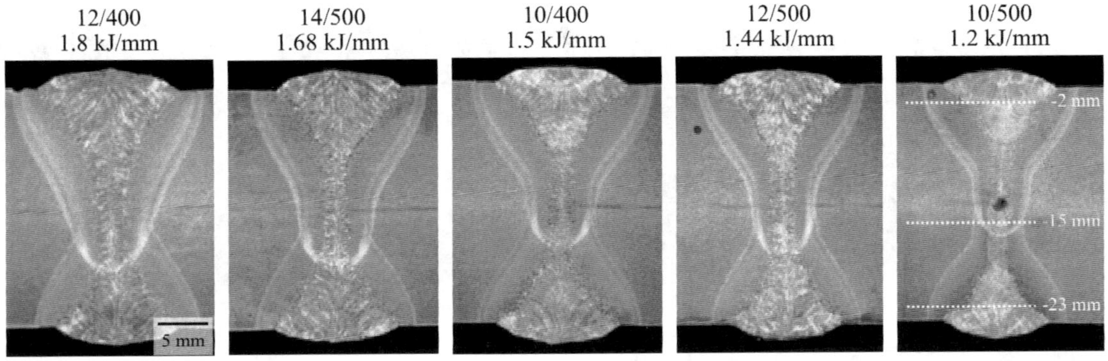

Figure 4: Macrographs of sections of the five remaining samples. The scale bar in *12/400* applies to all macrographs. *10/500* contains annotations to show the position of the hardness measurements presented in figure 5. The dark blemish in the center of *10/500* is a pore, acceptable according to DS/EN ISO 10675-1 level B. Note that the penetration decreases from left to right, in correspondence with the laser line energy from figure 3.

Tensile tests were performed in a Zwick/Roell 100 kN universal testing machine, according to DS/EN ISO 6892-1:2019, to determine the yield strength, ultimate tensile strength and A5 elongation, results shown in table 5. Samples *12/500* and *10/500* is qualified by all tensile test criteria in the base material standard set in DS/EN ISO 10025-6:2004+A1. The tensile test specimens were made to include the root zone from both welds in the center of the 12 mm diameter of the specimen.

Table 5: Results from tensile tests. Results that does not conform to the standard for S690QL are shaded red.

Sample	*12/400*	*14/500*	*10/400*	*12/500*	*10/500*	S690QL
Yield strength (MPa)	608.7	724.7	817.6	846.4	845.5	>690
Ultimate tensile strength (MPa)	756.7	801.2	820.4	847.7	852	770-940
A5 elongation (%)	5.6	5.28	8.1	17	18.18	>14

All macrographic samples are tested for Vickers hardness with a Struers Duramin 40 hardness

IOP Conf. Series: Materials Science and Engineering 1135 (2021) 012004 doi:10.1088/1757-899X/1135/1/012004

tester. Linearly spaced samples are made across the HAZ in three depths, referenced to the surface of the second weld: 2 mm below, 15 mm below (corresponding to the root of the second weld), and 23 mm below to allow comparison to the hardness of the first weld. The positions are noted in figure 4, in pane *10/500*.

The sampled hardness line scans are shown in figure 5, for the five samples. It is observed that the hardness generally increases with decreasing line energy, and that the Vickers hardness near the surface is higher for the first weld than the second weld. Higher line energies provides a more significant preheating, resulting in lower cooling rates and subsequently lower hardness. The first weld is performed on metal with ambient temperature, and experiences greater cooling rates, leading to increased hardness. DS/EN ISO 10025-6:2004+A1 require a hardness below 450HV10, and all weld samples pass this criterion.

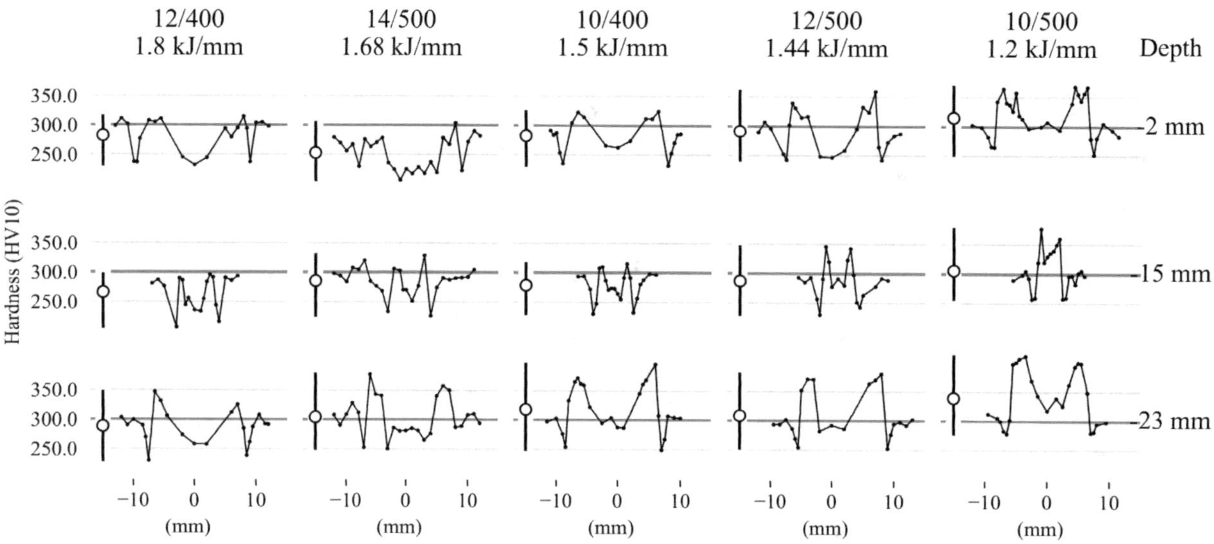

Figure 5: Linearly sampled hardness measurements on the Vickers scale (HV10). The vertical line in each graph spans the maximum and minimum hardness, and the circle show the mean. Note that the peak hardness is higher for samples welded with lower laser line energies (towards the right) and on the first weld surface. Sampled lines are shown in figure 4, pane *10/500*.

Impact toughness at decreased temperatures was tested according to DS/EN ISO 148-1:2006, in a Charpy V-notch apparatus. Samples were prepared with the notch in the HAZ and in the centerline of the weld. While DS/EN ISO 10025-6:2004+A1 specifies a minimum energy absorption of 27 J at −40 °C, SSAB[19] specifies a minimum energy absorption of 69 J for the base material in the transverse direction. Two samples of each configuration (HAZ/weld centre line) were tested.

All samples from the HAZ absorbed more than the instrument's full-scale reading (294 J) and was thus qualifiable. Results from the samples with notches in the weld centerline are shown in figure 6, where the average of the two samples all qualify according to DS/EN ISO 10025-6:2004+A1, but only one weld sample meets the toughness criteria set by SSAB in the base material.

4. Estimation of interpass temperature

A numerical model has been developed to estimate the samples' internal interpass temperature, specifically for the volume relevant to the second weld pass. The manufacturer specifies an

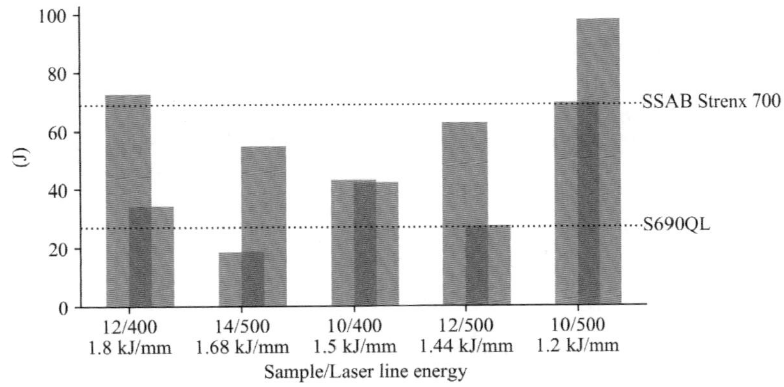

Figure 6: Charpy V-notch impact toughness of the two prepared samples from the weld centerline. All samples except *14/500* meet DS/EN ISO 10025-6:2004+A1, but only *10/500* also meet the base material criteria set by SSAB.

allowable maximum interpass temperature between room temperature and 300 °C[21]

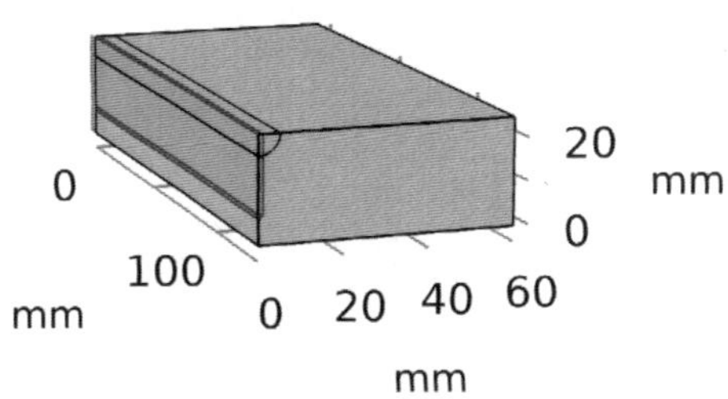

Figure 7: The simulation domains, with the laser deposition volume outlined in blue.

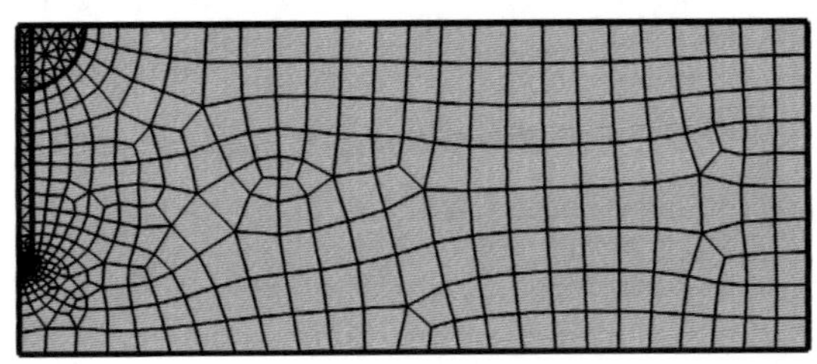

Figure 8: Transverse section of the simulation mesh; it is swept through the longitudinal direction. The third dimension (normal to the paper) is used to implement radiative and convective losses.

The sample volume (excluding the lead in and -out plates) was modelled in the Finite Element software COMSOL Multiphysics 5.4, with geometry as seen in figure 7. Two domains were set up to contain the accumulated energy from the laser source and the GMAW source, each with a thermal efficiency of 80%. At the simulated time $t = 0$, the temperatures in the domains were set according to:

$$T_{\text{arc}}\Big|_{t=0} = T_{\text{amb}} + \frac{E_{\text{arc}}\eta}{C_{\text{p}}\,\rho\,V_{\text{arc}}} \quad \text{and} \quad T_{\text{laser}}\Big|_{t=0} = T_{\text{amb}} + \frac{E_{\text{laser}}\eta}{C_{\text{p}}\,\rho\,V_{\text{laser}}}$$

where T_{arc}, E_{arc}, V_{arc} and T_{laser}, E_{laser}, V_{laser} is the resulting temperature, the accumulated energy (e.g. $E_{arc} = Q_{\text{GMAW}} \cdot 130\,\text{mm}$) and volume of the GMAW and laser domains respectively. T_{amb} is the ambient temperature (20 °C), C_{p} is the thermal capacity of steel ($500\,\text{J}\,\text{kg}^{-1}\,\text{K}^{-1}$), η is the thermal efficiency (80%) and ρ is the density of steel ($7880\,\text{kg}\,\text{m}^{-3}$). For a first order approximation, the temperature dependence of the density and specific heat is not included. Phenomena related to phase transitions in the melt are likewise disregarded. The external boundary conditions model natural convection losses from all exterior sides, as well as surface-ambient radiative losses. Conductive losses from fixturing are not included, and this overestimates the temperature for longer timescales.

The simulated results are not useful for short timescales due to the large gradients set by the initial boundary conditions. Still, as the sample temperature is renormalized, the simulation converges towards an upper bound of the internal material temperature.

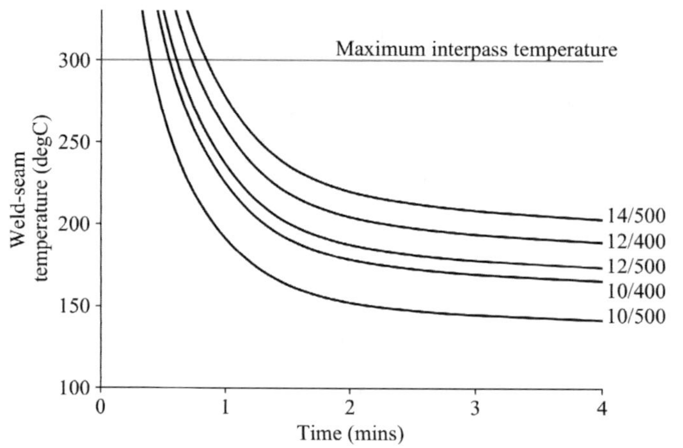

Figure 9: Average temperature of the internal laser weld volume, as a function of time. The five samples are outlined, along with the maximum interpass temperature according to SSAB guidelines [21].

The decay in temperature due to normalization and subsequent losses are shown in figure 9, for boundary conditions corresponding to the five samples' total line energy (laser and arc in their respective domains). It is observed that the temperature after 2 minutes decays less than $20\,^{\circ}\mathrm{C\,min^{-1}}$, only due to convective and radiative losses. At $t = 2$ minutes, the temperatures are between $150\,^{\circ}\mathrm{C}$ to $230\,^{\circ}\mathrm{C}$, and thus well below the SSAB specified allowable interpass temperature of $300\,^{\circ}\mathrm{C}$.

5. Conclusion

Double-sided welding is shown to be successful in joining thick section high strength steel. Samples were tested to fulfil DS/EN ISO 10025-6:2004+A1 and furthermore compared to the higher requirements for the base material from the manufacturer. Two sets of parameters gave weld samples that fulfilled the base material norm, with one parameter set also fulfilling the increased toughness specification from the supplier datasheet.

It was observed in figure 3 that high line energies were necessary to avoid root cracks, that are generally attributed to excessive cooling rates. However, a further increase in line energies resulted in a decrease of material properties like yield strength and elongation (table 5), but also a reduction in peak hardness (figure 5).

The numerical model indicates that the brief duration between the first and second side welds resulted in a non-negligible increase in interpass temperature. The estimated interpass temperature was below the manufacturer's specifications. It is observed that the peak hardness near the surface of the first weld (base material at room temperature) is significantly higher than that of the second weld, performed at elevated temperatures. This work predicts beneficial effects from introducing interpass heating with a double-sided welding process, thus not relying on external equipment such as inductive or flame heating. The remaining heat input that would otherwise be perceived as waste heat acts to improve the total weld quality. This method could be attractive for industries with a strong focus on environmental optimization.

Acknowledgments
The authors extend their gratitude for funding from Innovation Fund Denmark, project 9066-00007B, "Hybrid Laser-Arc Welding for Offshore Turbine Structures". The authors declare no conflicts of interest. The numerical model is available upon request.

ORCID iDs

C Sørensen	https://orcid.org/0000-0002-3473-0904
A Nissen	https://orcid.org/0000-0002-7602-9008
C Brynning	https://orcid.org/0000-0002-5547-3038
J Nielsen	https://orcid.org/0000-0001-5274-2979
R Schøn	https://orcid.org/0000-0001-5964-7333
R Malefijts	https://orcid.org/0000-0003-1552-7633
M Kristiansen	https://orcid.org/0000-0001-9652-3348

References

[1] Webster S, Kristensen J K and Petring D 2008 *Ironmaking & Steelmaking* **35** 496–504
[2] Silva R G N, de Paço C M M, Rodrigues M B, de Sousa J M S, Pereira M, Ramos B B, Schwedersky M B and Gonçalves e Silva R H 2020 *The International Journal of Advanced Manufacturing Technology* **110** 2801–2814
[3] Bakir N, Üstündağ Ö, Gumenyuk A and Rethmeier M 2020 *Welding in the World* **64** 501–511
[4] Lahdo R, Seffer O, Springer A, Kaierle S and Overmeyer L 2014 *Physics Procedia* **56** 637–645
[5] Lahdo R, Seffer O, Springer A, Kaierle S, Collmann M and Schaumann P 2015 *Stahlbau* **84** 1016–1022
[6] Üstündağ Ö, Gook S, Gumenyuk A and Rethmeier M 2019 *Procedia Manufacturing* **36** 112–120
[7] Üstündağ Ö, Avilov V, Gumenyuk A and Rethmeier M 2019 *Metals* **9** 594
[8] Meng X, Bachmann M, Artinov A and Rethmeier M 2019 *Journal of Manufacturing Processes* **45** 408–418
[9] Li L, Mi G, Zhang X, Xiong L, Zhu Z and Wang C 2019 *Optics & Laser Technology* **119** 105606
[10] Li L, Mi G and Wang C 2019 *Journal of Manufacturing Processes* **43** 276–291
[11] Farrokhi F, Siltanen J and Salminen A 2015 *Journal of Manufacturing Science and Engineering* **137** 061012
[12] Gebhardt M O, Gumenyuk A and Rethmeier M 2013 *Advances in Materials Science and Engineering* **2013** 1–8
[13] Farrokhi F, Endelt B and Kristiansen M 2019 *Optics & Laser Technology* **111** 671–686
[14] Bunaziv I, Akselsen O M, Frostevarg J and Kaplan A F H 2019 *The International Journal of Advanced Manufacturing Technology* **102** 2601–2613
[15] Wahba M, Mizutani M and Katayama S 2016 *Materials & Design* **97** 1–6
[16] Jiang Z, Hua X, Huang L, Wu D, Li F and Zhang Y 2018 *The International Journal of Advanced Manufacturing Technology* **97** 903–913
[17] Lei Z, Zhang K, Hu X, Yang Y, Chen Y and Wu Y 2015 *Journal of Materials Engineering and Performance* **24** 4518–4526
[18] Wiklund G, Akselsen O, Sørgjerd A J and Kaplan A F H 2014 *Journal of Laser Applications* **26** 012003
[19] SSAB 2019 Data sheet Strenx® 700 E/F
[20] ESAB 2021 OK Aristorod 69 datasheet
[21] SSAB 2017 SSAB Strenx Welding Brochure

IOP Conf. Series: Materials Science and Engineering 1135 (2021) 012005 doi:10.1088/1757-899X/1135/1/012005

Post-processing of additively manufactured metal parts by ultrashort laser pulses for high-quality net shape geometries and advanced functionality

Daniel Holder[1], Matthias Buser[1], Artur Leis[1,2], Rudolf Weber[1], Thomas Graf[1]

[1]Institut für Strahlwerkzeuge (IFSW), University of Stuttgart, Pfaffenwaldring 43, 70569 Stuttgart, Germany
[2]Graduate School of Excellence advanced Manufacturing Engineering (GSaME), University of Stuttgart, Nobelstraße 12, 70569 Stuttgart, Germany

E-mail: daniel.holder@ifsw.uni-stuttgart.de

Abstract. Additive manufacturing by means of laser-based powder bed fusion (LPBF) offers high flexibility with respect to the generation of individualized and light-weight metal parts. However, the produced parts are typically attached to support structures and deviate a few tens of micrometers from the targeted final component in geometrical net shape and surface roughness due to the melt-based fusion process. Therefore, different post-processing techniques were examined in the past to resolve the mentioned quality drawbacks. In our work, we investigated the potential of post-processing of LPBF-generated Ti6Al4V parts with ultrashort pulse laser ablation. As a result, the support structures were effectively removed, the surface roughness was reduced by 81% and complex geometries with high shape accuracy were fabricated. Furthermore, the LBPF-generated parts were laser surface structured to investigate the potential of post-processing with ultrashort laser pulses for advanced functionality, such as water-repellent surfaces. The generation of surface structures on the LPBF-generated Ti6Al4V part changed the wetting behaviour from hydrophilic to hydrophobic with an increased contact angle from 73° up to 130°.

1 Introduction

During the last decade additive manufacturing processes have made their way from rapid prototyping into industrial series production. Additive manufacturing by means of laser-based powder bed fusion (LPBF) offers high flexibility with respect to the generation of individualized and light-weight metal parts. With LPBF complex-shaped, metallic parts are generated from slices of selectively molten powder in the powder bed [1].

However, for most industrial applications the LPBF process does not allow to produce parts that are directly ready to use and thus post-processing is required [2]. For example, support structures are necessary to enable the generation of overhang structures which have to be removed from the part after the LPBF process [3]. Additionally, the surface quality of LPBF-generated parts is reduced by partially fused powder particles that adhere to the surface [4]. Furthermore, macroscopic deviations from the targeted geometry typically result from material shrinkage, build orientation and the staircase effect [5]. One promising approach to remove excess material such as support structures with minimized mechanical load is laser micromachining with short and ultrashort laser pulses. Excess material can be detected and measured in the processing area during micromachining using on-axis depth measurement based on optical coherence tomography (OCT). The online depth measurements can be used within a control system with a defined target depth for each location in the processing area to further process only locations that have not reached the target depth yet. OCT-controlled laser micromachining enabled the machining of highly accurate geometries in inhomogeneous materials, such as wood, bone [6] and carbon-fiber reinforced plastics (CFRP) [7]. In a previous work, we presented OCT-controlled laser micromachining with ultrashort laser pulses for high-quality net shape geometries in LPBF-generated aluminum parts with increased geometry accuracy and reduced surface roughness [8]. Surface structuring with ultrashort laser pulses can create advanced functionalities, e.g. preventing counterfeiting or tampering by encoding numerical information [4] or passive antibacterial behavior [9].

In this work, OCT-controlled laser micromachining with ultrashort laser pulses is demonstrated for post-processing of LPBF-generated parts manufactured from a Ti6Al4V powder. First, OCT-controlled laser micromachining is used for the removal of the support structures attached to the manufactured part. Then, OCT-controlled laser micromachining is used to smooth the underlying surface and to create complex geometries that are difficult or impossible to manufacture with the LPBF process. Finally, an additional laser process for advanced functionality is presented with the fabrication of hydrophobic surfaces by surface structuring of additively manufactured parts with ultrashort laser pulses.

2 Methods and material

The ultrafast laser system Pharos from Light Conversion was used for material processing and emitted pulses at a wavelength of λ_{Laser} = 1030 nm with a pulse duration of 260 femtoseconds (fs). The laser beam was linearly polarized and had a beam quality factor of $M^2 < 1.3$. The maximum average power on the sample surface was 13 W, which corresponds to a maximum pulse energy of E_p = 260 µJ at a pulse repetition rate of f_p = 50 kHz. The OCT-based depth measurement system CHRocodile 2 from Precitec was used for on-axis optical depth measurements [7], [8]. The superluminescence diode of the measurement system is a broadband light source that emits at a wavelength of λ_{OCT} = 1080±20 nm and with a beam quality factor of $M^2 < 1.1$. The measuring rate was set to 50 kHz to correspond to the pulse repetition rate of the processing laser of f_p = 50 kHz. The depth measurement system provided an axial measurement range of up to 6 mm with an axial measurement accuracy of ± 1 µm. The processing laser beam and OCT measurement beam were superimposed by means of a dichroic mirror as sketched in Figure 1.

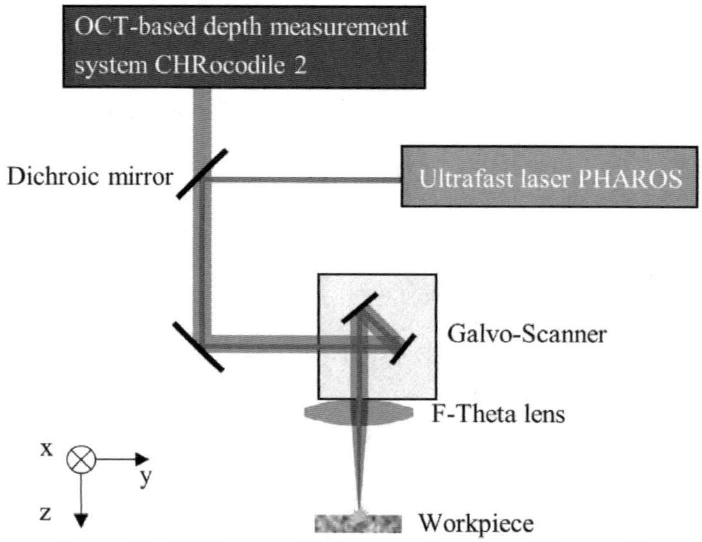

Figure 1. Experimental setup for OCT-controlled laser micromachining and laser surface structuring with ultrashort laser pulses.

Both beams were guided through a Galvo-Scanner for deflection and focused by an F-Theta lens. The experiments regarding OCT-controlled laser micromachining (cf. section 3) were performed using an F-Theta lens with a focal length of 163 mm, resulting in focal diameters of $d_{f,163mm}$ = 60±5 µm for the processing laser beam and 15±5 µm for the OCT measurement beam. The experiments regarding surface structuring (cf. section 4) were performed using an F-Theta lens with a focal length of 340 mm, resulting in a focal diameter of $d_{f,340mm}$ = 120±5 µm for the processing laser beam. The smaller beam diameter in micromachining allowed for a high lateral precision in the x-y-plane, whereas a larger beam diameter in surface structuring allowed for the usage of high pulse energies at a fixed repetition rate and fluence, which corresponds to a higher processed area per unit time.
The principle of OCT-controlled laser micromachining is described in [7]. In short, the sample surface is measured using the depth measurement system to receive the depth value (z-direction in Figure 1) while the measurement beam is deflected across the sample surface (x-y-plane in Figure 1) in a rasterized manner by the Galvo-Scanner. After each pass across the sample surface, a 3D-measurement of the current surface topography is generated and compared to the target topography. Locations in the x-y-plane with excess material are processed with a fixed pulse energy and locations where the target depth has already been reached are skipped in the following pass, until the measured topography fits to the target topography within a tolerance band of ±5 µm.
The samples were manufactured with a Ti6Al4V powder with a grain size distribution between 15 µm and 45 µm on an LPBF-system according to ASTM 52911-1 (ASTM 52911). A single-mode fibre laser with a wavelength of 1075 nm, an average power of 155 W and a beam diameter of 30 µm was used to generate powder layers with a thickness of 20 µm in an inert gas atmosphere of argon without additionally heated substrate. The part was generated with lateral dimensions of 8 mm and a part thickness of approximately 1.5 mm by hatching with a feed rate of 1200 mm/s and a hatching distance of 60 µm. Block support structures were generated so the parts were fixed metallurgically to the substrate. In other applications, the support structures are often required to enable the generation of overhang structures. The support structures, the manufactured sample and the resulting surface structure are shown in Figure 2.

Figure 2. a) SEM image of block support structures. b) Image of the sample generated from Ti6Al4V powder by LPBF with support structures on top side and the part on the bottom side. c) Surface structure after the manufacturing process.

The block support structures were built with a wall distance of 700 µm and wall height of 800 µm. Excess material adhered locally between the support structures from the mechanical separation of the sample from the substrate plate of the powder bed (Figure 2 a). Powder grains and partially molten material remained on the sample surface after the manufacturing process (Figure 2 c).
The 3D-measurements generated during OCT-controlled micromachining were used to evaluate the resulting quality and productivity of the micromachining process. The quality was quantified with the shape deviation D_a and surface roughness S_a as defined in [8]. The productivity was quantified by calculation of the ablation rate V_t, which is the removed material volume per processing time including auxiliary process times, e.g. calculations of the depth-control system. The micromachined surfaces and the surface structures were additionally characterized using a laser scanning microscope (LSM) and scanning-electron-microscopy (SEM). The wetting behaviour on different surfaces was characterized by measurement of the contact angle θ of deionised water with a camera-based contact angle measuring device.

3 Removal of support structures and fabrication of smooth surfaces with high accuracy

The support structures on the parts were removed using OCT-controlled laser micromachining with the setup as described in Figure 1. The target depth was set to 100 µm below the surface of the actual part in order to completely remove the excess material from the support structures. The maximum average power of 13 W was used for fast removal of the support structures at a peak fluence of $\Phi_0 = 20.2$ J/cm². The pulse distance in and perpendicular to the scanning direction was kept constant at 15 µm, which corresponds to a pulse overlap of 74%. This was achieved by using a feed rate of $v_x = 750$ mm/s and a hatching distance of $p_y = 15$ µm. The axial focus position of both beams was set on the top surface of the support structures. The moving average of 50 measurements of the ablation rate V_t as a function of the number of passes over the sample surface is shown in Figure 3. Two distinct regimes can be identified. The first regime lasts until approximately 67 passes. Within the first regime, the excess material adhering to the support structures and parts of the support structures were removed, which led to a drop of the ablation rate from 3.2 mm³/min to 0.6 mm³/min after 67 passes. As there was no material volume between the excess material and the part underneath, the ablation rate in this regime was artificially enhanced. In the second regime from 68 to 400 passes, the ablation rate gradually decreased, as more and more locations reached the target depth and were not further processed.

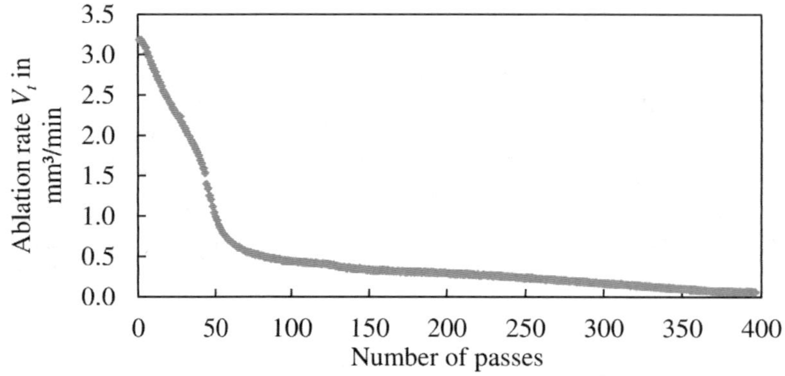

Figure 3. Ablation rate V_t as function of the number of passes for OCT-controlled laser micromachining with $\Phi_0 = 20.2$ J/cm².

After OCT-controlled micromachining for 400 passes over the sample surface, the support structures were completely removed from the sample surface. An SEM image of the resulting surface (left) and the deviation of the OCT-measurement to the targeted flat topography (right) are shown in Figure 4.

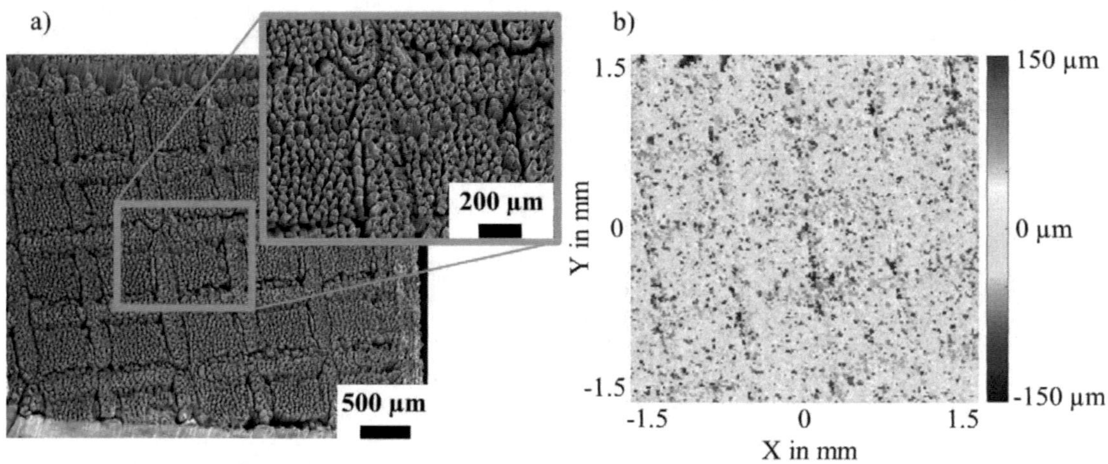

Figure 4. a) SEM image of the surface topography and b) deviation of OCT-measurement to the flat target topography after OCT-controlled laser micromachining with $\Phi_0 = 20.2$ J/cm² and 400 passes.

Large cones and holes can be seen in the SEM images in the inset of Figure 4 a), which is a well-known outcome of micromachining of metals with high peak fluences [4]. Furthermore, a lattice-shaped surface structure is visible on the sample surface with cavities of up to 100 µm depth between former support structure and samples surface. The cavities are deeper than the targeted ablation depth and presumably result from the reflection of the processing laser beam on the sidewalls of support structures. The formation of large cones and holes on the surface as well as the formation of the cavities on the sidewalls of the support structures yield a rather high surface roughness of $S_a = 33.7$ µm.
After removal of the support structures, the rough surface topography with large cones and holes was smoothed by additional micromachining with a lower peak fluence of $\Phi_0 = 1.9$ J/cm² for 2400 passes. The axial focus position was shifted on the surface of the part. The moving average of the ablation rate V_t as a function of the number of passes over the sample surface is shown in Figure 5.

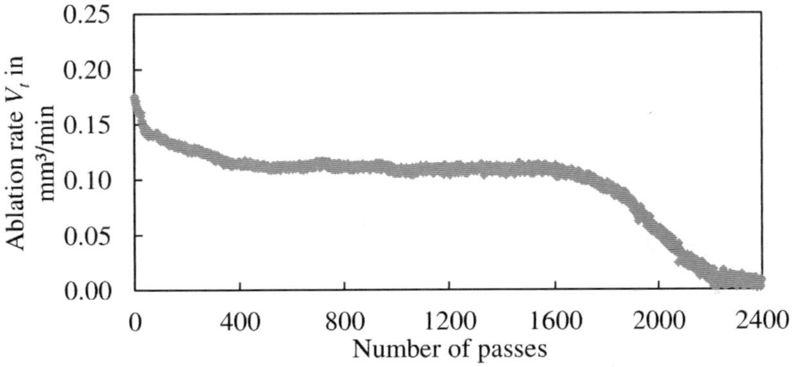

Figure 5. Ablation rate V_t as function of the number of passes for OCT-controlled laser micromachining with $\Phi_0 = 1.9$ J/cm².

Again, a decrease of the ablation rate can be found in the beginning of the micromachining process until about 400 passes. From 400 to 1500 passes, the entire surface was machined as no location had reached the target depth yet and the ablation rate remained constant at $V_t = 0.12 \pm 0.1$ mm³/min. From 1500 to 2400 passes, the ablation rate gradually decreased, as more and more locations had reached the targeted depth of 600 μm below the initial surface and were not further processed. The cause of the enhanced ablation from 0 to 400 passes is currently unknown and subject to future investigation.
A SEM image of the resulting surface after OCT-controlled micromachining for 2400 passes over the sample surface (left) and the deviation of the OCT-measurement to the targeted flat topography (right) are shown in Figure 6. Even after smoothing the surface for 2400 passes, a fine lattice-shaped surface structure is still visible on the sample surface with cavities of up to 30 μm depth. The measured surface roughness reached an 81% reduced value of $S_a = 6.3$ μm compared to the roughness $S_a = 33.7$ μm measured after removal of the support structures.

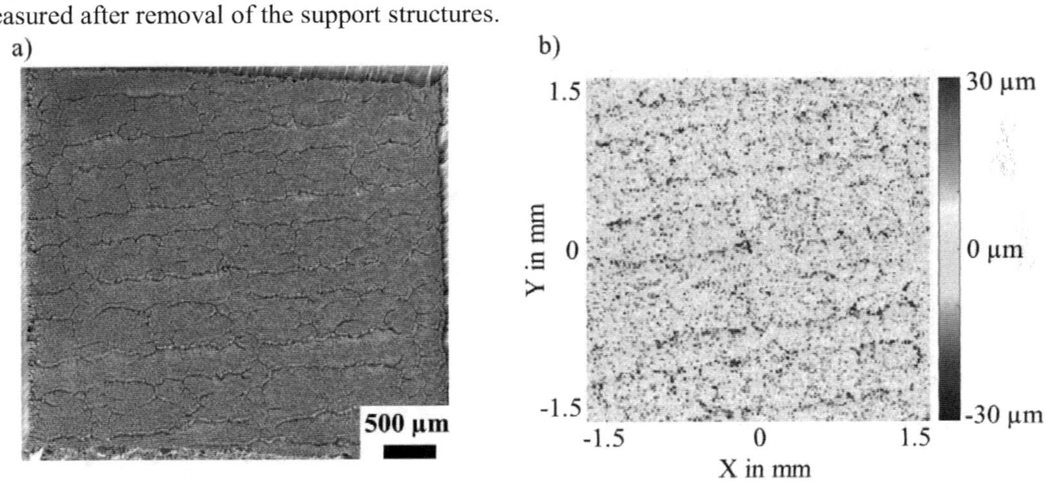

Figure 6. a) SEM image of the surface topography and b) deviation of OCT-measurement to the flat target topography after OCT-controlled laser micromachining with $\Phi_0 = 1.9$ J/cm² and 2400 passes.

In addition, OCT-controlled micromachining enables the fabrication of geometries that are difficult to manufacture by LPBF, such as sharp contours, thin tips, and inclined surfaces. The complex target geometry used in this work and the corresponding challenges of each geometry are described in detail in [8]. After removal of the support structures as described in the beginning of section 3, the target geometry was fabricated using OCT-controlled laser micromachining with a peak fluence of $\Phi_0 = 1.9$ J/cm² and 3700 passes. A SEM image of the resulting surface (left) and the deviation of the OCT-measurement to the targeted complex geometry (right) are shown in Figure 7.

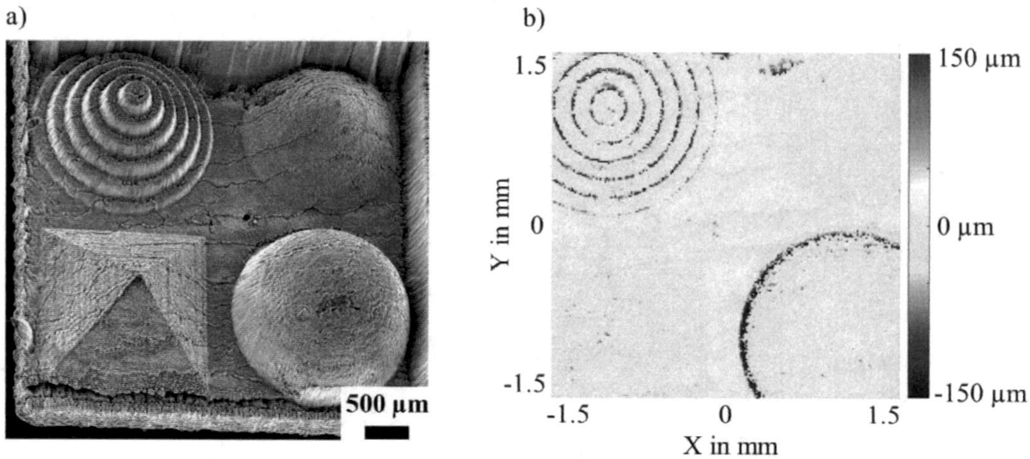

Figure 7. a) SEM image of the surface topography and b) deviation of OCT-measurement to the complex shaped target topography after OCT-controlled laser micromachining with $\Phi_0 = 1.9$ J/cm² and 3700 passes.

High-quality geometries with the features of each shape well pronounced and a rather smooth surface for an inhomogeneous material such as LPBF-generated Ti6Al4V can be seen in the SEM-image in Figure 7 a). As seen in Figure 7 b), deviations of the OCT-measurement to the targeted complex geometry are caused by the steep edges of the stepped geometry in the top left and by the steep area of the half sphere in the bottom right. A mean shape deviation of less than 22 µm was calculated for each of the geometries. These small deviations can be caused by the micromachining process or measurement errors. The pyramid and cone in the bottom left and top right were fabricated with higher accuracy and a shape deviation of less than 12 µm. The mean shape deviation calculated across all geometries is low with $D_a = 16.3$ µm.

The investigations confirm the high potential of post-processing of additively manufactured metal parts by ultrashort laser pulses for removal of the support structures, smoothing the surface and fabrication of high-quality net shape geometries. In this work, the approach was demonstrated for parts made from Ti6Al4V powder, but can be transferred to other metals and surface conditions, as long as the OCT-based depth measurement system is able to detect the excess material and as long as the used peak fluence is higher than the material-specific ablation threshold fluence. For the application in an industrial environment, the productivity and therefore the ablation rate must be significantly enhanced to remove the support structures within a reasonable period of time. This can be achieved with the power scaling of ultrashort pulse lasers into the kW range and the development of micromachining strategies for high machining productivity while maintaining high surface quality [10]. Furthermore, the ablation rate during selective removal in the later stage of the OCT-controlled ablation process can be increased when the auxiliary times are reduced, e.g. with a scanning path optimization as shown in [11].

4 Defined wetting behavior by surface structuring

The same laser system and processing setup can be used for additional manufacturing processes, e.g., functionalization of the sample surface with a defined wetting behavior. To demonstrate this capability, different original surface states were fabricated before the surface structuring process by mechanical polishing one half of the parts. The other half of parts without any mechanical pretreatment was used "raw" as a reference. Mechanical polishing significantly reduced the roughness from $S_a = 12.1$ µm of the raw surface shown in Figure 2 c) to $S_a = 0.1$ µm. Subsequently, the parts were laser surface structured in one pass with a peak fluence of $\Phi_0 = 0.8$ J/cm² and a hatching distance of $p_y = 24$ µm with different feed rates of $v_x = 1000$ mm/s and $v_x = 100$ mm/s. SEM images of the resulting surfaces are shown in Figure 8.

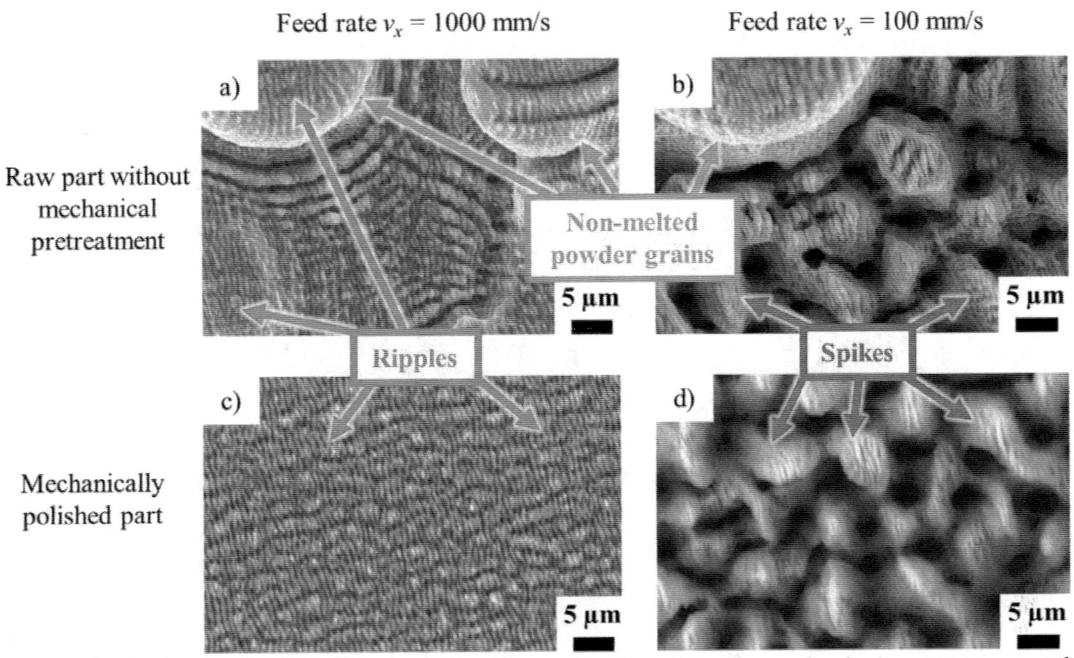

Figure 8. SEM images of resulting surface topography after different mechanical pretreatment and surface structuring with different feed rates.

Structuring with a feed rate of v_x = 1000 mm/s yielded a surface covered with fine ripples. The ripples were superimposed on the rough surface topography of the raw part, even on the non-melted powder grains (Figure 8 a). The roughness after structuring remained in the range of the raw part with S_a = 9.2 µm. Mechanical polishing of the part before structuring with v_x = 1000 mm/s yields a homogeneous surface structure (Figure 8 c) and low roughness S_a = 0.6 µm. Structuring with a lower feed rate of v_x = 100 mm/s resulted in a surface covered with rough spikes. The spikes were partially superimposed on the initially rough surface topography of the raw part. The non-melted powder grains were again covered with ripples (Figure 8 b). By superimposing the rough spikes on the rough surface topography of the raw part, the surface roughness was increased to S_a = 14.5 µm. Mechanical polishing of the part before structuring with v_x = 100 mm/s resulted in a homogeneous distribution of the spikes (Figure 8 d) and increased roughness S_a = 3.3 µm compared to the roughness of the polished surface of S_a = 0.1 µm.

The wetting behavior on the different surfaces was characterized with the sessile drop technique by measurement of the contact angle θ of deionised water after storing the samples in ambient atmosphere for 180 days. A commercial, camera-based measuring device with a frame rate of 25 Hz was used to record the time-dependent wetting behaviour of water droplets with a volume of 0.5 µL over a period of 20 s. The measured contact angle as a function of the time after drop application is shown in Figure 9. The mechanically polished part showed to an initial contact angle of θ = 73° that slightly decreased over time to θ =65° (grey triangles). The initially hydrophilic wetting state of additively manufactured parts was observed before [12], however, without consideration of the short-term behavior after application of the drop. The structured polished parts with fine ripples (green triangles) showed a hydrophobic wetting state within the investigated time with a maximum contact angle of θ =114° after 20 s. The surface covered rough spikes (red triangles) showed an even higher contact angle of θ =128° after 20 s.

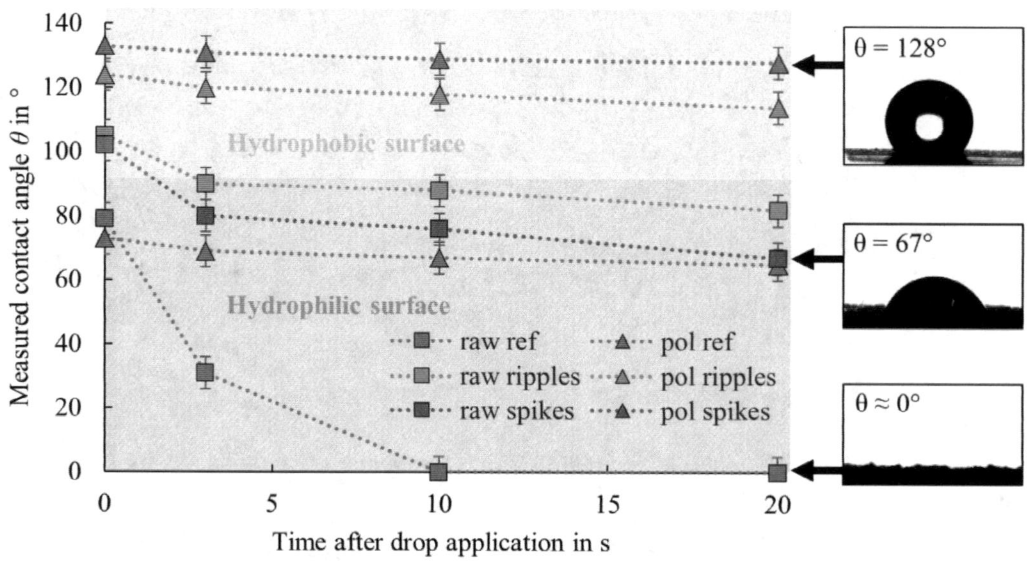

Figure 9. Measured contact angle θ as a function of time after application of a droplet with a volume of 0.5 μL for different surface topographies.

Further increase of the contact angle for a superhydrophobic behavior of additively manufactured parts can be achieved, e.g. with laser processing in ethanol as shown in [12]. The change of the wetting behavior from hydrophilic without surface structures to hydrophobic with surface structures is a result from the combined effect of surface morphology and surface chemistry and is explained in detail in [13]. The raw part without surface structures showed a time-dependent wetting behavior with a contact angle of $\theta = 79°$ directly after application of the drop (grey squares). Over time, the fluid spreaded over the sample surface with a superhydrophilic wetting state of $\theta \approx 0°$ after about 10 s. The structured raw parts with ripples (green squares) and spikes (red squares) showed a hydrophobic wetting state directly after application of the drop with contact angles of $\theta = 105°$ and $\theta = 102°$, respectively. The contact angles of both surfaces then quickly decreased in the hydrophilic range <90°. The time-dependent major decrease of the measured contact angle of the raw part is presumably caused by the large initial roughness of the raw part and is subject to future investigation.

5 Conclusion

In summary, we presented post-processing methods for LPBF-generated Ti6Al4V parts based on OCT-controlled laser micromachining and surface structuring with ultrashort laser pulses that can be used for removal of support structures, smoothing the surface, fabrication of complex geometries and generation of a hydrophobic wetting behavior.

By OCT-controlled laser micromachining the block support structures with a wall distance of 700 μm and wall height of 800 μm were completely removed using the maximum average power of 13 W, which corresponds to a peak fluence of $\Phi_0 = 20.2$ J/cm². Large cones and holes resulted in a high surface roughness of $S_a = 33.7$ μm after removal of the support structures, which was subsequently smoothed with OCT-controlled micromachining at a peak fluence of $\Phi_0 = 1.9$ J/cm² for 2400 passes, resulting in a reduced roughness of $S_a = 6.3$ μm. The fabrication of very small complex geometries that are difficult to manufacture by LPBF, such as sharp contours, thin tips, and inclined surfaces was achieved by OCT-controlled micromachining with a low mean shape deviation of $D_a = 16.3$ μm. Finally, the generation of advanced functionalities on LPBF-generated parts was demonstrated using surface structuring with ultrashort laser pulses, which resulted in the formation of ripples and spikes on the surface. Structuring at a low feed rate of $v_x = 100$ mm/s of mechanically polished parts led to the homogenous formation of spikes on the surface, resulting in a hydrophobic wetting state with a contact angle of $\theta = 128°$.

Future work will include further smoothing of the surface after OCT-controlled micromachining, e.g. by laser polishing [14], to be able to omit the mechanical polishing step before surface structuring for subsequent hydrophobic wetting behavior.

Acknowledgements
The authors would like to thank the Ministry of Science, Research and Arts of the State of Baden-Württemberg for the financial support of the InnovationsCampus Mobility of the Future in the frame of the project “AddSub”. Furthermore, the authors would like to thank Ulrich Höchner and Light Conversion for providing the USP laser Pharos and Dr. Markus Kogel-Hollacher and Precitec for providing the OCT measurement system CHRocodile 2. Finally, the authors thank Clemens Maucher and the Institute of Machine Tools (IfW) for providing the Ti6Al4V samples and Liane Hoster and Dimitra Stanigloudi for their support with the surface preparation and the SEM images.

ORCID-IDs
D Holder https://orcid.org/0000-0002-9810-2747
M Buser https://orcid.org/0000-0002-9320-8569
A Leis https://orcid.org/0000-0002-2065-6651
R Weber https://orcid.org/0000-0001-8779-2343
T Graf https://orcid.org/0000-0002-8466-073X

References
[1] Emmelmann C, Petersen M, Kranz J and Wycisk E 2011 Bionic lightweight design by laser additive manufacturing (LAM) for aircraft industry *SPIE Eco-Photonics 2011: Sustainable Design, Manufacturing, and Engineering Workforce Education for a Green Future* 80650L 1-12
[2] Flynn J M, Shokrani A, Newman S T and Dhokia V 2016 *International Journal of Machine Tools and Manufacture* **101** 79–101
[3] Jiang J, Xu X and Stringer J 2018 *JMMP* **2** 64 1-23
[4] Worts N, Jones J and Squier J 2019 *Optics Communications* **430** 352–7
[5] Zhu Z, Keimasi S, Anwer N, Mathieu L and Qiao L 2017 Review of Shape Deviation Modeling for Additive Manufacturing *Advances on Mechanics, Design Engineering and Manufacturing (Lecture Notes in Mechanical Engineering)* ed B Eynard *et al* (Cham: Springer International Publishing) 241–50
[6] Webster P J L, Wright L G, Ji Y, Galbraith C M, Kinross A W, van Vlack C and Fraser J M 2014 *Optics letters* **39** 6217–20
[7] Holder D, Buser M, Boley S, Weber R and Graf T 2021 *Materials & Design* **203** 109567
[8] Holder D, Leis A, Buser M, Weber R and Graf T 2020 *Advanced Optical Technologies* **9** 101–10
[9] Lutey A H A, Gemini L, Romoli L, Lazzini G, Fuso F, Faucon M and Kling R 2018 *Scientific reports* **8** 10112 1-10
[10] Holder D, Weber R, Röcker C, Kunz G, Bruneel D, Delaigue M, Graf T and Ahmed M A 2021 *Optics letters* **46** 384–7
[11] Buser M, Onuseit V and Graf T 2021 *Optics and Lasers in Engineering* **138** 106412 1-7
[12] Jiao L, Chua Z Y, Moon S K, Song J, Bi G and Zheng H 2018 *Nanomaterials (Basel, Switzerland)* **8** 601 1-10
[13] Kietzig A-M, Hatzikiriakos S G and Englezos P 2009 *Langmuir: the ACS journal of surfaces and colloids* **25** 4821–7
[14] Bhaduri D, Penchev P, Batal A, Dimov S, Soo S L, Sten S, Harrysson U, Zhang Z and Dong H 2017 *Applied Surface Science* **405** 29–46

IOP Conf. Series: Materials Science and Engineering 1135 (2021) 012006 doi:10.1088/1757-899X/1135/1/012006

Hardening in laser forming under the temperature gradient mechanism

Georgi Nikolaev Nikolov[1], Anders Noel Thomsen[1], Morten Kristiansen[1]

[1]Department of Materials and Production, Aalborg University, Fibigerstræde 16, 9220 Aalborg, Denmark

gnn@mp.aau.dk

Abstract. Laser forming is a contactless thermal forming process that can be applied for both single and double-curved geometries. When it comes to prototyping and small batch production, laser forming has the potential to compete with conventional sheet-metal forming processes; however, an investigation of the relationship between process parameters, hardness distribution and the bend rate is lacking. This study examines the influence of using different sets of processing parameters on the bend rate and the hardness distribution. ANSI 304 stainless steel samples of 1 and 3 mm thickness are laser formed up to 90° with a bend radius equal to their thickness. A theoretical discussion of the material's hardening kinetics is used to generalize the results. Micro-Vickers hardness test is used to measure the hardness distribution along the 3 mm samples to support the theoretical discussion. The results show that the bend rate increases when using different sets of process parameters; furthermore, the bend arc length has shown to have a significant influence over the bend rate. An increase of hardness is observed on the bottom side of the laser formed samples, indicating potential strain hardening.

1. Introduction

Laser forming is a contactless incremental thermo-mechanical forming process that uses a defocused laser beam to induce thermal stresses and strain into a workpiece. The process is highly flexible and can be used both for single and double-curved geometries. There are three main mechanisms in play during laser forming: temperature gradient mechanism (TGM), upsetting mechanism (UM) and buckling mechanism (BM) [1]. TGM is characterized with a steep thermal gradient along the material's thickness resulting in the material bending towards the laser beam [2,3]. The bend is a result of the initial material expansion during heating and the subsequent cooling of the surface layer which causes thermal contraction. TGM is one of the most reported mechanisms in laser forming, with an achievable bend angle up to 3° in a single pass [4,5]. BM is active when a negligibly small temperature gradient is developed through the material's thickness and the diameter of the beam is significantly larger than the material's thickness, resulting in a bend that can be either towards or away from the laser beam [4,6]. UM examines the same negligible thermal gradient as BM but with a laser diameter smaller than the material's thickness which results in material in-plane shortening [4]. In practice, laser forming is a complex interaction between the various mechanisms.

As stated, laser forming is an incremental process, meaning that if a bend larger than 3° is to be achieved using TGM, more than one laser passes will have to be utilized, this is called multi-pass laser forming. However, the issue with multi-pass laser forming is the bend rate decay, that is, with each laser pass the rate at which the bend develops decreases [3,7]. There are two major review studies on laser forming done in the recent years [4,8], and both reviews show an abundance of parameter studies, microstructural and control studies.

Shen et al. [9] performed a numerical study of two-scan laser forming with varying time intervals between laser passes along the same forming path; the lower the time interval between the passes the higher the peak temperature leading to a steeper thermal gradient which in terms results in a higher deformation. That said, the buildup of the peak temperature with every subsequent laser pass in multi-pass laser forming which may result in burning or melting the sample, needs to be taken into consideration. Lambiase et al. [10] examined both the productivity and necessity of multi-pass laser forming. The authors observe that even under same processing conditions, samples that are laser formed normal to the rolling direction tend to bend more. Moreover, Lambiase et al. [10] note that for passive air cooling the velocity of the laser is highly influential for the cooling time, with higher velocity resulting in highly reduced cooling time. The studies show both an influence of the process parameters over each other and over the bending rate.

An increase of hardness in the heat affected zone (HAZ) with each subsequent laser pass, either attributed due to strain hardening or phase transformation, has been observed in a multitude of studies [4,8] and cited as an influencing factor for the bend rate decay [3]. Edwardson et al. [3] made an overview of the factors influencing the bend rate in multi-pass laser forming using TGM. The investigated factors were strain hardening, section thickening, thermal effects, variation in absorption and geometrical effects. The authors observed that an increase of laser passes will result in an increase of micro-hardness in the HAZ; furthermore, with the increase of laser passes, the peak hardness will shift deeper into the materials' thickness. The geometrical factor, proposed by Edwardson et al. [7] and examined again by Edwardson et al. [3], refers to the geometrical distortion of the laser beam incident due to bend angle increase. Furthermore, at higher number of passes and higher bend angles, both strain hardening and the geometrical factor are dominant influencers for the bend rate decay [3]. Yang et al. [11] investigate the relation between the surface behaviors of the HAZ and the process parameters of pulse laser forming. The results show that both the depth of the HAZ together with the micro-hardness relationship between the HAZ and the hardening depth is dependent on process parameters. Moreover, the study shows that the micro-hardness decreases when moving away from the laser beam center, giving it a Gaussian distribution. However, a holistic investigation of the relationship between process parameters, hardness distribution and the bend rate seem to be missing.

Studies thus far have examined either the individual influence of parameters over the bending rate and hardness or finding optimal parameters for specific materials and sample thickness. This study attempts to investigate the correlation between the process settings and the hardening of the material and its influence on the bend behavior and bend rate decay. A theoretical analysis of the hardening kinetics under TGM, which is used to devise three hypotheses for the bend behavior will be presented in this study. Samples of 1 mm and 3 mm thick AISI 304 steel are bent up to 90° using different sets of parameters, hardness tests are made before switching parameters for the 3 mm samples.

2. Theoretical analysis

Laser forming is a highly dynamic process where thermal, mechanical and spatial aspects are closely coupled to create unique thermal and mechanical profiles all along the laser scan path [12]. The theory presented here is meant to give an overview of the possible paths of the transverse stress, perpendicular to the scanning direction, in relation to temperature and to demonstrate the complexity of the problem. The goal of the model is to make qualitative predictions on correlations. The theory has the following constraints:

1. Only considers TGM.
2. Neglect the influence of longitudinal, normal and shear stresses. This is purely for simplification as they are all bound to occur due to the non-uniformity of the transverse stresses and will change the deformation and hardening kinetics.

3. Neglect the influence of position along the laser scan path and other geometric factors such as width and length of the part. The results of Thomsen et al. [12] showed that laser forming under TGM is suspect to differences in deformation and temperature profiles along the laser scan path.
4. Considers only the transverse stress and temperature through the thickness.

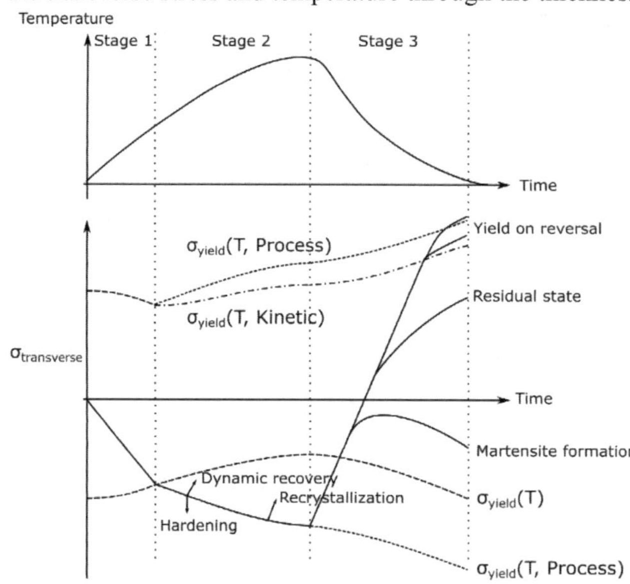

Figure 1. The temperature and transverse stress over time at a single material point. Note that a part can be in all three stages at different positions through the thickness at the same time.

The laser heats the material and as it is turned off, the material cools down, before the process repeats itself. This behavior is similar in shape albeit not in size throughout the thickness of the material. The peak temperature will be reached at different times through the thickness [13]. Figure 1 also shows the possibilities of the transverse stress over time and can be divided into several stages. Note that a part can be in all three stages at different positions during laser forming. Figure 2 shows how varying the speed will change the temperature profiles and their influence through the thickness.

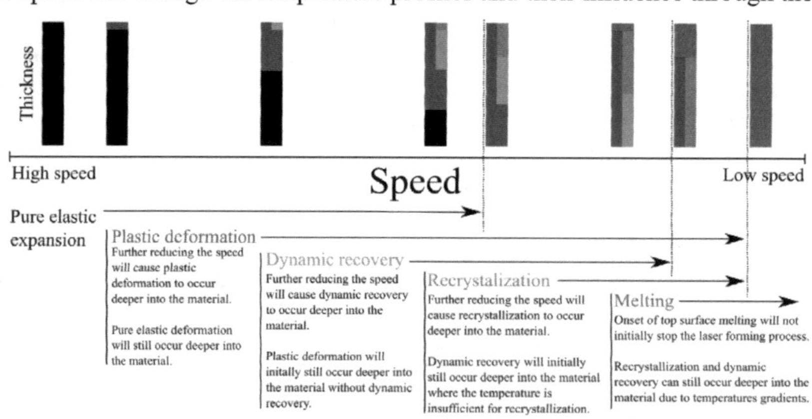

Figure 2. A one-dimensional model of the influence of the speed. Note that the position and occurrence of overlap between different factors is dependent on other process settings and the total thickness of the part. Similar models can be made for other process parameters as well.

Stage 1

The increasing temperature drives the increase in tensile thermal strains. As the thermal expansion is resisted by the surrounding material, the material experiences an elastic compressive response causing compressive elastic stresses and strains. No yielding occurs during stage 1.

Stage 2

As the thermal strains increases, the compressive transverse stresses increase to beyond the thermally reduced yield stress, resulting in plastic compressive stresses and strains. During this stage, depending on the temperature two mechanisms influence the hardening:

1. Increasing thermal strains cause increasing compressive stresses which leads to further work hardening of the material.
2. Increasing temperature leads to dynamic recovery and possible recrystallization, thus reducing hardening.

Stage 3

As the temperature starts to decrease, the thermal strains start to decrease as well. This leads to contraction, which in turn changes direction of the stresses to tensile. As the material contracts, the tensile stresses may increase to surpass the tensile yield strength. If one considers the kinetic hardening, the tensile yield strength may have been reduced by the compressive yield flow only moments before.

Another consideration is that the rate of cooling can lead to martensite formation. As martensite has a larger volume, the residual stresses can even remain compressive.

Hypotheses

The model in Figure 1 leads to several interesting hypotheses:

Firstly, traditional bending theory with an anti-symmetrical distribution of tensile and compressive strains around a neutral axis positioned halfway through the thickness cannot be applied. The position of the neutral axis is dependent on the depth of the heat affected zone i.e., what portion of the material is under stage 1-3. Furthermore, most of the plasticity occurs at elevated temperatures at the surface. This means that anti-symmetry cannot be expected. From a macroscopic point of view, limited stretching of the bottom surface may occur during TGM as part of the equilibrium during the cooling of the upper surface. An interesting consequence of this is that laser forming is primarily driven by thickening of the part even under TGM. The first hypothesis can be stated as:

1. There is a significant top surface thickening in laser forming, especially when using smaller bend radii.

Secondly, the most significant hardening will not necessarily occur at the surface despite this being the position of the most plastic deformation, because increasing temperatures can lead to dynamic recovery and recrystallization during stage 2. This in turn creates a question of where the limiting hardness does occur. As the temperature and especially the temperature profile, meaning gradients, can be modified through the process settings. It becomes apparent that the hardness distribution can be altered through the process settings. This leads to the second hypothesis:

2. The through-thickness hardness distribution can be influenced by process parameters.

Hardening of the material is known for influencing the bend rate decay [3]. Considering that the most plastic deformation occurs at the top surface, this would, without dynamic recovery, increase the top surface hardness; thereby, causing the bend rate decay. However, with dynamic recovery, the increase in hardness occurs at a shifted position through the thickness. As the plastic deformation is smaller at the shifted position; therefore, the effect on the bend rate decay is reduced. This leads to the third hypothesis:

3. Shifting the hardness distribution affects the laser forming bend behavior.

3. Experimental setup

The experiments were designed and conducted at Aalborg University's laser cell, the setup can be seen in Figure 3. The material used was AISI 304/ EN 1.4301 with thickness of 1 and 3 mm. All samples were 100 mm in length and 50 mm in width with clamping length of 20 mm and free length of 80 mm, as depicted in Figure 3.

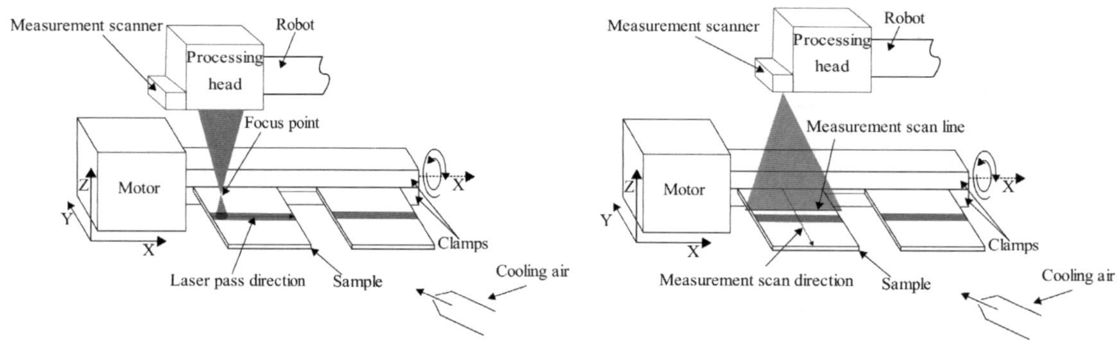

Figure 3. Experimental setup of the laser forming process. There are two moving parts in the setup, a Beckhoff AM8112 motor, and a KUKA KR 120. The motor provides rotation along the X axis with the intent of tilting the samples. The HighYang (modified) processing head is mounted on the robot which controls the motion of the laser beam. A Wenglor MLWL153 measurement scanner is mounted on the side of the processing head. The samples are air cooled from the bottom. The laser used is IPG YLS-3000 SM.

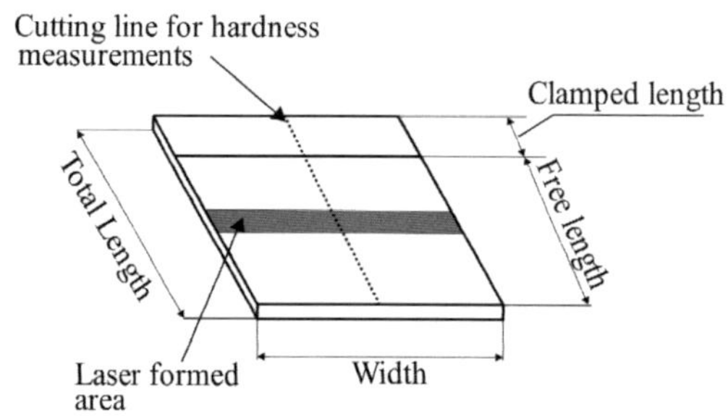

Figure 4. Dimensions for the samples. Total length is 100 mm, width is 50 mm. Clamped length is 20 mm and the free length is 80 mm. Sample's thickness is 1 and 5 mm.

All samples were laser formed, with the direction of the laser beam perpendicular to the sheet's rolling direction. Beam diameter is equal to the sample's thickness. For each thickness, several parameter sets were generated, the parameter sets can be seen in Table 1. Each parameter set consists of 30 laser passes and is repeated five times, after each iteration of the parameter set a measurement scan of the surface is performed to note down the bend angle. To reduce the influence of the geometrical distortion of the laser beam incident due to the developing bend angle, the samples are pre-tilted at 7.5°. For every 30° of bending, the sample will be tilted with an additional 15°; hence, from 0 to 30° bend angle the sample is tilted at 7.5°, from 30 to 60° the sample will be tilted to 22.5° and from 60 to 90° the sample will be at 37.5°.

After laser forming, three samples with 3 mm thickness will be cut midsection along the length, as seen in Figure 4, and prepared for micro-hardness testing. Three hardness measurements per sample were performed using Struers Duramin-40 and HV0.1 values were obtained with dwell time of 15 s and a load force of 0.9806 N. For reference measurements, the midsection of a non-laser formed sample was used. The midsection thickening of each sample was measured using a Wild Heerbrugg stereo microscope together with a AxioCam ICc 5 camera.

Table 1. Process parameter sets. The naming convention is [sample thickness]-[parameter set number], 1-1 will be parameter set 1 for thickness of 1 mm.

Parameter set	Laser diameter [mm]	Power [W]	Scan speed [mm/min]	Number of passes per iteration	Dwell time between passes [s]
1-1	1	120	3461	30	30
1-2	1	90	1442	30	30
3-1	3	270	1154	30	60
3-2	3	180	320	30	60
3-3	3	180	240	30	60

4. Results and Discussion

In this section the results from the experimental part of the study will be presented and discussed. The bending rate for 1 and 3 mm samples can be seen in Figure 5. The results show that even though there is an increase in the overall bend, the more a parameter set is used, the lower the bend rate becomes. From Figure 5 it can also be seen that every time the parameter set is changed there is an initial increase in the bending rate. To investigate the change in the bending rate, the hardness distribution of the samples is investigated.

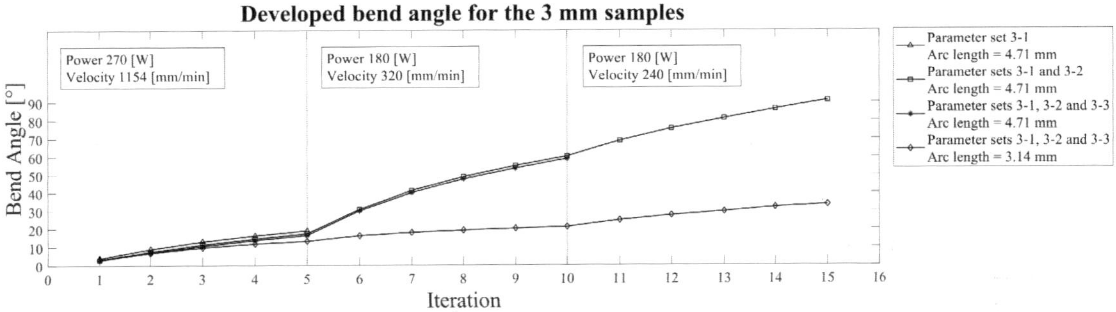

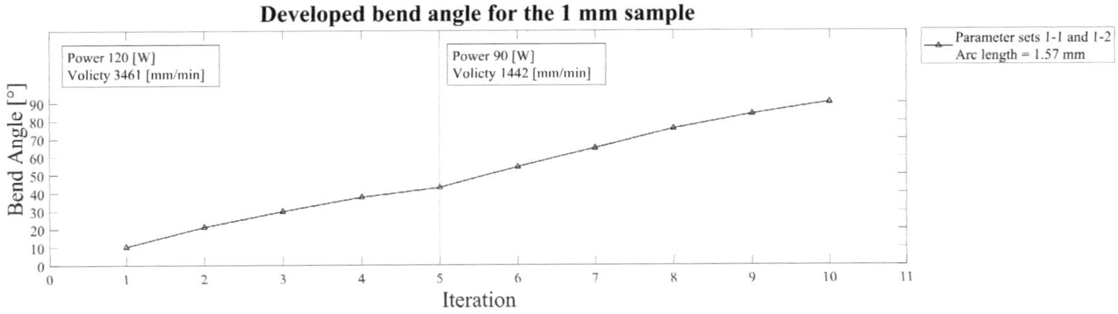

Figure 5. Developed bend for the 1 and 3 mm plates. The red vertical lines are separators for the parameter sets.

Furthermore, it is observed that the arch length has a significant influence over the bend rate and the maximum obtainable bend. When examining Figure 5, there are two samples that are being laser formed with parameter sets 3-1, 3-2 and 3-3; however, they exhibit vastly different final bend angles. The first sample has a final bend angle of 91.1°and the second has 33.4°. The difference between the two samples is the dwell time, 60 seconds and 30 seconds and the arch length, 4.71 and 3.14 mm, respectively. Of the two differences, only the decrease of 1.57 mm in arch length is deemed influential for the bend reduction. The only potential influences the lower dwell time might have is increased peak temperature with each subsequent laser pass. The increase of peak temperature will result in an increased bend rate or potentially burning the material, neither of which was observed for the sample with the 33.4° bend.

Figures 6 and 7 show the hardness distribution of both the reference and laser formed samples. There is good correlation between the hardness results and the theoretical section of the study, as the most significant hardening does not occur at the top surface, close to the laser incident despite the region having the most plastic deformation. Figures 6 and 7 show that the hardness distribution shifts deeper into the material, when changing process settings. This can explain the increased bend rate, the change of settings will result in modifying the temperature profile through the thickness, as it can be seen in Figure 8. The increased temperature in the upper midsection of the sample will lead into expanding the HAZ deeper into the sample's thickness which causes an increased bend rate.

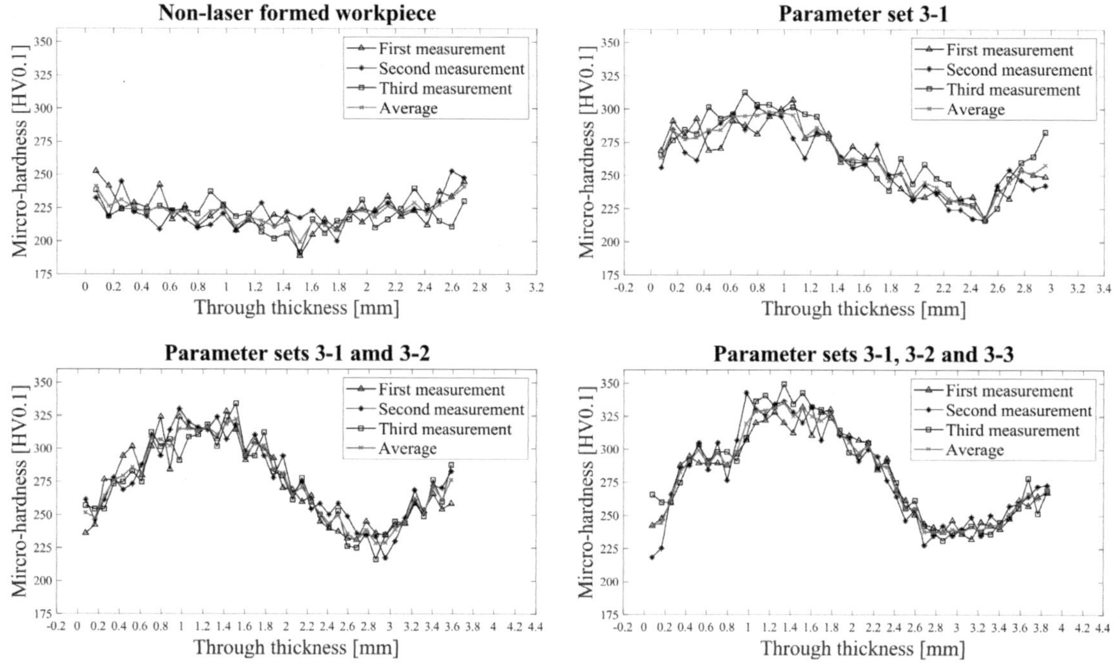

Figure 6. Micro-hardness measurements of the 3 mm samples with arch length of 4.71 mm. Parameter set 3-1 power 270 W, scan speed 1154 mm/min. Parameter set 3-2 power 190 W, scan speed 320 mm/min. Parameter set 3-3 power 190 W, scan speed 240 mm/min.

In both Figure 6 and Figure 7, all samples seem to follow the same hardness distribution trend. From the top surface, the hardness increases and peaks away from it, it then drops to levels similar to the reference material before slightly increasing again towards the bottom surface. The drop in hardness can be an indication of a neutral axis, that seems to shift deeper into the material when changing process settings. Thereby, the increase in hardness at the bottom surface would be from strain hardening, induced by the bending moment during laser forming. This bending moment is believed to occur during cooling of the HAZ, when the material contracts and the modulus of elasticity and yield strength of the HAZ begins to increase again. If the flow stresses have surpassed the yield strength of the material, flow may begin to occur on the bottom side. This would mean that the yielding was tensile in nature. None of the samples exhibit any significant oxidation on the bottom surface that might indicate elevated temperatures which could result in hardening due to either phase transformation or grain refinement.

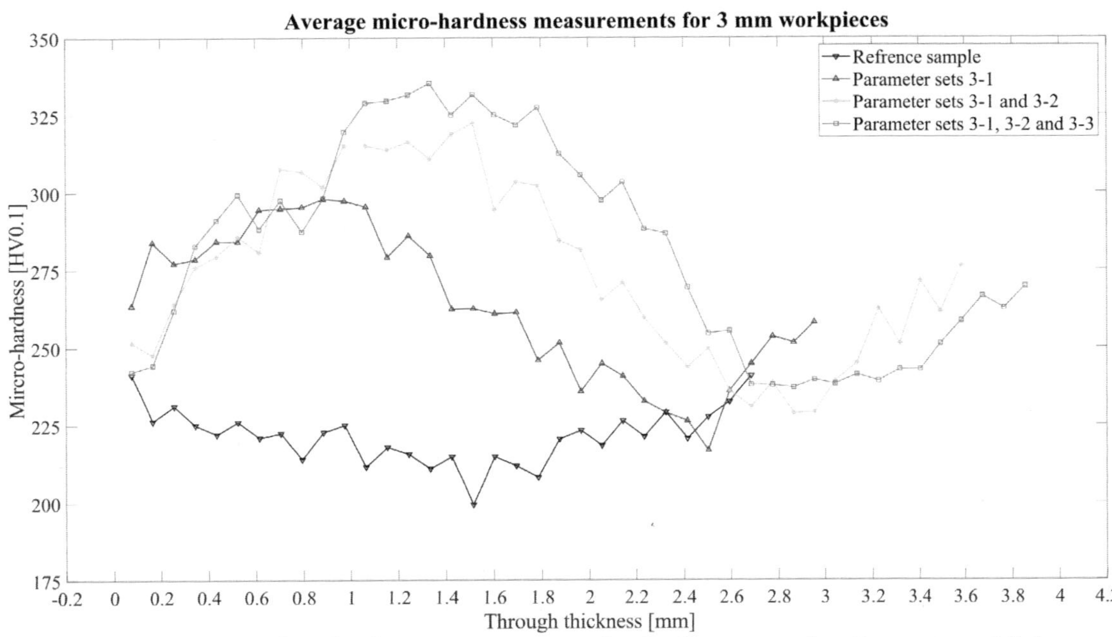

Figure 7. Average micro-hardness measurements for the 3 mm samples. Parameter set 3-1 power 270 W, scan speed 1154 mm/min. Parameter set 3-2 power 190 W, scan speed 320 mm/min. Parameter set 3-3 power 190 W, scan speed 240 mm/min.

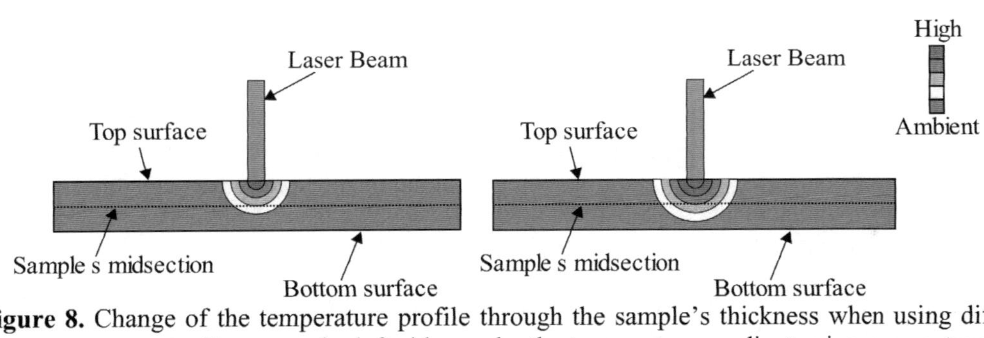

Figure 8. Change of the temperature profile through the sample's thickness when using different process parameter sets. For example, left side can be the temperature gradient using parameter set 3-1 and on the right parameter set 3-2.

Significant section thickening is observed in the laser formed 3 mm samples that were cut for the hardness measurements, with thickness increasing by 34% in the sample formed with parameter sets 3-1, 3-2 and 3-3. The section thickening can be seen in Figure 9 and the measurements can be seen in Table 2. The increased section thickening needs to be taken into account when considering process parameters, as settings for a 3 mm thick section may not work as intended for a 4 mm thick section. Furthermore, this can lead to some unintended consequences when designing part tolerances and may require machining.

Table 2. Section thickening of the HAZ for 3 mm samples.

Bend Angle [°]	Section thickness	Increase compared to nominal thickness [%]
19.00	3.27	9
57.40	3.75	25
91.10	4.04	34

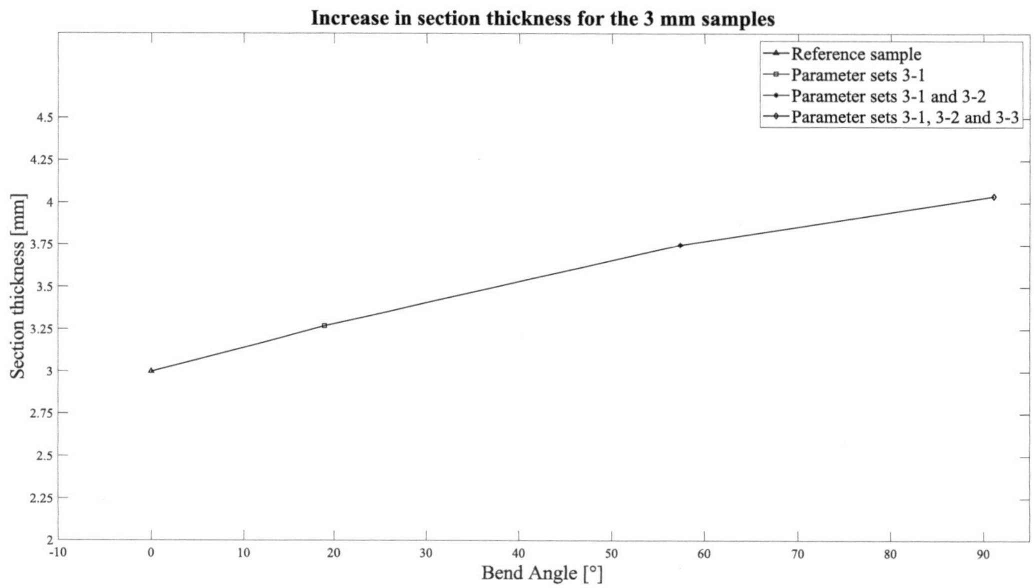

Figure. 9. Cross section thickness for the 3 mm samples laser formed using parameter sets 3-1, 3-2, 3-3. Parameter set 3-1 power 270 W, scan speed 1154 mm/min. Parameter set 3-2 power 190 W, scan speed 320 mm/min. Parameter set 3-3 power 190 W, scan speed 240 mm/min.

5. Conclusion

This study has examined the hardening kinetics of the laser forming process using TGM. ANSI 304 stainless steel samples of 1 and 3 mm thickness have been laser formed, using different sets of parameters, up to 90° and the cross-section hardness distribution of the 3 mm samples has been examined. From the results and discussion section of the study, several things can be concluded:

- The most significant hardening does not occur close to the top surface. The hardening distribution shifts deeper into the material due to the change of the temperature profile in result of changing the process parameter.
- Changing process parameters to gradually increase the amount of energy introduced to the sample will result in increase of the bend rate, which can translate into higher maximum bend angles. The hardness is responsible for the bend rate decay, by altering the hardness distribution, the bend rate decay is influenced. By reducing the bend rate decay, higher bend angles can be achieved.
- There have been a significant section thickening in the 3 mm samples when the bend radius is equal to the thickness. This needs to be taken into consideration when planning process parameters for higher bend angles, e.g., the for the 90° 3 mm sample, the section thickness has increased to 4 mm; hence, using the parameters designed for 3 mm cross section might not work as well for a 4 mm one due to the through thickness temperature gradient not being the same.
- Hardness increase have been observed close to the bottom surface, this can be an indication of strain hardening which can be a result of limited stretching occurring in the region while cooling.
- The hardness drop between the top and bottom surface can be an indication of a neutral axis, similar to the one described in conventional sheet metal forming.
- The arc length of the bend has shown a significant influence over the bend rate. Decreasing the arch length for the same bend angle and processing parameters, will result in increased local hardening, due to heating a smaller area.

This study has taken a more holistic approach to show that there is a correlation between the process settings and the hardening of the material and its influence on the bend behavior and bend rate decay. The study has shown that there is a potential of examining different processing schemes and their time cost effectiveness based on bend development.

For future work, it will be of interest to examine samples with varying arc lengths and its influence over the hardness distribution, section thickening and bending rate. Another topic of interest can be to investigate the bend development rate when using multiple parameter sets, as done in this study, and compare it to the bend rate of a sample that is only formed using the last parameter set. The focus here will be to see if the maximum obtainable bend is influenced by the chosen processing scheme. Lastly, the change in material strength, in the bend, due to section thickening compared to conventional sheet-metal forming could a topic of investigation.

Acknowledgments

This study was supported by the Innovation Fund Denmark through the INTERLASE project with project number 7050-00024B. The experimental equipment used for the experiments was partly supported by the Poul Due Jensen Foundation.

ORCID IDs

G N Nikolov https://orcid.org/0000-0001-5161-9223
A N Thomsen https://orcid.org/0000-0003-4865-1032
M Kristiansen https://orcid.org/0000-0001-9652-3348

References

[1] Geiger M and Vollertsen F 1993 The Mechanisms of Laser Forming *CIRP Ann. - Manuf. Technol.* **42** 301–4
[2] Shi Y, Yao Z, Shen H and Hu J 2006 Research on the mechanisms of laser forming for the metal plate *Int. J. Mach. Tools Manuf.* **46** 1689–97
[3] Edwardson S P, Griffiths J, Dearden G and Watkins K G 2010 Temperature Gradient Mechanism: Overview of the multiple pass controlling factors *Phys. Procedia* **5** 53–63
[4] Dixit U S, Joshi S N and Kant R 2015 Laser forming systems: A review *Int. J. Mechatronics Manuf. Syst.* **8** 160–205
[5] Lawrence J, Schmidt M J J and Li L 2001 The forming of mild steel plates with a 2.5 kW high power diode laser *Int. J. Mach. Tools Manuf.* **41** 967–77
[6] Hu Z, Kovacevic R and Labudovic M 2002 Experimental and numerical modeling of buckling instability of laser sheet forming *Int. J. Mach. Tools Manuf.* **42** 1427–39
[7] Edwardson S P, Abed E, Bartkowiak K, Dearden G and Watkins K G 2006 Geometrical influences on multi-pass laser forming *J. Phys. D. Appl. Phys.* **39** 382–9
[8] Safari M, de Sousa R A and Joudaki J 2020 Recent advances in the laser forming process: A review *Metals (Basel).* **10** 1–19
[9] Shen H, Shi Y and Yao Z 2006 Laser forming of plates using two sequent scans of different intervals *Proc. Inst. Mech. Eng. Part C J. Mech. Eng. Sci.* **220** 507–11
[10] Lambiase F, Di Ilio A and Paoletti A 2016 Productivity in multi-pass laser forming of thin AISI 304 stainless steel sheets *Int. J. Adv. Manuf. Technol.* **86** 259–68
[11] Yang L J, Tang J, Wang M L, Wang Y and Chen Y B 2010 Surface characteristic of stainless steel sheet after pulsed laser forming *Appl. Surf. Sci.* **256** 7018–26
[12] Thomsen A N, Kristiansen M, Kristiansen E and Endelt B 2020 Online measurement of the surface during laser forming *Int. J. Adv. Manuf. Technol.* **107** 1569–79
[13] McBride R, Bardin F, Gross M, Hand D P, Jones J D C and Moore A J 2005 Modelling and calibration of bending strains for iterative laser forming *J. Phys. D. Appl. Phys.* **38** 4027–36

IOP Conf. Series: Materials Science and Engineering 1135 (2021) 012007 doi:10.1088/1757-899X/1135/1/012007

Laser-TIG welding of galvanized steel – numerical and experimental assessment of the effect of arc in various setups

Hana Šebestová[1], Petr Horník[1], Jan Novotný[1], Libor Mrňa[1]

[1]Institute of Scientific Instruments of the Czech Academy of Sciences, Královopolská 147, 612 64 Brno, Czech Republic

sebestova@isibrno.cz

Abstract. The technology of laser-TIG welding utilizes the arc as a secondary heat source during laser welding. In TIG-leading configuration, the low-current arc precedes the beam to preheat the material. The numerical simulations representing various setups combining laser and arc were performed to study the changes of thermal cycles on the interface of thin metal sheets of overlap joint. The relations between the position of the arc towards the beam, additional heat input, and temperature gradients are discussed. The technology of laser-TIG welding of zinc-coated deep-drawing steel was experimentally applied in the same joint configuration. A good agreement between the calculated and experimental welds was achieved. The arc current less than 40 A did not cause the vaporization, neither oxidation of zinc coating on the interface surface of metal sheets. Nevertheless, the quality of laser-TIG welds was better compared to laser welds. The 40A arc current increased the heat input by about 50% and led to an almost 60% decrease in cooling rate compared to autonomous laser welding. Prolonged heating and cooling time are the key factors of improving the weld quality.

1. Introduction

Galvanized steel is widely used in the automotive industry for body-in-white production. However, laser welding of galvanized steel is still challenging. Welding defects like spatter, blowholes, or inner porosity occurs. Kim et al. [1] demonstrated that the problematics of defect formation during laser welding concerns especially overlap joints. In the case of bead-on-plate welding, good-quality welds were produced. The problems originate from a low boiling point of zinc. The vaporized zinc escape from the free material surface to the surroundings before the melting point of steel is reached. During the overlap welding, the zinc vaporizes also from the interface surfaces. The vapor has very limited space and time to exhaust during laser welding in a gap-free configuration. Therefore, some vapor is trapped by the melt pool when the beam penetrates through the interface of metal sheets. As the melt pool heats up, the zinc vapor expands. The zinc bubbles explode through the melt pool and/or escape through the keyhole usually destabilizing it (Norman et al. [2], Pan and Richardson [3]). The liquid metal is ejected from the weld pool and produces spatter, blowholes, and other weld defects (Kim et al. [1], Yang et al. [4], Ma et al. [5]). Yang et al. [6] recommend shielding with the addition of active gas (O_2 and CO_2) to enlarge and stabilize the keyhole so as the zinc vapor could escape through the keyhole. However, this approach modifies weld metal chemical composition determining its mechanical properties. Similar cons appear when welding with interlayer metal foil or powder (Zhou et al. [7]).

Mechanical or chemical methods of removing the galvanized layer expense the production. Effective methods of a larger gap between the sheets (Bley et al. [8], Ozkat et al. [9]) require sheets pre-processing and they are often problematic to be implemented in industry. Iqbal et al. [10] proposed a laser-cut kerf for zinc vapor ventilation before the welding itself. However, the kerf surface oxides increase oxygen content in further weld metal (Frostevarg et al [11]). Many other alternative approaches are summed up in Yang et al. [4] or more recently in Wang et al. [12].

Methods of preheating seem more effective. Production time is not increased. Based on the preheating peak temperature, the zinc melts (420 °C) or even evaporates (907 °C) and escape before laser beam impact. The melted zinc oxidizes rapidly. The zinc oxide (ZnO) melting point (1975 °C) is higher than the melting point of steel (about 1500 °C). Thus, it does not negatively affect the melt pool, as proved by Yang et al. [4]. What is more, the coupling efficiency of laser power is higher on oxidized surfaces thanks to the decreased reflectivity (Steen [13]). Thus, a deeper penetration can be achieved.

Only a local preheating is required not to decrease the protective purpose of zinc coating. Tandem laser beams (generated separately or split from the same source) or a laser with a shaped beam (Xie and Denney [14], Gäbler [15]) can be applied. These technologies are very sophisticated and effective. However, special welding facilities and welding heads are required. Laser welding supplemented with arc preheating provides a much cheaper solution. The welding electrode can be fixed to the current laser welding head easily and the power source can be carried by the positioning robot (Šebestová et al. [16]).

Laser-TIG technology combines laser beam with tungsten inert gas (TIG) welding. Although the laser-TIG technology is already known for a few decades, it is still somewhat challenging since the effect of arc is strongly dependent on individual setup and processing parameters. It allows modifying the weld shape, its microstructure, and mechanical properties. The setup is variable. The arc helps to bridge the gap in butt welding. The increase in the total power of welding technology is also beneficial. Therefore, it is a suitable technology for many industries.

In our research, we focus on galvanized thin sheet overlap laser-TIG welding with a low-current arc. Based on the numerical simulations of low-carbon deep-drawing steel welding, we evaluate the effect of position and heat input of arc on thermal cycles on the interface of metal sheets and discuss the possible behavior of zinc coating. Numerical simulations are supplemented with experiments performed with equivalent material. The aim is to distinguish whether the positive effect of arc addition results from the modification of the galvanic layer or from the overall increase in heat input.

2. Method and experimental setup

The technology of laser and laser-TIG welding in TIG-leading configuration were experimentally investigated during overlap welding of cold-rolled galvanized extra deep-drawing steel grade A14 (WSS-M1A365-A14-50G/50G). The average thickness of the zinc coating measured by PosiTector 6000 was 12 μm. The scheme of laser-TIG method and experimental setup are displayed in figure 1.

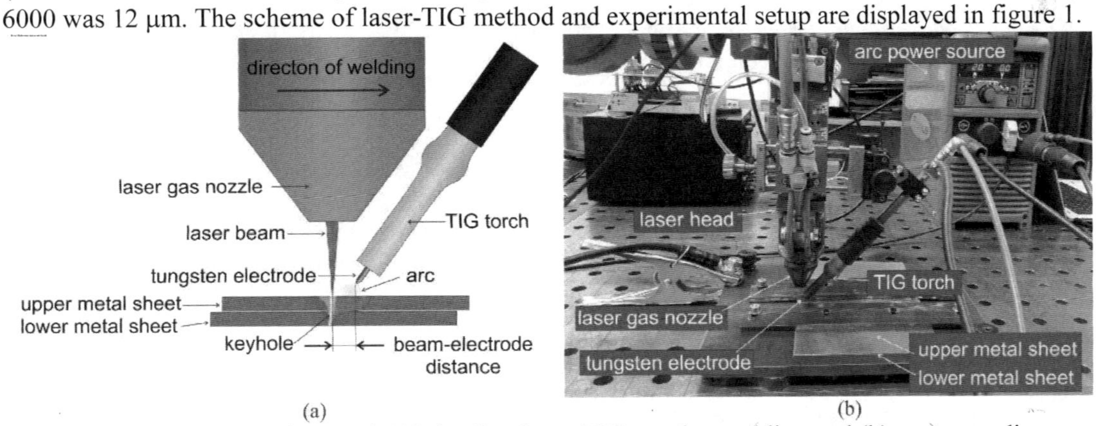

(a) (b)

Figure 1. (a) The scheme of TIG-leading laser-TIG overlap welding and (b) corresponding experimental setup.

The IPG YLS2000 fiber laser was used in our experiments. The laser beam was delivered by a 200μm optical fiber to the Precitec YW30 processing head. It was focused 1 mm under the metal sheet surface

IOP Conf. Series: Materials Science and Engineering 1135 (2021) 012007 doi:10.1088/1757-899X/1135/1/012007

and reached 0.4 mm spot diameter at the surface. The ABB IRB 2400 robot was used for beam positioning. The Fronius MagicWave 1700 Job TIG power source was employed.

The TIG torch was fixed with a special holder to the laser welding head. It was tilted off 45 degrees towards the beam axis. The cathode with a 2.4mm diameter was used. The tip of the cathode was placed 2 mm above the upper metal sheet surface and preceded the beam by 2 mm. The sheets with a thickness of 0.9 mm were fixed in the clamping jig in the gap-free configuration.

Argon 6.4 shielding gas, laser power 800 W, and welding speed 20 $mm \cdot s^{-1}$ were kept constant in the experiments. The applied arc current was 20 A, 30 A, and 40 A. Laser-TIG welds were compared to the laser weld prepared with the same laser power and welding speed (delivered heat per unit length 50 $J \cdot mm^{-1}$). Note that the absorbed energy (heat input) is lower than that delivered. It is dependent on the absorption efficiency which is generally lower for arc compared to the laser welding technology.

Further, the autonomous TIG was used in the experimental setup of laser-TIG welding described above. The arc current 40 A and 80 A were applied at welding speed 20 $mm \cdot s^{-1}$ to evaluate the effect of arc heat on zinc coating oxidation.

3. Numerical simulation of laser-TIG welding

Commercial software ESI SYSWELD was used for transient finite element analysis of laser and laser-TIG welding corresponding to the setup displayed in figure 1a. This software does not provide a complete physical simulation of the welding process. However, it respects the thermal dependency of the physical and mechanical properties of a welded alloy. Concerning moving heat source, it calculates a temperature field in every time step of the simulation. It integrates the metallurgical phase transformations during the heating and cooling and predicts also weld deformations and mechanical properties. Such simulation allows us to estimate the effect of preheating source which is often problematic to be evaluated experimentally, especially in lap joint configuration.

The pre-defined 3D conical model of heat source (figure 2) characterized by power density function of gaussian type (equation (1)) is well adapted to describe thermal load to the material during laser welding. The volumetric power density Q in a node with coordinates (x, y, z) can be expressed as follows:

$$Q(x,y,z) = Q_0 \exp\left[-\frac{r^2}{r_0^2(z)}\right] \quad (1)$$

where Q_0 is the maximum value of volumetric power density and

$$r^2 = x^2 + y^2, r_0 = r_e - \frac{r_e - r_i}{z_e - z_i}(z_e - z). \quad (2), (3)$$

The gaussian parameters r_e and r_i are upper and lower cone radius, respectively. The difference $z_e - z_i$ represents the depth of the melt pool. These parameters should be optimized to achieve the required shape of the weld metal.

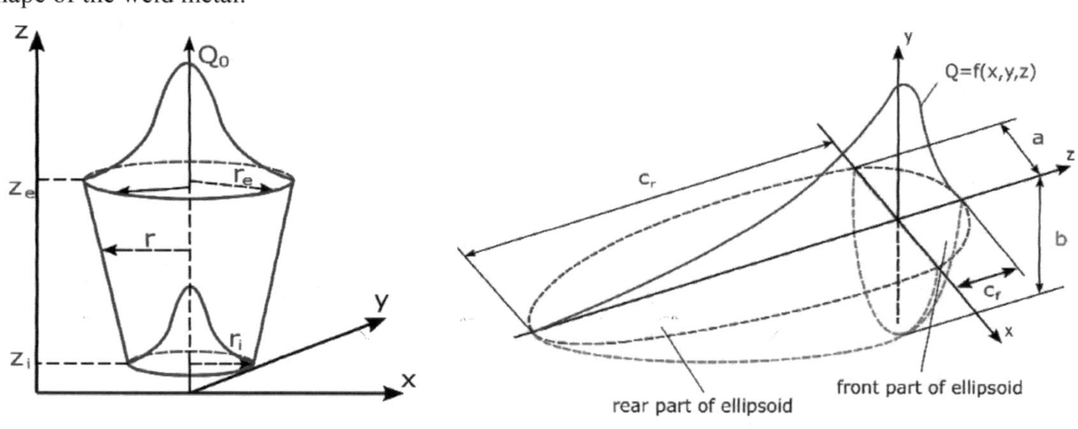

Figure 2. The 3D conical model of heat source suitable for laser welding simulation [18].

Figure 3. Goldak's double ellipsoidal model of heat source suitable for TIG welding simulation [18].

Goldak's double ellipsoidal heat source model (figure 3) is used for TIG welding simulation. This model is built of two ellipsoids placed perpendicular to each other. The model is described by two equations, separately for the front and rear part of the ellipsoid:

$$Q_{\mathrm{f}}(x,y,z)=\frac{6\sqrt{3}f_{\mathrm{f}}Q}{abc_{\mathrm{f}}\pi\sqrt{\pi}}\exp\left[-3\left(\frac{x^2}{a^2}+\frac{y^2}{b^2}+\frac{z^2}{c_{\mathrm{f}}^2}\right)\right] \tag{4}$$

$$Q_{\mathrm{r}}(x,y,z)=\frac{6\sqrt{3}f_{\mathrm{r}}Q}{abc_{\mathrm{r}}\pi\sqrt{\pi}}\exp\left[-3\left(\frac{x^2}{a^2}+\frac{y^2}{b^2}+\frac{z^2}{c_{\mathrm{r}}^2}\right)\right] \tag{5}$$

where Q_{f} and Q_{r} are the volumetric power density in the front and rear part of the model, respectively, Q is the total introduced power, and a, b, c_{f}, and c_{r} are the width, depth, and length of the front and rear part of the estimated melt pool, respectively. The parameters f_{f} and f_{r} characterize the portion of total heat transferred to the front and rear part of the ellipsoid. For more details, see [17] or [18].

3.1. Geometrical model, material, and simulated welding conditions

The geometrical model of the simulated overlap joint is shown in figure 4. The total width of the model is 95 mm. The weld length is 15 mm. The two 0.9mm thick sheets are separated by the 0.05mm gap in the model. It represents a possible air gap that is always present because of sheet surface roughness or and/or dust presence.

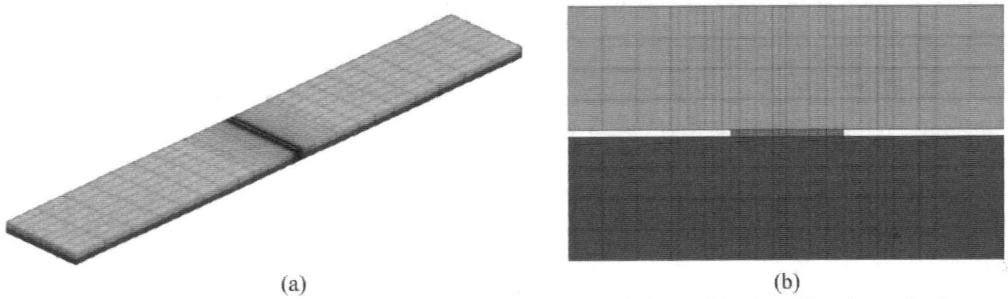

Figure 4. (a) Geometrical model of simulated overlap joint. (b) Detail of mesh in a plane perpendicular to the welding trajectory.

Based on the materials available in the database of SYSWELD, the low-carbon deep-drawing steel DC04 was chosen for the simulations. The prescribed chemical composition is concluded in table 1. This steel is equivalent to the alloy used in the experiment. The galvanic layer itself is not included in the model because of its very small thickness (typically up to few tens of microns). Therefore, its effect on the temperature field can be neglected. However, the behavior of possible galvanic layer can be discussed relative to the calculated sheet surface temperatures.

Table 1. Chemical composition of DC01 used in simulations (in weight percent).

C [%]	Mn [%]	Si [%]	S [%]	P [%]	Al [%]
≤ 0.08	≤ 0.4	≤ 0.1	≤ 0.025	≤ 0.025	≥ 0.02

The effect of beam-electrode distance (BED) and the effect of arc heat input were simulated at constant laser heat input per unit length 33 J·mm^{-1} and welding speed 20 mm·s^{-1}. Temperature field distribution and thermal cycles of laser-TIG welds are discussed relative to laser weld.

3.2. The effect of TIG position on temperature field distribution during laser-TIG welding

IOP Conf. Series: Materials Science and Engineering 1135 (2021) 012007 doi:10.1088/1757-899X/1135/1/012007

The three different positions (3 mm, 5 mm, and 7 mm) of the electrode towards the beam were simulated at constant arc heat input 17 J·mm^{-1} (arc current 40 A). The thermal analysis for autonomous TIG (17 J·mm^{-1}) was also calculated to compare corresponding thermal cycles. Figure 5 shows the temperature field distributions for different setups at the same time moment.

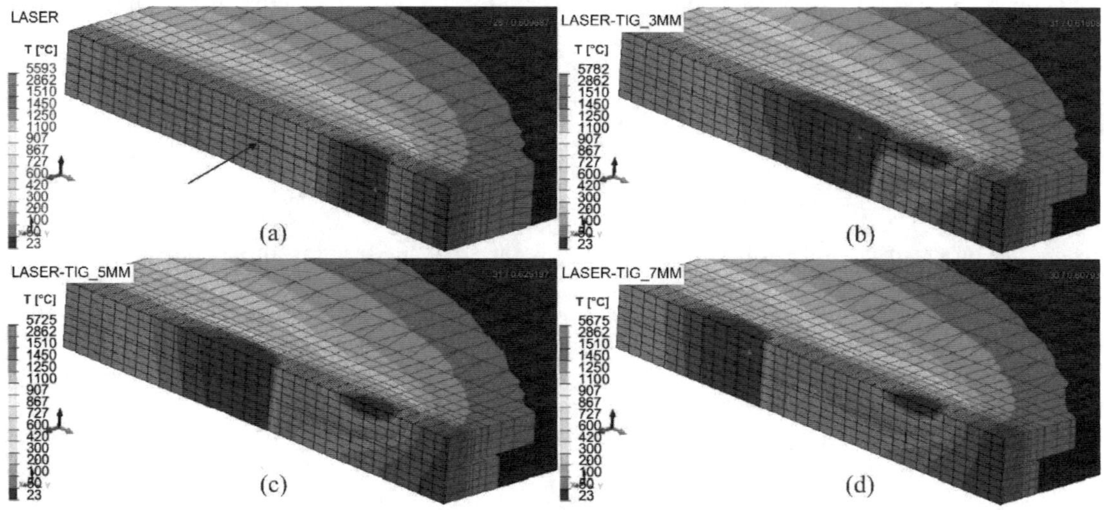

Figure 5. Temperature field distribution during (a) laser welding and laser-TIG welding with the electrode preceding the beam by (b) 3 mm, (c) 5 mm, and (d) 7 mm; clipping plane given by the beam axis and the trajectory of welding.

The thermal cycles in an axial node on the bottom sheet interface surface (marked with the black arrow in figure 5a) are displayed in figure 6 for laser, TIG, and laser-TIG welds. The autonomous TIG heats up the sheet interface up to 471 °C. The time above the melting point of zinc is 125 ms. The autonomous laser heats the interface from the melting to boiling point of zinc in 18 ms. The arc preheating with 3mm and 5mm BED allows prolonging this time to 74 ms and 171 ms, respectively. The 7mm BED would be sufficient for the zinc re-solidification before the laser passes. Increasing BED decreases both weld fusion and heat-affected zone since the welding processes become more separated.

Based on the thermal cycles, the cooling times $\Delta t_{8/5}$ (and corresponding cooling rates) in the range 800 °C to 500 °C were calculated. It reached 0.276 s (1086 °C·s^{-1}) for laser weld. The heat of the arc prolonged the $\Delta t_{8/5}$ to 0.672 s, 0.676 s, and 0.671 s for the BED of 3 mm, 5 mm, and 7 mm, respectively.

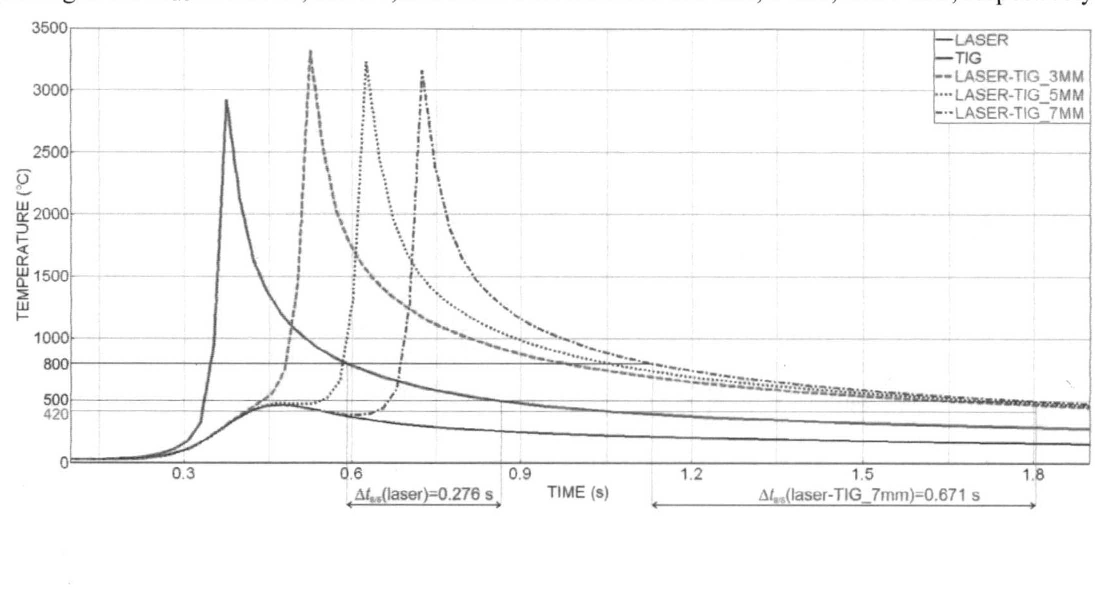

Figure 6. Thermal cycles of laser, TIG, and laser-TIG welding for different electrode-beam distances at the same arc heat input.

3.3. The effect of arc heat input on temperature field distribution during laser-TIG welding

The arc current 40 A, 60 A, and 80 A corresponding to the arc heat input 17 J·mm^{-1}, 26 J·mm^{-1}, and 34 J·mm^{-1}, respectively, were considered for the simulation of laser-TIG welding at constant BED (3 mm). Increasing heat input leads to the transformation of rather a combined laser-arc process into a hybrid process during which only one melt pool is formed (figure 7). The decrease of BED would have a similar effect.

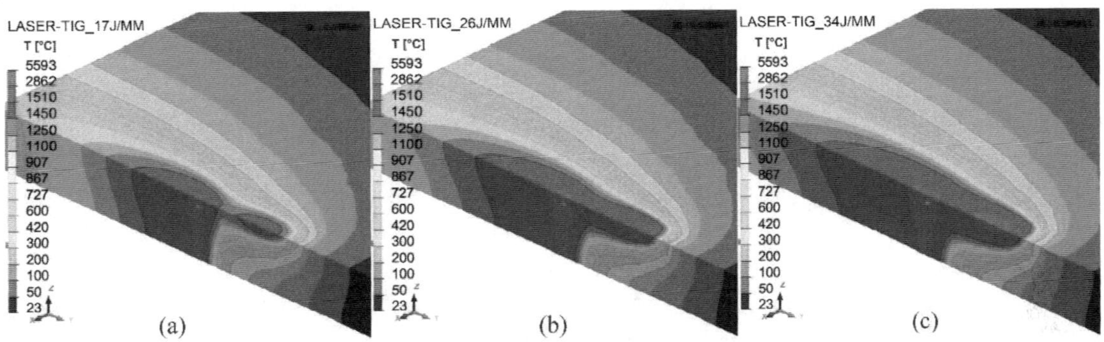

Figure 7. Temperature field distribution during laser-TIG welding with arc heat input (a) 17 J·mm^{-1}, (b) 26 J·mm^{-1}, and (c) 34 J·mm^{-1} at constant BED; clipping plane given by the beam axis and the trajectory of welding.

The width of a weld can be evaluated from the isotherms of peak temperatures on the weld cross-section (figure 8a–c). Increasing heat input increases both the fusion and the heat-affected zone (HAZ) width. The temperature range above the boiling point of steel (about 2862 °C) represents the keyhole formation in the simulation. The calculated keyhole opening is wider at higher arc heat inputs.

Figure 8d displays the effect of arc heat input on the thermal cycle of the node marked with the black arrow in figure 8a (the same node as marked in figure 5a). The cooling times $\Delta t_{8/5}$ (and corresponding cooling rates reached about 0.68 s (442 °C·s^{-1}), 0.90 s (334 °C·s^{-1}), and 1.16 s (260 °C·s^{-1}) for arc heat input 17 J·mm^{-1}, 26 J·mm^{-1}, and 34 J·mm^{-1}, respectively. The deeper penetration can also be expected since the peak temperature is higher at higher arc heat inputs. As well as the higher preheating temperatures can be reached at the same depth.

IOP Conf. Series: Materials Science and Engineering 1135 (2021) 012007 doi:10.1088/1757-899X/1135/1/012007

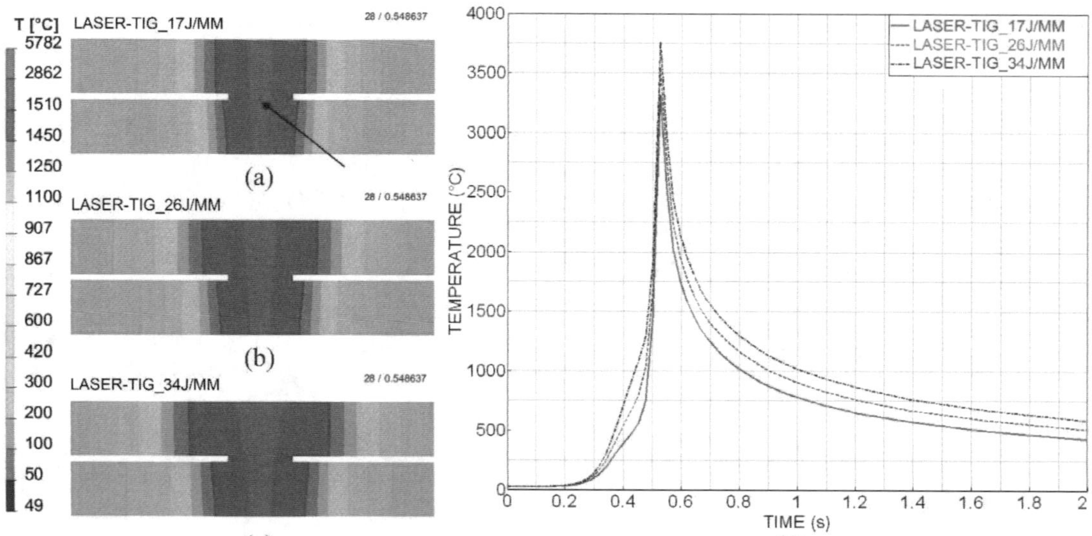

(c) (d)

Figure 8. The peak temperatures in the perpendicular cross-section in the middle of the weld length for arc heat input (a) 17 J·mm^{-1}, (b) 26 J·mm^{-1}, and (c) 34 J·mm^{-1}, and (d) thermal cycles of the same node for three different arc heat inputs at constant BED.

4. Experimental verification

Unacceptably large depressions and blowholes (up to a third of sheet thickness) of both face and root surface were observed in laser weld. Increasing arc current led to the smoothing of both surfaces of laser-TIG welds (figure 9). The face depression was present also in the case of laser-TIG welds. Nevertheless, it did not exceed 0.1 mm. Increasing heat input led to the widening of both the fusion zone and HAZ. The higher widening was detected in the upper metal sheet.

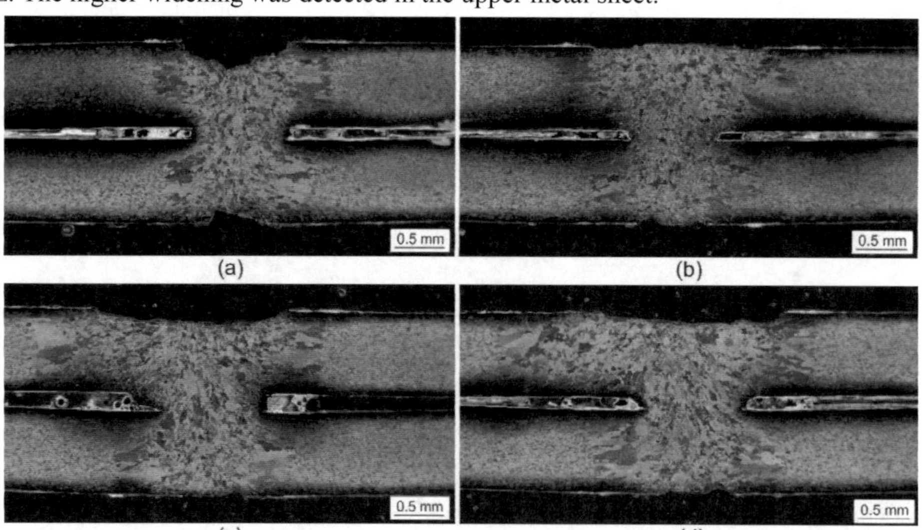

(c) (d)

Figure 9. The macrostructure of A14 overlap (a) laser weld and laser-TIG welds made with (b) 20 A, (c) 30 A, and (d) 40 A; cross-sections perpendicular to the direction of welding.

Laser-TIG welds were free of ZnO on the interface surfaces. However, the ZnO was present in the HAZ of both laser and laser-TIG welds on both the face and root side. Laser-TIG welds exhibited a narrower zone of oxides in the root region compared to laser weld.

Then, the autonomous TIG was applied in the same experimental setup with a disabled laser beam. The arc current 40 A was sufficient to melt the interface surface of the upper sheet, whereas the galvanized layer on the interface surface of the bottom sheet was only tempered. In the case of arc current 80 A, the steel sheets were fused by the melted zinc. Nevertheless, no eye-visible ZnO was present on the sheet interface after breaking the metal sheets apart (figure 10).

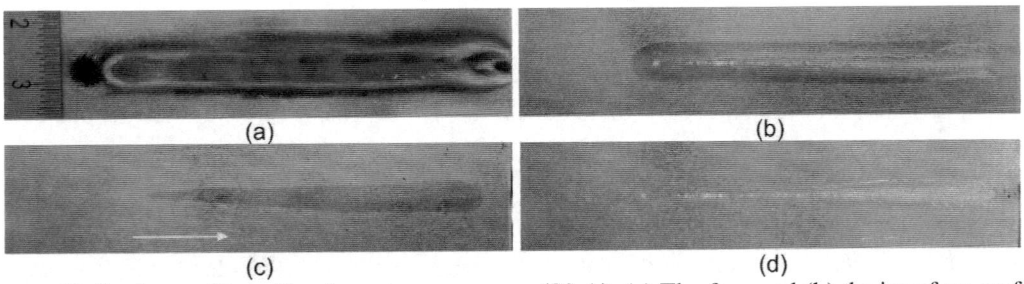

Figure 10. Surfaces affected by the autonomous arc (80 A). (a) The face and (b) the interface surface of the upper metal sheet, and (c) the root and (d) the interface surface of the bottom metal sheet. The yellow arrow represents the direction of welding.

5. Discussion

Laser welding is characterized by rapid heating and cooling. In our numerical simulation of laser overlap welding, the heating time (heating rate) and cooling time (cooling rate) in the temperature region 500–800 °C reached 0.014 s (22095 $°C \cdot s^{-1}$) and 0.276 s (1088 $°C \cdot s^{-1}$), respectively. The time interval between the melting and boiling point of zinc is only 18 ms on the interface of sheets. During the experiment, such rapid vaporization could destabilize the melt pool and promote the formation of welding defects. Laser-TIG welding in TIG-leading configuration was simulated to discuss the effect of local arc preheating on thermal cycles on the interface of metal sheets in lap joint.

First, the effect of BED was evaluated at constant both laser and arc heat input (33 $J \cdot mm^{-1}$ and 17 $J \cdot mm^{-1}$). Increasing BED decreased both weld fusion and heat-affected zone since the welding processes become more separated. The heating time between the melting and boiling point of zinc was calculated. It reached 74 ms and 171 ms for BED 3 mm and 5 mm, respectively. In the air environment, the melted zinc could oxidize producing the ZnO with a high melting point. Slower heating prolongs the time available for zinc oxidation or escape. The autonomous arc (17 $J \cdot mm^{-1}$) heated up the sheet interface up to 471 °C. The time above the melting point of zinc was 125 ms. According to the welding speed of 20 $mm \cdot s^{-1}$, the 7mm BED would lead to the re-solidification and/or oxidation of the melted zinc on the interface of metal sheets before the laser passes.

Second, the effect of arc heat input at constant BED (3 mm) is discussed. Increasing heat input increased both weld fusion zone and HAZ. The $\Delta t_{8/5}$ reached about 0.68 s, 0.90 s, and 1.16 s for arc heat input 17 $J \cdot mm^{-1}$, 26 $J \cdot mm^{-1}$, and 34 $J \cdot mm^{-1}$, respectively. Corresponding cooling rates were reduced to 442 $°C \cdot s^{-1}$, 334 $°C \cdot s^{-1}$, and 260 $°C \cdot s^{-1}$, respectively. The slower heating and cooling would promote a more fluent escape of the possible zinc vapor from the melt pool.

The isotherm representing the boiling point of steel could represent the keyhole width. However, the process of keyhole formation is more complex and this simulation does not cover complete physical phenomenon (material losses caused by evaporation, the pressure inside the keyhole, melt flow, the effect of shielding gas, etc.).

The values presented above were derived from the linear interpolation of calculated time-temperature data. According to the time step of simulation (23 ms) these values can be subject to relatively large error, especially in the heating period. Nevertheless, the found trends are unambiguous.

Finally, the experiment of overlap laser-TIG welding was performed. The BED was set to 2 mm. The arc was unstable at higher BEDs at low arc currents. Assuming 65% absorption efficiency, the experimental arc current 20 A, 30 A, and 40 A correspond to the arc heat input 9 $J{\cdot}mm^{-1}$, 13 $J{\cdot}mm^{-1}$, and 17 $J{\cdot}mm^{-1}$, respectively. The last mentioned was considered in the simulation. Laser-TIG welds reached a higher quality level compared to laser weld. On the other hand, the increasing arc current led to the widening of both the fusion zone and HAZ. The higher widening was observed in the upper sheet which corresponds to the results of simulations.

The simulation of autonomous TIG welding predicted that arc heat input 17 $J{\cdot}mm^{-1}$ should be sufficient to exceed the melting point of zinc on the interface surface of the bottom metal sheet. To validate this presumption, the corresponding experiment was performed. Nevertheless, 40 A did not lead to the melting of the zinc layer in this region. The calculated overheating is probably not sufficient for the phase transformation of zinc. Note that the latent heat of zinc melting is not included in the simulation.

The same experiment was performed at arc current 80 A which was already sufficient to join the metal sheets by the melted zinc. The metal sheets were broken apart to evaluate the formation of zinc oxide on their interface. No eye-visible ZnO (white powder) was present. We assume that the amount of oxygen between the sheets is not sufficient for the melted zinc oxidation, which is in contrast to the observations of Yang et al. [4]. The oxides were present only on open surfaces in the HAZ. The presence of ZnO on the face surface of the upper sheet indicates insufficient argon shielding.

6. Conclusions

Laser, TIG, and laser-TIG overlap welding of low-carbon deep-drawing steel were numerically investigated. We focused on the modification of thermal cycles on the interface of metal sheets by the variation of arc position and heat input and discussed the behavior of the possible zinc coating.

The autonomous low-current TIG does not vaporize the zinc on the interface of metal sheets. It only heats this region slightly above the melting point of zinc. The time for zinc oxidation is very limited. The TIG preceding the beam by 5 mm increases the heating time between the melting and boiling point of zinc almost ten times compared to autonomous laser welding.

Increasing arc heat input at the fixed beam-electrode distance modifies rather a combined process into a hybrid process with one melt pool. We showed that about a 50% increase in heat input leads to an almost 60% decrease in cooling rate compared to autonomous laser welding.

The effect of TIG heat input on galvanized steel A14 laser-TIG lap joints was experimentally verified. The surface defects were reduced compared to laser weld. A good agreement was found between the experimental and simulated weld cross-sections. Although the maximal applied pre-heating itself was sufficient for the zinc melting on the interface surfaces of metal sheets, no ZnO was found here.

The positive effect of TIG addition in our experiments is rather in increased heat input than in the modification of the zinc interface layer by the arc before the laser pass. Reduced both heating and cooling rates result in a less violent, smoother welding process.

Acknowledgments

This research was supported by the Technology Agency of the Czech Republic [TH04010366]. The institutional support of the Czech Academy of Sciences [RVO:68081731] is acknowledged.

ORCID iDs

H Šebestová https://orcid.org/0000-0003-4445-1324
P Horník https://orcid.org/0000-0002-9320-9561
L Mrňa https://orcid.org/0000-0002-3928-5619
J Novotný https://orcid.org/0000-0001-6175-869X

References

[1] Kim C, Choi W, Kim J and Rhee S 2008 Relationship between weldability and the process parameters for laser-TIG hybrid welding of galvanized steel sheet. *Mater. Trans.* **49** 179–86

[2] Norman P, Eriksson I and Kaplan A 2009 Monitoring laser beam welding of zinc coated sheet metal to analyze the defects occurring *Proc. 12th Nordic Laser Materials Processing Conference*, ed ED Mortensen (Kgs. Lyngby: ATV-SEMAPP)

[3] Pan Y and Richardson IM 2011 Keyhole behaviour during laser welding of zinc-coated steel *J. Phys.* D*: Appl. Phys*. **44** 045502

[4] Yang S and Kovacevic R. 2009 Laser welding of galvanized DP980 steel assisted by the GTAW preheating in a gap-free lap joint configuration *J. Laser Appl.* **21** 139–148

[5] Ma J, Kong F, Carlson B and Kovacevic R 2012 Mitigating zinc vapor induced weld defects in laser welding of galvanized high-strength steel by using different supplementary means *IntechOpen* chapter Welding processes, ed R Kovacevic

[6] Yang S, Carlson B and Kovacevic R 2011 Laser welding of high-strength galvanized steels in a gap-free lap joint configuration under different shielding conditions *Weld. J.* **90** 8–18

[7] Zhou S, Zhao Y, Peng Z and Ren F 2011 The investigation of laser lap welding process on high-strength galvanized steel sheets *ISRN Mech. Eng.* **2011** 923964

[8] Bley H, Weyand L, Luft A 2007 An alternative approach for the cost-efficient laser welding of zinc-coated sheet metal *CIRP Annals* **56** 17–20

[9] Ozkat EC, Franciosa P and Ceglarek D 2017 Laser dimpling process parameters selection and optimization using surrogate-driven process capability space *Opt. Laser. Technol.* **93** 149–64

[10] Iqbal S, Gualini MMS and Grassi F 2007 Laser welding of zinc-coated steel with tandem beams: Analysis and comparison *J. Mater. Proc. Technol.* **184** 12–18

[11] Frostevarg J, Qinglong P, Mizutani M, Kawahito Y and Katayama S 2018 Effects of edge oxides from laser cutting in laser-arc hybrid welding *J. Laser Appl*. **30** 012014

[12] Wang C, Lei B, Jiang P, Xu X and Mi G. 2018 Numerical and experimental investigation of vacuum-assisted laser welding for DP590 galvanized steel lap joint without prescribed gap *Int. J. Adv. Manuf. Technol.* **94** 4177–85

[13] Steen WM 2003 *Laser Material Processing* (London: Springer) 230

[14] Xie J and Denney P 2001 Galvanized steel joined with lasers *Weld. J.* **80** 59–61

[15] Gäbler F 2009 Optimizing automotive welding applications with CleanWeld: A unique approach to fiber laser welding *PhotonicsViews* **4** 52–5

[16] Šebestová H, Horník P, Mrňa L, Jambor M, Horník V, Pokorný P, Hutař P, Ambrož A and Doležal P 2020 Fatigue properties of laser and hybrid laser-TIG welds of thermo-mechanically rolled steels *Mater. Sci. Eng. A* **772** 138780

[17] Kik T 2020 Heat source models in numerical simulations of laser welding *Materials* **13** 2653

[18] Kik T and Górka 2019 J Numerical simulations of laser and hybrid S700MC T-joint welding *Materials* **12** 516

IOP Conf. Series: Materials Science and Engineering 1135 (2021) 012008 doi:10.1088/1757-899X/1135/1/012008

Influence of mill scale on oxygen laser cutting processes

Handika Sandra Dewi[1], Joerg Volpp[1], Jan Frostevarg[1], Jukka Siltanen[2]

[1]Department of Engineering Sciences and Mathematics, Luleå University of Technology, 971 87 Luleå, Sweden
[2]SSAB Europe Oy, Finland

contact@hsdewi.com

Abstract. Mill scale formed on the surface of hot rolled steels consists of magnetite (Fe_3O_4), hematite (Fe_2O_3) and wustite (FeO) layers, which can protect the steels from corrosion and other atmospheric effects. Existence of mill scale on the specimens' surface has shown to be able to decrease the cut edge quality. Since the mechanism behind influence of mill scale on the laser cutting process is unknown, this work performs direct observation of oxygen laser cutting processes on specimens with and without removed mill scale layers. Oxygen laser cutting processes were carried out using Ytterbium fibre laser 1070 nm along the edge of 20-mm-thick-steel specimens which were attached to a borosilicate glass. Focal point of the laser beam was positioned to be 0.7 mm below the specimens' surface. A high speed imaging system was arranged to face the glass, recording the cut front and kerf dynamics during cutting processes. It was found that cut front inclination angle increase when the mill scale was removed from the specimens' surface. This implies that mill scale on the specimens' surface seem to contribute in increasing the exothermal energy during laser cutting processes.

1. Introduction

Laser cutting is a common and well-established manufacturing technique. The process involves focused laser beam to melt the materials and assisting gas to blow out the molten materials. Oxygen is the most common assist gas in laser cutting due to its ability in initiating exothermal reaction which adds energy to the process [1]. Although there is an increased trend of nitrogen usage in laser cutting, the oxygen laser cutting still offers the best cut edge quality in thicker specimens. Laser cut edge quality is determined by the striations mark on the cut edge, which is parallel grooves appearing as periodic lines on the cut edge surface. A good cut edge quality has minimum striations mark on the cut edge surface.

The laser cutting process itself is a complex process which can be influenced by the chemical composition and surface conditions of the specimens [2,3]. For instance, mill scale is oxide layers formed on the surface of hot-rolled steel products after reheating processes, hot rolling and subsequent controlled cooling [4]. Wustite FeO, magnetite Fe_3O_4, and hematite Fe_2O_3 are the main types of mill scale with its own chemical and physical properties [5]. Mill scales has benefit of protecting the steel from corrosion and environmental impact [6]. However, it was found to influence the laser processes. For instance, mill scale causes CO porosity in laser welding of mild steel which indicates a lack of deoxidizing element and increase of oxygen content in the fusion zone [7,8]. In laser cutting, a study tested varieties of hot rolled steels found that hot-rolled pickled and oil steels produce much more consistent cut and better cut quality than steels cut straight-from-the-mill [9]. The study concluded that a smooth and homogenous surface condition is necessary for a consistent and good cut quality. Furthermore, SSAB and ESAB have seen that removing mill scale on the surface of thick specimens can reduce the cut edge quality. Studies on surface tension and viscosity of molten iron showed that surface tension and viscosity of molten iron are lower for higher oxygen content [10,11]. Lower viscosity and surface tension can affect the melt dynamics in laser cutting. As mill scale consists of oxides, it has potential to influence the surface tension and viscosity of molten steel during oxygen laser cutting, affecting the cut quality. Therefore, it is necessary to study the influence of mill scale on the oxygen laser cutting for process optimisation.

Observations of the laser cutting processes have been carried out using high speed imaging in many ways, such as having a high speed camera facing a cut intersection to observe the cut front morphology or replacing the specimen's edge with a glass to enable a side view of the process [12,13]. Both methods enable recording of dynamics in the cut kerf during the process and improve our knowledge in laser cutting. For instance, low inclination angle of the cut front can reduce the striation marks on the cut edge [14]. The inclined cut front change the melt flow direction and dynamics, affecting the cut edge quality [1,15]. This study uses a glass as replacement edge to record the cut front and kerf dynamics during the processes. Analysis is thus focused on the cut front inclination angle and melt flow direction to understand the influence of mill scale removal on the oxygen laser cutting processes.

2. Methodology

2.1. Experimental set-up

To answer the research question, oxygen assisted laser cutting processes using an IPG Ytterbium fibre laser 1070 nm were carried out on two different types of materials with chemical compositions listed in Table 1. The specimens were prepared in three conditions; (1) without treatment, (2) removed mill scale on the top side, and (3) removed mill scale on both top and bottom side. Mill scale removals were performed using laser oxide removal at SSAB.

Table 1. Chemical composition of the two steel grades (in wt% and balanced Fe).

Materials	Thickness	C	Si	Mn	Cr	Ni	Mo	Al
Steel A	20 mm	0.07	0.008	0.81	0.029	0.02	0.004	0.049
Steel B	20 mm	0.19	0.21	0.71	0.8	0.06	0.022	0.055

In situ high speed diagnosis technique was used to record the laser cutting processes [13]. The technique employed borosilicate glass as a replacement edge allowing a direct view on the cut kerf during the process. A high speed imaging equipment and a continuous illumination laser were placed facing the specimen edge as shown in Figure 1(a). The laser beam was placed at the specimens' edge and moved along the positive X-axis. The edge quality of specimens cut using glass is comparable to the specimens cut under normal laser cutting condition. Accordingly, the dynamics seen through the glass can represent the dynamics during actual laser cutting condition. The laser cutting parameters and recording frequency were kept constant for all specimens as shown in Table 2.

Table 2. List of laser cutting parameters.

Laser power	4500 W
Focal position of the laser beam	0.7 mm below specimens' surface
Cutting speed	0.7 m/min
Gas	Oxygen
Gas pressure	1.75 bar
Nozzle diameter	1.5 mm
Distance between specimen surface and nozzle	0.3 mm
Camera recording frequency	10240 Hz

2.2. Analysis

Observation was carried out on the specimens' cut edge and cut kerf using macroscopic pictures and recorded high-speed videos as shown in Figure 1(b) and (c). Figure 1(b) shows an example of a cut edge where three zones (top, middle, bottom) can be distinguished according to the visible striations. The amount of striations per centimetre was counted for each zone respectively.

The recorded videos were also observed according to the distinguishable zones on Figure 1(b). Figure 1(c) shows a frame taken during a laser cutting process. The frame reveals an inclined cut front and melt pool direction during the process. The inclination angle was measured for each zone by comparing the cut front to the normal axis, while the melt pool width was measured along the visible melt on the cut edge up to the cut front. The melt pool width was measured at five different heights within each zone on three frames from a laser cutting process. The standard deviation of the melt pool width was calculated for each zone based on the measurement from three frames to show how big the melt pool width varies along the height.

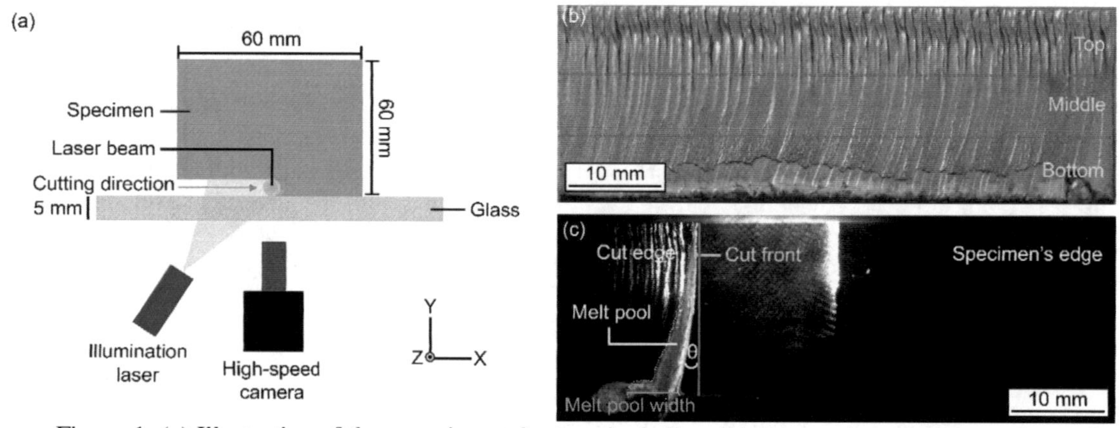

Figure 1. (a) Illustration of the experimental set-up including (b) observation on the cut edge and (c) recorded videos.

3. Results

Untreated specimens (1A and 1B) have flat cut edge for both materials as seen in Figure 2(a) and (b). Specimen A with removed mill scale on the top surface (2A) shows a concave cut edge while specimen B (2B) shows a rather flat edge (Figure 2(c) and (d)). Removing the mill scale on both top and bottom side results in a flattened cut edge (Figure 2(e) and (f)) compared to specimens with removed mill scale on the top side. Dross can be seen on all specimens and specimens with removed mill scale show visible heat affected zone close to the cut edge as seen in Figure 2(o) – (r), (w), (x).

Figure 3 shows examples of the frames taken during laser cutting processes. The cut front inclination angle and melt pool width seen on the cut edge increase at the bottom zone for all processes. Specimens with removed mill scale on the top side (2A and 2B) show a gentle slope in their general cut front appearance compared to the other specimens.

In Figure 4(a), the top zone shows more striations compared to the other zones and removing mill scale leads to reduced amount of striation per cm for both steel A and B. The cut edge of steel A with removed mill scale on the top side (2A) has the smallest amount of striations among the examined specimens due to wider groves shown in Figure 2(i).

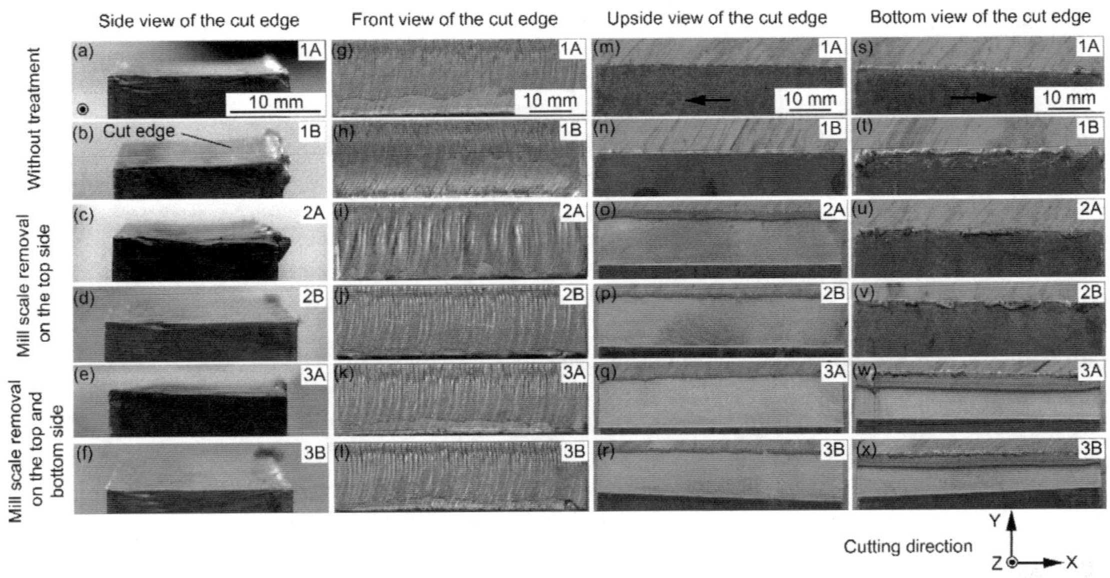

Figure 2. Macroscopic images of the specimens showing the side view (a) – (f), cut edge surface (g) – (l), top (m) – (r) and bottom view (s) – (x) of the specimen, including cutting direction shown in arrows and dots.

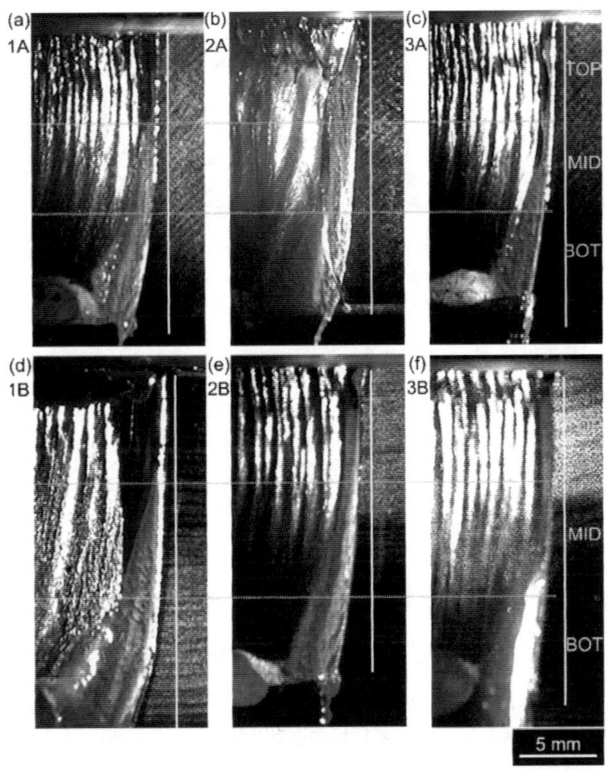

Figure 3. Frames taken during laser cutting processes for steel A and B at three mill scale conditions; (a) (d) with mill scale, (b) (e) removed mill scale on the top side, and (c) (f) removed mill scale on both the top and bottom side.

The tendency of having a high inclination angle at the bottom zone shown in Figure 3 is confirmed in Figure 4(b). Steel A with removed mill scale on the top side (2A) shows an increased inclination angle at the top and middle zones, portraying a gentle slope cut front in general dissimilar to the reference case (1A). This behaviour is also shown in Figure 3(b). However, steel A with removed mill scale on both the top and bottom side (3A) show similar inclination angle behaviour to its reference (1A). Significant change in the inclination angle due to mill scale removal is not found for the steel B specimens.

During the process, the melt pool seen on the bottom zone was the widest among the top and middle zones for all specimens while in general, steel A with removed mill scale on the top side (2A) shows the widest melt pool (Figure 4(c)). Steel B with removed mill scale on the top side (2B) shows an increased width at the top zone, but decreased width at the bottom zone. In addition, the specimens with removed mill scale on both the top and bottom side (3A and 3B) have similar melt pool width values compared to the untreated specimens (1A and 1B).

Melt pool seen on the cut edge widens at the middle and bottom zone where a significant widening happens at the bottom zone. This can be seen on the standard deviations of the melt pool widths for all specimens (Figure 4(c)). At the bottom zone, the standard deviation is higher for steel A than B. Mill scale removal does not seem to affect the standard deviation of the melt pool width.

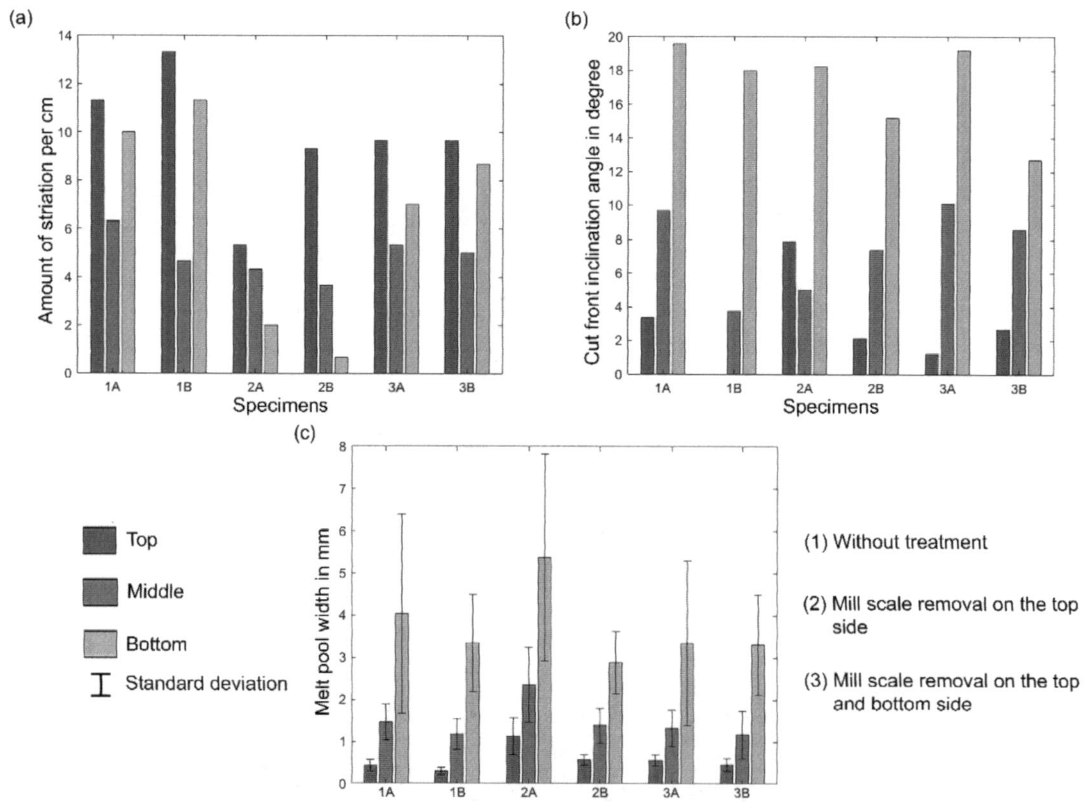

Figure 4. Observation results on the amount of striations per cm (a), cut front inclination angle (b), melt pool width seen on the cut edge during the process and standard deviation of the melt pool width (c) at the top, middle and bottom zones.

IOP Conf. Series: Materials Science and Engineering 1135 (2021) 012008 doi:10.1088/1757-899X/1135/1/012008

4. Discussion

The observed tendencies of increasing cut front inclination angle and melt pool width at the bottom zone will be discussed in the following paragraphs. The discussion will come to explain the influence of mill scale on laser cutting processes.

The inclined cut front is known to be due to a lack of energy input to melt the materials [16]. A high inclination angle at the bottom zone implies that this zone experienced less energy input during the process compared to the top and bottom zones. Since the oxygen laser cutting process is driven by energy from the laser and exothermal reaction of Iron-Oxygen [1], the inclined cut front at the bottom zone can be related to the properties of the laser beam and local oxygen concentration. The bottom zone is located about 10 mm away from the focal position of the laser beam, meaning that the laser beam hitting this zone is defocused. The defocused laser beam then delivers a small power density to this zone. Additionally, the position of the bottom zone is far away from the nozzle opening, causing less oxygen concentration as the oxygen is contaminated by ambient air [17] . Accordingly, the bottom zones experience the least energy input from both the laser and exothermal reaction of Iron-Oxygen.

This phenomenon impacts the melt pool behaviour at the lower zones of the specimens. Less energy input can lead to a low temperature on the cut kerf. Since the energy input decreases towards the bottom zone, the temperature might also decrease towards the bottom zone. The top, middle and bottom zones are most likely experiencing temperature gradients where the bottom zone possesses the lowest temperature respectively. The temperature can influence viscosity of the molten material where low temperature leads to a high viscosity [18,19]. As the temperature is decreasing towards the bottom zone, the viscosity of the molten materials is expected to increase at the bottom zone. An increase of viscosity causes a widening of the melt pool seen on the cut edge (Figure 1(c) and 3(c)). Molten material with a high viscosity can have difficulties to flow away from the specimen and thus sticks on the cut edge.

The bottom zones have more molten material sticking on the cut edge compared to the other zones. It means that heat conduction from the molten material to the cut edge occurs longer at the bottom zone than the other zones. This situation can cause widening of the heat affected zone at the bottom side of the specimens shown in Figure 2(w) and (x). Steel A with removed mill scale on the top side (2A) also shows a wider heat affected zone than the other specimens (Figure 2(o)). The melt pool width at the top zone (Figure 3(c)) of this specimen is also the highest among the other specimens. These are indications that the appearance of the heat affected zone close to the cut edge correlates with the molten material sticking on the cut edge during the process. However, quite large dross was created in the process, which can also cause the widening of the heat affected zone. Therefore, the widening of the heat affected zone at the bottom zone can be due to both dross formation and molten material sticking on the cut edge.

Mill scale seems to influence the exothermal reaction of Iron-Oxygen during the cutting process. High cut front angles and wide melt pool widths seen on steel A with removed mill scale on the top (2A) indicate that the process experienced lower energy than the untreated specimen (1A). This means that the mill scale might contribute in increasing the exothermal reaction of Iron-Oxygen. However, removed mill scale on the bottom side of the specimen (3A) seems to reduce the impact of having high cut front angles and wide melt pool during the process as the values of the cut front angles and the melt pool widths are similar to the reference (1A). Steel B does not show significant change in the inclination angle and melt pool width after the mill scale removal. This can be due to difference in chemical composition of the two materials. Previous investigation on similar steel grade revealed that increased carbon content can reduce the exothermal energy and result in more stable cutting process with reduced risk of rough cut edge [3]. While silicon content does not seem to influence the cut edge quality, reduce amount of manganese content can improve the cut edge quality [3]. These indicate that influence of mill scale on the oxygen laser cutting processes also depends on the alloying element of the materials although more observation in this topic needs to be carried out separately.

5. Conclusion

Oxygen assisted laser cutting processes of thick materials possess varied energy input along the specimen thickness due to the characteristics of the laser beam caustic and gradual reduction of oxygen concentration. Varied energy input can lead into temperature and viscosity variation in the molten material along the specimen thickness where molten material at the bottom zone has the lowest temperature and highest viscosity. High-viscous molten material has difficulties to flow away from the specimens resulting in wider melt pool width seen on the bottom zone of the cut edge.

The existence of mill scale seems to contribute to increasing the exothermal reaction of Iron-Oxygen during the process. Accordingly, the energy input to the process zone is reduced when the mill scale is removed from the top side of the specimen. Reduced energy input causes an increase of the cut front inclination angle and viscosity of the molten material during the process which results in increased amount of striations on the cut edge. However, changing the chemical composition of the specimens or pre-treating specimens' surfaces before cutting to have a homogenous surface condition will possibly improve the consistency of oxygen laser process and thus cut edge quality.

Acknowledgement
The authors gratefully acknowledge the funding from VINNOVA, Swedish Innovation Agency for the project ACCEL no. 2019-00781_VINNOVA.

ORCID
H.S. Dewi https://orcid.org/0000-0002-3060-5831
J. Volpp https://orcid.org/0000-0003-0194-9018
J. Frostevarg https://orcid.org/0000-0003-4265-1541
J. Siltanen https://orcid.org/0000-0003-0058-3480

References

[1] A. Riveiro, F. Quintero, M. Boutinguiza, J. Del Val, R. Comesaña, F. Lusquiños, J. Pou, Laser Cutting: A Review on the Influence of Assist Gas, Materials (Basel, Switzerland). 12 (2019) 157. https://doi.org/10.3390/ma12010157.
[2] J. Hu, X. Qiu, Z. Zhang, H. Shen, Influence of oxidation on flow structure in laser-oxygen cutting, Journal of Applied Physics. 112 (2012). https://doi.org/10.1063/1.4751455.
[3] A. Ivarson, J. Powell, J. Siltanen, Influence of Alloying Elements on the Laser Cutting Process, Physics Procedia. 78 (2015) 84–88. https://doi.org/10.1016/j.phpro.2015.11.020.
[4] X.J. Hu, B.M. Zhang, S.H. Chen, F. Fang, J.Q. Jiang, Oxide scale growth on high carbon steel at high temperatures, Journal of Iron and Steel Research International. 20 (2013) 47–52. https://doi.org/10.1016/S1006-706X(13)60043-6.
[5] R.M. Cornell, U. Schwertmann, Introduction to the Iron Oxides, The Iron Oxides. (2003) 1–7. https://doi.org/https://doi.org/10.1002/3527602097.ch1.
[6] J. Ahlström, J. Tidblad, L. Tang, B. Sederholm, S. Leijonmarck, Electrochemical Properties of Oxide Scale on Steel Exposed in Saturated Calcium Hydroxide Solutions with or without Chlorides, International Journal of Corrosion. 2018 (2018) 5623504. https://doi.org/10.1155/2018/5623504.
[7] J. Frostevarg, P. Qinglong, M. Mizutani, Y. Kawahito, S. Katayama, Effects of edge oxides from laser cutting in laser-arc hybrid welding, Journal of Laser Applications. 30 (2018) 012014. https://doi.org/10.2351/1.5009792.

[8] K. Ono, K. Adachi, Y. Matsumoto, I. Miyamoto, T. Inoue, R. Narita, Influence of oxide film on weld characteristics of mild steel in CO2 laser welding, Journal of Laser Application. 73 (2002) D83–D91. https://doi.org/10.2351/1.5059237.

[9] Steve Pugh, Better material surface, better laser cutting, The Fabricator. (2018). https://www.thefabricator.com/thefabricator/article/lasercutting/better-material-surface-better-laser-cutting (accessed June 7, 2021).

[10] K. 'Uchikosi, M. 'Isshiki, M. 'Fukuyama, H. 'Morohoshi, Surface tension of liquid iron as functions of oxygen activity and temperature, ISIJ International. 51 (2011) 1580–1586.

[11] D.B. Dingwell, D. Virgo, The effect of oxidation state on the viscosity of melts in the system Na2O-FeO-Fe2O3-SiO2, Geochimica et Cosmochimica Acta. 51 (1987) 195–205. https://doi.org/https://doi.org/10.1016/0016-7037(87)90231-6.

[12] J. Pocorni, J. Powell, E. Deichsel, J. Frostevarg, A.F.H. Kaplan, Fibre laser cutting stainless steel: Fluid dynamics and cut front morphology, Optics & Laser Technology. 87 (2017) 87–93. https://doi.org/https://doi.org/10.1016/j.optlastec.2016.08.002.

[13] D. Arntz, D. Petring, F. Schneider, R. Poprawe, In situ high speed diagnosis—A quantitative analysis of melt flow dynamics inside cutting kerfs during laser fusion cutting with 1 μ m wavelength , Journal of Laser Applications. 31 (2019) 022206. https://doi.org/10.2351/1.5096091.

[14] J. Powell, S.O. Al-Mashikhi, A.F.H. Kaplan, K.T. Voisey, Fibre laser cutting of thin section mild steel: An explanation of the "striation free" effect, Optics and Lasers in Engineering. 49 (2011) 1069–1075. https://doi.org/10.1016/j.optlaseng.2011.03.011.

[15] A. Riveiro, F. Quintero, F. Lusquiños, R. Comesaña, J. Pou, Study of melt flow dynamics and influence on quality for CO2laser fusion cutting, Journal of Physics D: Applied Physics. 44 (2011) 135501. https://doi.org/10.1088/0022-3727/44/13/135501.

[16] A. Mahrle, E. Beyer, Theoretical aspects of fibre laser cutting, Journal of Physics D: Applied Physics. 42 (2009). https://doi.org/10.1088/0022-3727/42/17/175507.

[17] K. Chen, Y.L. Yao, V. Modi, Numerical simulation of oxidation effects in the laser cutting process, International Journal of Advanced Manufacturing Technology. 15 (1999) 835–842. https://doi.org/10.1007/s001700050140.

[18] Y. Sato, Representation of the viscosity of molten alloy as a function of the composition and temperature, Japanese Journal of Applied Physics. 50 (2011). https://doi.org/10.1143/JJAP.50.11RD01.

[19] M. Hirai, Estimation of Viscosities of Liquid Alloys, Isij International. 33 (1993) 251–258. https://doi.org/10.2355/isijinternational.33.251.

Influence of the laser cutting front geometry on the striation formation analysed with high-speed synchrotron X-ray imaging

Jannik Lind[1,2], Christian Hagenlocher[1], David Blazquez-Sanchez[2], Marc Hummel[3], A. Olowinsky[4], Rudolf Weber[1], Thomas Graf[1]

[1] Institut für Strahlwerkzeuge, University of Stuttgart, Pfaffenwaldring 43, 70569 Stuttgart, Germany

[2] Precitec GmbH & Co. KG, Draisstraße 1, 76571 Gaggenau, Germany

[3] Chair for Laser Technology LLT, RWTH Aachen University, Steinbachstraße 15,52074 Aachen, Deutschland

[4] Fraunhofer Institute for Laser Technology ILT, Steinbachstr. 15, 52074 Aachen, Germany

jannik.lind@ifsw.uni-stuttgart.de

Abstract. The generation of low surface roughness of the cut edge during laser beam cutting is a challenge. The striation pattern, which determines the surface roughness, can be distinguished into regular and interrupted striations, the latter resulting in an increased surface roughness. In order to analyse their formation, the space- and time-resolved cutting front geometry and melt film thickness were captured during laser beam fusion cutting of aluminium sheets with a framerate of 1000 Hz by means of high-speed synchrotron X-ray imaging. The comparison of the contours of the cutting fronts for a cut result with regular und interrupted striations shows that the contour fluctuates significantly more in case of interrupted striations. This leads to a strong fluctuation of the local angle of incidence. In addition, the average angle of incidence decreases, which results in an increase of the average absorbed irradiance. Both phenomena, local increase of absorbed irradiance and its dynamic fluctuation, result in a local increase of the melt film thickness at the cutting front which is responsible for the formation of the interrupted striations.

1. Introduction

The generation of low surface roughness of the cut edge during laser beam fusion cutting is challenging, especially when cutting thick metal sheets. In [1, 2] it was shown, that a low surface roughness coincides with a regular striation pattern, while a high surface roughness coincides with an interrupted striation pattern. An example of a cut edge in case of a regular striation pattern a) and for an interrupted striation pattern b) is shown in Figure 1. The origin of the coordinate system (x_c, y_c, z_c) was set at the intersection point of the beginning of the cut and the top surface of the cut edge. The relative movement between the laser beam and the workpiece was performed by moving the workpiece against the x_c-direction, as highlighted by the black arrow. The difference of these patterns is that the striations do not run continuously from the top to the bottom surface of the sample in the case of interrupted striations, as indicated by the black dashed rectangle in Figure 1 b). Recent publications [3–6] identified local temperature maxima, which exceeded evaporation temperature, as a reason for the formation of the interrupted striation pattern. The local exceed of the evaporation temperature coincides with an increase of molten material, which flows not only in z_c-direction downward the cutting front, but also horizontally against the x_c-direction [7]. Furthermore, the thickness of the melt film and its flow exhibits strong instabilities [8–10]. Since the geometry of the cutting front influences the absorbed irradiance and therefore the temperature and melt film thickness, it is mandatory to know about the dynamic changes of the geometry of the cutting front and the melt film thickness during the transition from the process regime of regular striation to the regime of interrupted striation.

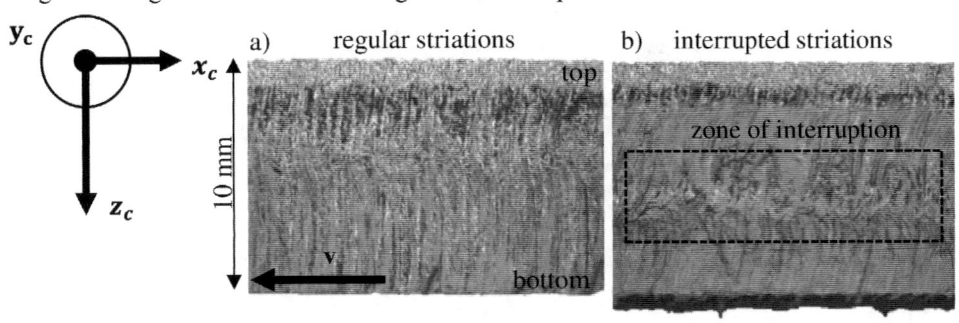

Figure 1 Regular striation pattern on the surface of the cut edge at a feed rate of $v = 2.3$ m/min a) and pattern with interrupted striations at an increased feed rate of $v = 2.6$ m/min b); $P = 6$ kW.

The space- and time-resolved in-process observation of the geometry of the cutting front was demonstrated in [11–13] by means of high-speed X-ray imaging. However, the presented approaches do not allow to observe the thickness of the melt film due to the used incoherent X-rays. By using coherent X-rays, it has been shown that the transition from the solid to the liquid physical state became visible during laser beam welding [14–20].

This paper presents for the first time a space- and time-resolved experimental determination of the geometry of the cutting front and melt film thickness during the transition from regular striation formation to interrupted striation formation within the same sample by means of high-speed synchrotron X-ray imaging.

2. Experimental Setup

Cutting of the aluminium alloy AlMg3 was investigated using a disk laser with a wavelength of 1.03 µm in combination with a *Precitec-ProCutterZoom2.0ET* cutting head. The beam-delivery fiber had a core diameter of 50 µm and the beam was focused to a waist diameter of approximately 150 µm to a top-hat intensity distribution with a Rayleigh-length of about 2.5 mm as measured with a Primes FocusMonitor. Figure 2 sketches the experimental setup, which was realized in cooperation with the RWTH in Aachen. The X-ray imaging system consisted of the DESY-synchrotron [21] providing the X-ray source (green), a scintillator (blue), and a high-speed camera (purple).

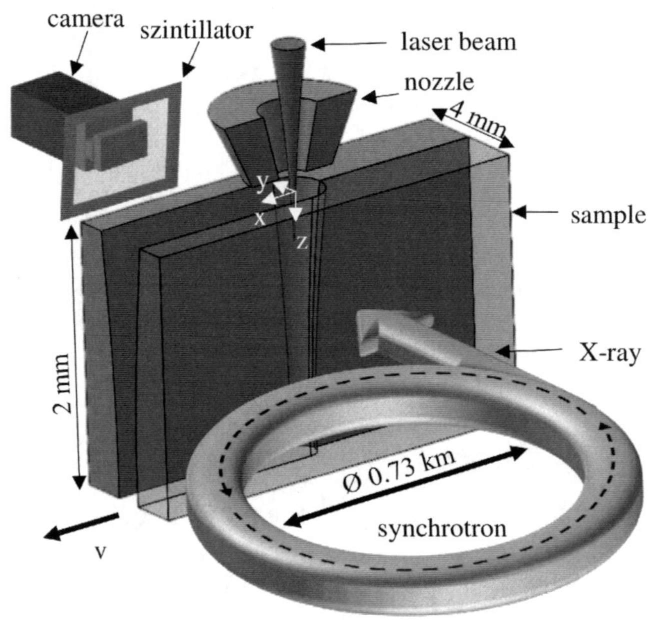

Figure 2 Sketch of the experimental setup.

The monochromatic X-ray beam from the electron accelerator ring DESY in Hamburg with an energy of 37.7 keV and an average beam diameter of approximately 3 mm transmitted the samples during the cutting process. Locally varying thickness due to the cutting kerf and locally varying attenuation coefficients depending on the physical state of the sample yielded a local attenuation of the beam. This locally attenuated X-ray projection behind the sample was converted by the scintillator to visible light which was recorded by the high-speed camera, in the following called "X-ray video". A similar system was previously used to study the laser cutting process using a microfocus tube as the X-ray source [11, 12]. However, the herein presented cutting process was observed by high-speed X-ray imaging in side-view at a frame rate of 1000 fps, with a spatial resolution of 256 pixels/mm and an image size of 920 x 920 pixels, resulting in an exposed area of approximately 3 mm in diameter. Due to the small divergence of the X-ray beam between 0.0055 mrad and 0.027 mrad, the impact of the penumbra to the observed blurring is negligible. The origin of the coordinate system (x, y, z) was set at the intersection point of the laser beam axis and the surface of the sample. The X-ray videos were post-processed with a flat-field correction. To be able to suitably observe the cutting process, aluminum was chosen as the sample material and the width of the samples (in the direction of the X-rays) was chosen to be 4 mm. Aluminum absorbs less X-rays than steel due to its lower attenuation coefficient, which allows for shorter exposure times of the X-ray images.

During the cutting process the nozzle (orange) with an outlet diameter of 3 mm was positioned 1 mm above the 2 mm thick samples. In order to observe the transition from regular to interrupted striation formation and to compare their formation differences, a critical process parameter regime was chosen for the synchrotron experiments at which regular and interrupted striations were generated within the same sample. This critical parameter were determined in previous work. The beam waist was positioned 2 mm above the sample's surface, the laser power was set to $P = 1$ kW, the feed rate was set to $v = 1.75$ m/min, and nitrogen with a pressure of 2 bar was used as the processing gas. The experiment within this critical process regime was repeated three times.

Figure 3 (Multimedia view [22]) shows an averaged image of 500 frames of an X-ray video recorded during one of the three repeated cutting processes. The frames were taken 500 ms after the beginning of the cutting process. The image shows the color-coded local transmittance of the X-ray radiation through the sample. A clear contrast between the solid sample material (blue, high absorption of X-rays) and the cutting kerf (red and orange, low absorption of X-rays) is visible in the images. Bright areas indicate a low local absorption of the X-rays and therefore a high local width of the cutting kerf. From these images, the geometry of the cutting front (line at the center of the cutting front) can be identified, as highlighted by the green dashed line in Figure 3. The position of the beam in the X-ray images was determined by applying a couple of laser pulses (0.8 ms at 1 kW and 588 Hz) on the sample. The symmetry axis of the resulting hole was defined as the position of the beam axis, which is represented by the white dash-dotted line in Figure 3. The caustic of the laser beam which results for the measured beam characteristics is represented by the white dashed line in Figure 3.

The transition from the solid to the liquid physical state causes interference effects of the transmitted coherent X-ray beam on the scintillator [23]. This interference effect results in a clearly visible transition from the solid to the liquid physical state, i.e. the liquidus isotherm, as marked by the black dashed line in Figure 3. Therefore, the width of the melt film in the x-direction can be derived.

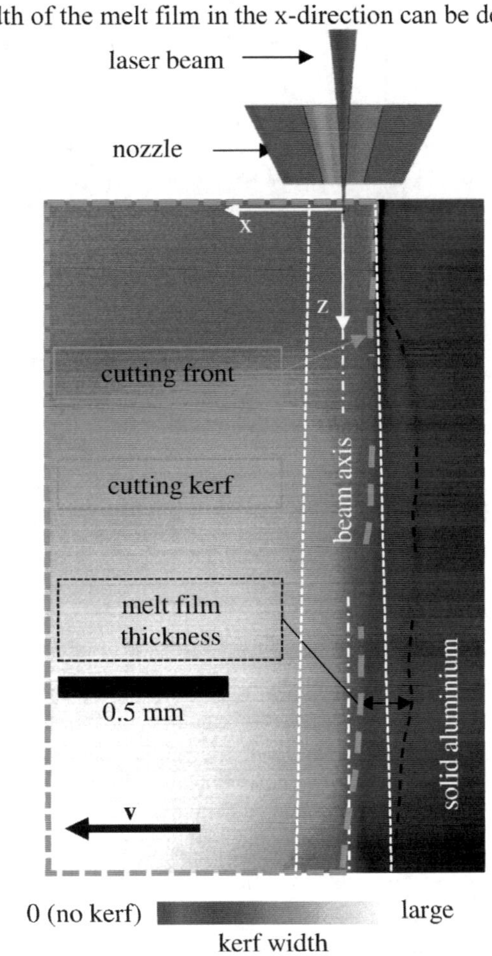

Figure 3 Average image of 500 frames from an X-ray video showing the cutting process in a 2 mm thick aluminium AlMg3 sheet. Red, yellow and white colored areas show the shape of the cutting front and cutting kerf. The black dashed line marks the liquidus isotherm; $P = 1$ kW, $v = 1.75$ m/min, Multimedia view [22].

3. Analysis of the transition from regular to irregular striations

3.1. Topography of the surface of the cut edge for regular and interrupted striations

In order to distinguish between regular and interrupted striations, Figure 4 presents the color-coded topography of the surface of the cut edge at the location of the transition from regular to interrupted striations within the same sample. The topography of the surface was measured with a scanner-based optical coherence tomography (OCT) system. Its setup and specifications are described in [24]. The dross at the bottom of the cut edge was not measured and is therefore not shown in Figure 4. The origin of the coordinate system (x_c, y_c, z_c) was set at the intersection point of the beginning of the cut and the top surface of the cut edge. The lower horizontal axis represents the time $t = x_c/v$ when the local position x_c of the presented cut edge was processed. Three lines parallel to the top surface of the cut edge were extracted for $z_c = 0.25$ mm (1), $z_c = 1.00$ mm (2), and $z_c = 1.75$ mm (3), as shown by the curves in Figure 4 bottom.

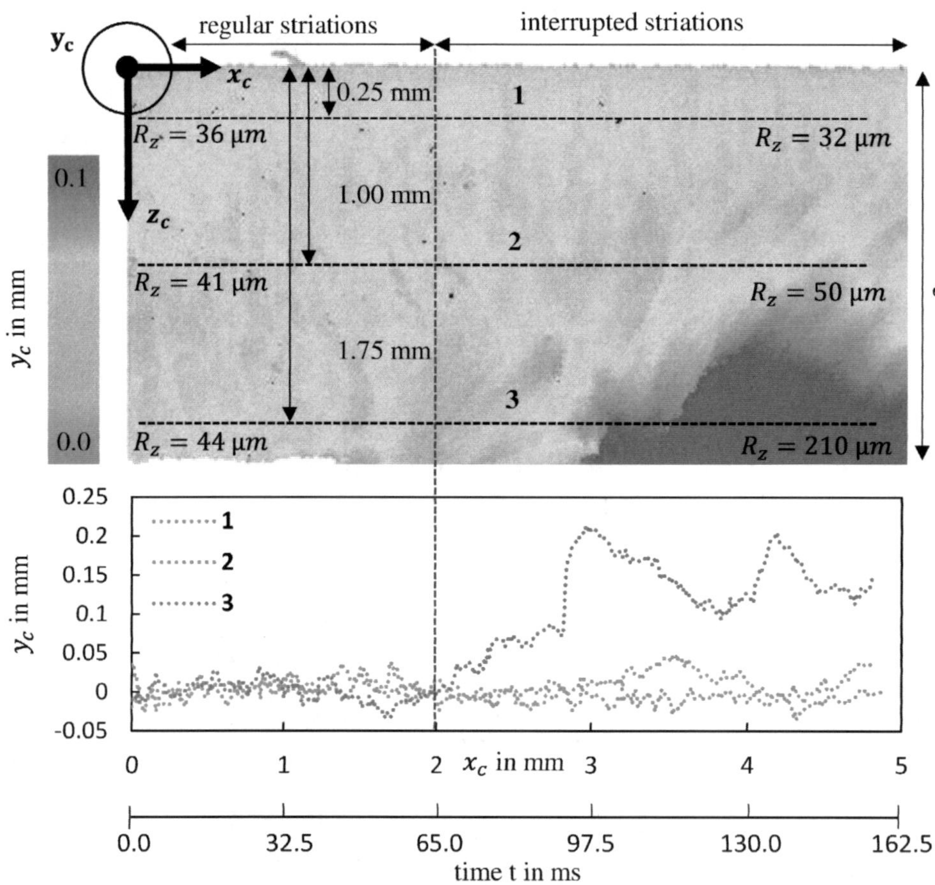

Figure 4 Topography image of the surface of the cut edge (top) and height profiles derived at three different z_c-positions (bottom). The vertical dashed purple line marks the transition from regular to interrupted striations within the same sample; P = 1 kW, v = 1.75 m/min.

The results show, that for x_c<2 mm, the height profile for all three measured positions is y_c<0.05 mm. For x_c>2 mm the height profile increases especially in the lower part (z_c = 1.75 mm) of the cut edge. The surface roughness R_z increases for x_c<2 mm from R_z=44 µm to R_z=210 µm for

x_c>2 mm. The vertical dashed purple line marks the transition from low to high surface roughness i.e. indication of the transition from regular to interrupted striations.

3.2. Geometry of the cutting front for regular and interrupted striations
During the cutting process the axis of the laser beam passed the position of the purple line in Figure 4 at the point in time t=65 ms, as indicated by the lower horizontal axis in Figure 4. Regular striations were generated on the surface of the cut edge for t<65 ms.

The contours of the cutting fronts during the formation of regular striations from 20 ms to 65 ms are represented with the curves in Figure 5a. The contours of the cutting fronts during the formation of interrupted striations at t>65 ms are represented with the curves in Figure 5b.

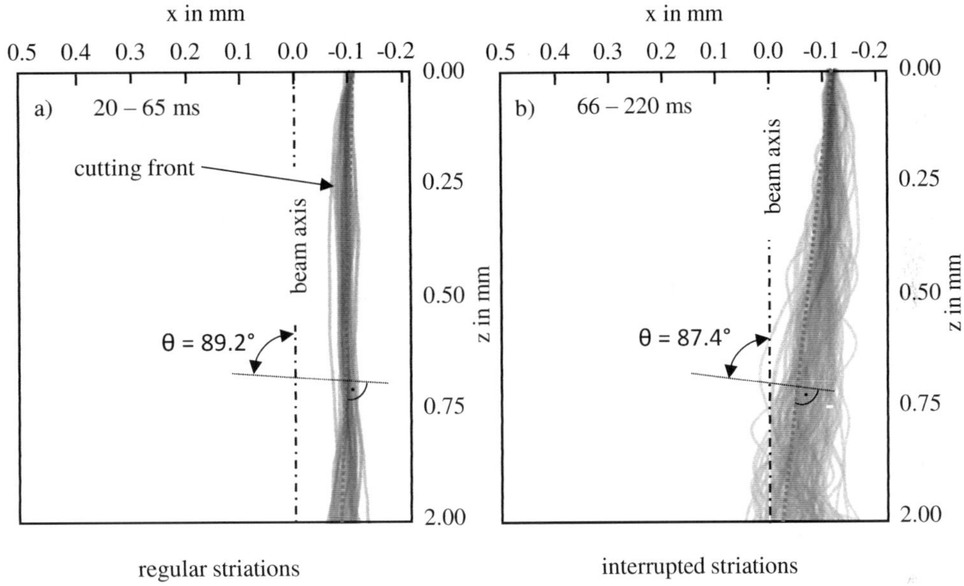

Figure 5 Contours of the center line of the cutting front along the cutting depth for cutting conditions with regular a) and interrupted b) striation formation; P = 1 kW, v = 1.75 m/min.

In the case of regular striations, the average angle of incidence θ on the cutting front with respect to the laser beam was 89.2°, as highlighted by the red dashed line in Figure 5. In the case of interrupted striations, the average angle of incidence θ decreases to 87.4° compared to the case of regular striations. When comparing the contours of the cutting fronts of the two cases, it can be seen that the fluctuation of the contour significantly increases in case of interrupted striations and therefore the local angle of incidence also changes more frequently. The absorbed irradiance of the laser beam increases with decreasing angle of incidence, as quantified in [11]. It is concluded that in case of interrupted striations, due to the smaller angles of incidence and the higher dynamic changes of the contour of the cutting front, the absorbed irradiance increases and changes more frequently.

3.3. Melt film thickness for regular and interrupted striations
In order to investigate the effect of the increase of the absorbed irradiance and of the fluctuations of the geometry of the cutting front, the melt film thickness at the cutting front was analyzed.

Figure 6 (Multimedia view [22]) shows six single images of an X-ray video of the cutting process in a 2 mm thick sheet of aluminium AlMg3 at different points in time (noted below). After 1359 ms the laser beam and the cutting gas were switched off. The good agreement of the liquidus isotherm (visible in the last frame at 1359 ms) and the recast layer in the longitudinal section verifies the capability to determine the thickness of the melt film with high temporal and spatial resolution. For t<65 ms, regular striations were generated on the surface of the cut edge as indicated in Figure 4. During this period, the liquidus isotherm is not visible in the X-ray image. This indicates, that the melt film thickness is below the spatial resolution and thus smaller than approximately 30 µm in x-direction.

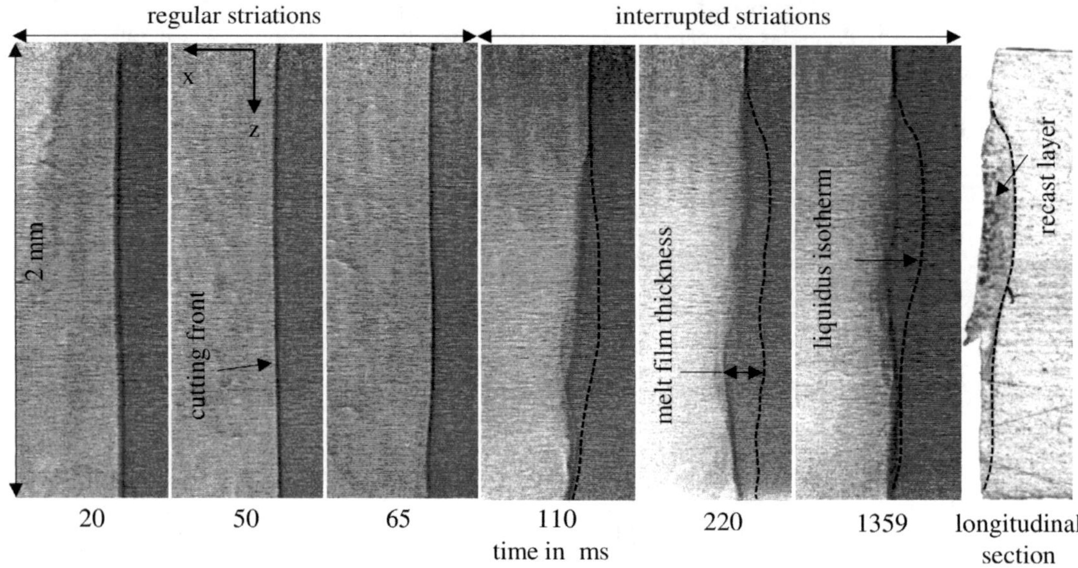

Figure 6 Single frames of an X-ray video of the cutting process in a 2 mm thick sheet of aluminium AlMg3 at six points in time (left) and longitudinal section of the cut front (right). Red, yellow and white colored areas show the shape of the cutting front and cutting kerf; P = 1 kW, v = 1.75 m/min, Multimedia view [22].

For t>65 ms, interrupted striations were generated on the surface of the cut edge and the liquidus isotherm can be identified as highlighted by the black dashed line. For z<0.25 mm, the melt film thickness is below the spatial resolution. For z>0.25 mm, the melt film thickness increases and temporal fluctuations occur. Since the frames contain motion blur at the position of the cutting front, it can be concluded that the geometry fluctuates with more than 1 kHz. It is concluded that in case of interrupted striations an increase of the melt film thickness coincides with the increase of the absorbed irradiance for t>65 ms.

4. Conclusion

This paper presents for the first time a space- and time-resolved experimental determination of the geometry of the cutting front and melt film thickness during the transition from regular striation formation to interrupted striation formation by means of high-speed synchrotron X-ray imaging.

The comparison of the contours of the cutting fronts for a cut result with regular und interrupted striations shows that the contour fluctuates significantly more in case of interrupted striations. This leads to a strong fluctuation of the local angle of incidence. In addition, the average angle of incidence decreases, which results in an increase of the average absorbed irradiance. Both phenomena, local increase of absorbed irradiance and its dynamic fluctuation, result in a local increase of the melt film thickness at the cutting front which is responsible for the formation of the interrupted striations.

Future work will focus on strategies to prevent the formation of interrupted striations based on the presented results.

5. Acknowledgement

This research was supported by Precitec GmbH & Co. KG and Trumpf GmbH & Co. KG. The presented investigations were carried out in cooperation with DESY in Hamburg and with RWTH Aachen University within the framework of the Collaborative Research Centre SFB1120-236616214 "Bauteilpräzision durch Beherrschung von Schmelze und Erstarrung in Produktionsprozessen" and funded by the Deutsche Forschungsgemeinschaft e.V. (DFG, German Research Foundation). We acknowledge DESY (Hamburg, Germany), a member of the Helmholtz Association HGF, for the provision of experimental facilities. Parts of this research were carried out at PETRA III and we would like to thank F. Beckmann and J. Moosmann for assistance in using P07 EH4. Beamtime was allocated for proposal I-20191140. The sponsorship and support is gratefully acknowledged.

6. ORCID iDs

J. Lind https://orcid.org/0000-0002-2196-5616
C. Hagenlocher https://orcid.org/0000-0003-2929-9723
D. Blazquez-Sanchez https://orcid.org/0000-0002-3504-229X
M. Hummel https://orcid.org/0000-0002-9097-9843
A. Olowinsky https://orcid.org/0000-0002-5231-8251
R. Weber https://orcid.org/0000-0001-8779-2343
T. Graf https://orcid.org/0000-0002-8466-073X

7. References

[1] Wandera C, Salminen A and Kujanpaa V 2009 *Journal of Laser Applications* **21** 154–61
[2] Bocksrocker O, Berger P, Fetzer F, Rominger V and Graf T 2019 *Lasers Manuf. Mater. Process.* **6** 1–13
[3] Bocksrocker O, Berger P, Kessler S, Hesse T, Rominger V and Graf T 2020 *Lasers Manuf. Mater. Process.* **7** 190–206
[4] Onuseit V, Ahmed M A, Weber R and Graf T 2011 *Physics Procedia* **12** 584–90
[5] Petring D 2016 *Proc of JLPS 84th Laser Materials Processing Conf Nagoya Japan*
[6] Onuseit V, Michael J, Weber R and Graf T 2011 *Proc. of Icaleo Paper 104*
[7] Olsen F 2006 *Laser materials processing conference ICALEO* 401
[8] Levichev N, Staudt T, Schmidt M and Duflou J R 2021 *CIRP Annals*
[9] Kaplan A FH 2015 *Optics and Lasers in Engineering* **68** 35–41
[10] Hirano K and Fabbro R 2011 *J. Phys. D: Appl. Phys.* **44** 105502
[11] Lind J, Fetzer F, Blazquez-Sanchez D, Weidensdörfer J, Weber R and Graf T 2020 *Journal of Laser Applications* **32** 32015
[12] Lind J, Fetzer F, Hagenlocher C, Blazquez-Sanchez D, Weber R and Graf T 2020 *Journal of Manufacturing Processes* **60** 470–80
[13] Ozaki H, Le M Q, Kawakami H, Suzuki J, Uemura Y, Doi Y, Mizutani M and Kawahito Y 2016 *Journal of Materials Processing Technology* **237** 181–7
[14] Simonds B J, Tanner J, Artusio-Glimpse A, Williams P A, Parab N, Zhao C and Sun T 2021 *Applied Materials Today* **23** 101049
[15] Wolff S J, Wu H, Parab N, Zhao C, Ehmann K F, Sun T and Cao J 2019 *Scientific reports* **9** 962
[16] Börner S, Dittrich D, Mohlau P, Leyens C, García-Moreno F, Kamm P H, Neu T R and Schlepütz C M 2021 *Journal of Laser Applications* **33** 12026
[17] Shevchik S, Le-Quang T, Meylan B, Farahani F V, Olbinado M P, Rack A, Masinelli G, Leinenbach C and Wasmer K 2020 *Scientific reports* **10** 3389
[18] Cunningham R, Zhao C, Parab N, Kantzos C, Pauza J, Fezzaa K, Sun T and Rollett A D 2019 *Science (New York, N.Y.)* **363** 849–52

IOP Conf. Series: Materials Science and Engineering 1135 (2021) 012009 doi:10.1088/1757-899X/1135/1/012009

[19] Miyagi M, Kawahito Y, Kawakami H and Shoubu T 2017 *Journal of Materials Processing Technology* **250** 9–15
[20] Wagner J, Hagenlocher C, Hummel M, Olowinsky A, Weber R and Graf T 2021 *Metals* **11** 797
[21] Schell N, King A, Beckmann F, Fischer T, Müller M and Schreyer A 2013 *MSF* **772** 57–61
[22] Lind J, Hagenlocher C, Blazquez-Sanchez D, Hummel M, Olowinsky A, Weber R and Graf T 2021 *High-speed synchrotron X-ray video of the laser beam cutting process in a 2 mm thick sheet of aluminium: https://doi.org/10.18419/darus-1824* 1st edn (DaRUS)
[23] van der Veen F and Pfeiffer F 2004 *J. Phys.: Condens. Matter* **16** 5003–30
[24] Holder D, Buser M, Boley S, Weber R and Graf T 2021 *Materials & Design* **203** 109567

IOP Conf. Series: Materials Science and Engineering 1135 (2021) 012010 doi:10.1088/1757-899X/1135/1/012010

Reduced finite-volume model for the fast numerical calculation of the fluid flow in the melt pool in laser beam welding

Jonas Wagner[1], Peter Berger[1], Philipp He[1], Florian Fetzer[1], Rudolf Weber[1], Thomas Graf[1]

[1]Institut für Strahlwerkzeuge, Universität Stuttgart, Pfaffenwaldring 43, 70569 Stuttgart, Germany

jonas.wagner@ifsw.uni-stuttgart.de

Abstract. In this paper we propose a reduced two-dimensional finite-volume model for the fast calculation of the melt flow. This model was used to determine the influence of the welding speed, viscosity in the melt and vapour flow inside of the keyhole on the fluid flow field, the temperature distribution, and the resulting weld-pool geometry for laser beam welding of aluminium. The reduced computational time resulting from this approach allows the fast qualitative investigation of different aspects of the melt flow over a wide range of parameters. It was found that the effect of viscosity within the melt is more pronounced for lower welding speeds whereas the effect of friction at the keyhole walls is more pronounced for higher welding speeds. The weld-pool geometry mainly depends on the welding speed.

1. Introduction

The fluid flow in the weld pool is related to many characteristics of the laser beam welding process such as the temperature distribution [1], quality aspects like humping [2], and the distribution of alloying elements [3, 4]. Due to the interdependence of pressure, velocity and temperature and the varying boundary conditions like the time-dependent shape of the keyhole, a direct analytical calculation is only possible to a limited extent. Numerical models have long proven to be a valuable tool for gaining insight into the physical processes involved in laser beam welding [1, 5–11]. Recently impressive multi-physical models have been shown, including multiple reflection models for the absorptance of the laser beam [12, 13], 3D-flow calculation and free surface flow [6, 7, 14–16], evaporation models [6, 7, 14–16] and models for the calculation of the resulting microstructure [17–20] and solidification cracking [21–23].

While these models give great insight into the welding process, the large number of influencing variables makes it difficult to determine their impact on the process result. Additionally, new beam shaping technologies [24–26] offer the potential to influence the shape of the keyhole. Knowledge about the interaction between keyhole and melt is therefore needed for adapting the suitable strategy for each process. In the present paper, we propose a reduced two-dimensional finite-volume model with the intention that the reduced complexity and the reduced simulation time resulting from this approach allow the fast qualitative investigation of different aspects of the melt flow over a wide range of parameters. With this model, the influence of viscosity in the melt and of friction at the wall of the keyhole is analysed and compared for different welding speeds.

2. Numerical Model

The reduced numerical model was implemented for the case of a moving sheet under a static laser beam using the open source software *OpenFOAM®* v2012 (ESI-OpenCFD) [27, 28]. For each time step at first the basic heat conduction equation [29–31]

$$\frac{\partial(\rho c_p T)}{\partial t} + \nabla \cdot (\rho u c_p T) = \nabla \cdot (\lambda_{\text{th}} \cdot \nabla T) + S_h \tag{1}$$

is solved, where T is the temperature, t is the process time, u is the vector of the velocity, λ_{th} is the heat conductivity, c_p is the specific heat capacity and ρ is the density. As introduced by Rösler and Brüggemann [30], the equation is extended by the source term

$$S_h = -\frac{\Delta h_f \left(4 \cdot \exp\left(\left(\frac{4(T - T_{\text{m}})}{T_{\text{liquid}} - T_{\text{solid}}}\right)^2\right)\right)}{(T_{\text{liquid}} - T_{\text{solid}})\sqrt{\pi}} \cdot \frac{\partial T}{\partial t} \tag{2}$$

in order to account for the latent heat Δh_f of the phase change between liquid and solid, where T_{liquid} is the liquidus temperature, T_{solid} is the solidus temperature and T_{m} is the average of liquidus and solidus temperature.

This calculation of the temperature distribution is coupled to the calculation of the melt flow, which is based on the solver *icoFoam* from *OpenFOAM®*. With this the incompressible laminar Navier-Stokes equations are solved for each discrete time step by applying the PISO algorithm [32–34]. In order to distinguish between the fluid flow in the weld pool and the constant velocity of the solid sheet, which is equal to the welding speed, a source term

$$S_m = -\frac{C\left(1 - (1 - f_s)\right)^2}{(1 - f_s)^3 + b} \cdot u_{\text{rel}} \tag{3}$$

based on the Darcy law for the flow in porous media is added to the momentum equation

$$\frac{\partial u}{\partial t} + (u \cdot \nabla)u - \nabla \cdot (\nu \nabla u) = -\nabla p + S_m \tag{4}$$

as proposed in [35], where C is the mushy zone constant, b is a small value which is introduced to avoid division by zero, u_{rel} describes the velocity vector relative to the feed rate of the solid part, ν is the kinematic viscosity, p is the pressure and f_s is the solid fraction which describes the phase state. A solid fraction of $f_s = 0$ represents the liquid state and a solid fraction of $f_s = 1$ represents the solid state. During the solidification between T_{liquid} and T_{solid} the solid fraction increases from $f_s = 0$ to $f_s = 1$. In this model, the course of this increase of the solid fraction is described by a sigmoid function according to [30]. During melting the decrease of the solid fraction from $f_s = 1$ to $f_s = 0$ is described by the same sigmoid function.

As shown by Rösler and Brüggemann [8], for each material property one value for the solid state β_{solid} and one for the liquid state β_{liquid} is specified and β is calculated as

$$\beta = f_s \cdot \beta_{\text{solid}} + (1 - f_s) \cdot \beta_{\text{liquid}}. \tag{5}$$

For each state the material parameters are considered as constant values. In order to drastically reduce the simulation time, the model is applied to a two-dimensional mesh. It is assumed that the process is symmetric with respect to the weld's centreline. In order to model this, the rectangular calculation area is limited to one half of the weld seam and a symmetry boundary condition is applied. To further reduce the complexity, the keyhole is assumed to be of a constant cylindrical shape. This is represented in the mesh by a constant geometrical boundary with a cylindrical shape with a diameter equal to the beam diameter on the surface d_f. Different boundary conditions can be assigned to the front and the back of the keyhole. In order to account for the feed rate of the sheet, a constant velocity is applied at the boundary of the mesh with constant temperature T_{amb} and $f_s = 1$. This results in the simulation of a sheet which is moved at constant feed rate $-u_{weld}$, while the laser beam is resting with constant position and orientation. The calculation area with all applied boundary conditions is shown in Figure 1. The coordinate system is placed at the centre of the keyhole with x in direction of the sheet's feed rate and y transverse to the feed rate. The detail view in Figure 1 shows the keyhole and a representation of the local mesh.

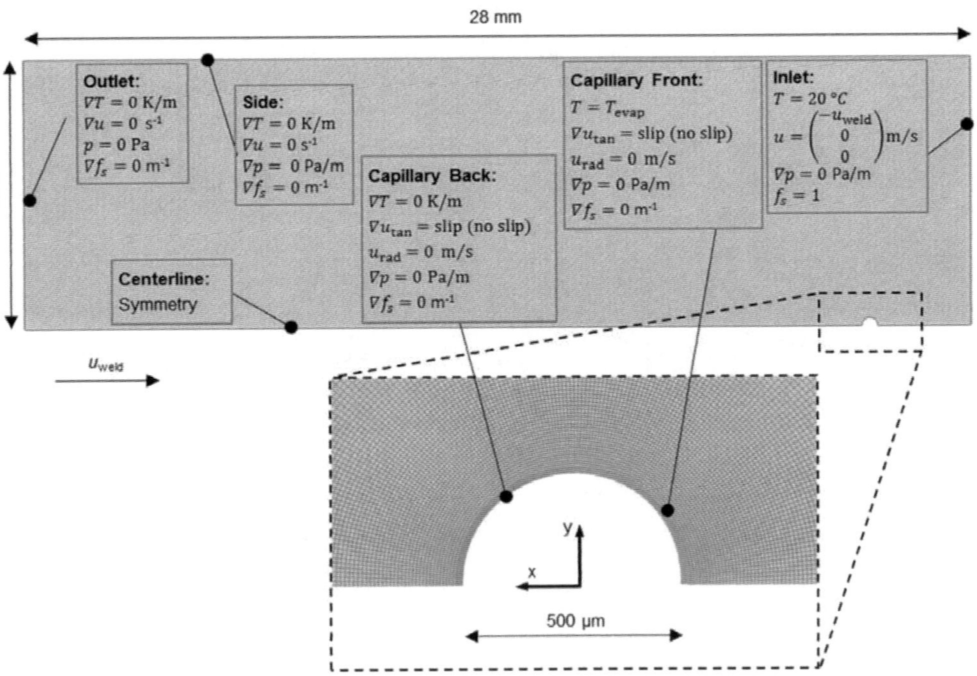

Figure 1. Two-dimensional calculation area with used boundary conditions. Top: Total calculation area. Bottom: Detail view of the keyhole with indication of the local mesh.

It has been shown that most of the laser beam power is absorbed at the front wall of the keyhole [36]. This is taken into account by using an adequate heat source: A constant temperature at the front side of the keyhole (orange in Figure 1) which is equal to the material's evaporation temperature. This results in a source term for each adjacent cell i

$$\dot{Q}_i = -\lambda_{\mathrm{th}} \cdot A_i \cdot \nabla \mathrm{T}_{\mathrm{rad}} \tag{6}$$

where A_i is the interface area at the keyhole wall for each cell and $\nabla\mathrm{T}_{\mathrm{rad}}$ is the corresponding radial temperature gradient. Consequently, the total absorbed power per penetration depth $P_{\mathrm{depth}} = \sum_i \dot{Q}_i$ varies for each simulated case and is provided with the results. At the keyhole's rear (blue in Figure 1) a *zeroGradient* boundary condition is applied for the temperature field. In order to gain insight into the interaction of the melt flow with the wall of the keyhole, the boundary conditions for the tangential flow u_{tan} at the wall were varied during the investigations. No mass transfer through the keyhole wall was assumed e.g. $u_{\mathrm{rad}} = 0\ \mathrm{m/s}$ for all cases.

The model can be considered as the top layer of a deep penetration process. With this model, laser beam welding of the aluminium alloy AA6016 was investigated. The material properties were chosen as listed in Table 1. In order to achieve faster convergence of the solution for all cases, at first the temperature field with a homogenous flow field with $u = u_{weld}$ was calculated. This allowed for a convergence to a steady state flow in the subsequent flow simulation after $t = 0.1$ s of simulated process time.

Due to the two-dimensional approach and the restriction of influencing factors, the model enables reduced simulation times. This provides the possibility to vary a large number of process parameters in short time.

Table 1. Constant properties used in the simulation of welding of aluminium alloy AA6016 [37, 38].

Parameter	Unit	Value
Ambient temperature T_{amb}	°C	20
Solidus temperature T_{solid}	°C	610
Liquidus temperature T_{liquid}	°C	660
Evaporation temperature T_{evap}	°C	2450
Enthalpy of fusion Δh_f	J/kg	390000
Heat capacity (solid) $c_{p,\text{solid}}$	J/(kg K)	900
Heat capacity (liquid) $c_{p,\text{liquid}}$	J/(kg K)	1126
Thermal conductivity (solid) $\lambda_{\text{th,solid}}$	W/(m K)	210
Thermal conductivity (liquid) $\lambda_{\text{th,liquid}}$	W/(m K)	105
Density (solid) ρ_{solid}	kg/m³	2700
Density (liquid) ρ_{liquid}	kg/m³	2220
Kinematic viscosity ν	m²/s	$4.08 \cdot 10^{-7}$
Mushy zone constant C	kg /(m³·s)	10^{7}
Small value (Darcy) b	-	10^{-7}
Focal / Keyhole diameter d_f	µm	500

3. Determination of factors influencing the melt flow

By separate modification of the boundary conditions of the model, the effect of different phenomena on the resulting melt flow can be investigated. In this section, the influence of viscosity in the melt and friction at the keyhole wall is analysed.

3.1. Influence of viscosity in the melt

The influence of the viscosity in the melt was investigated by comparison of the fluid flow with a kinematic viscosity ν of zero to the fluid flow with $\nu = \nu_{\text{aluminium}} = 4.08 \cdot 10^{-7}\ \text{m}^2/\text{s}$.

The flow fields resulting from the simulation with $\nu = 0\ \text{m}^2/\text{s}$ are given in Figure 2a with $u_{\text{weld}} = 6$ m/min (top) and $u_{\text{weld}} = 24$ m/min (bottom). Figure 2b shows the results for $\nu = 4.08 \cdot 10^{-7}\ \text{m}^2/\text{s}$. All cases were simulated without friction at the keyhole wall, i.e. a *slip* boundary condition. The local absolute values of the velocity vector u_{magn} are indicated according to the colour-coded scale. The upper and the lower limits of the scale indicate the minimum and maximum of the local velocity of the melt u_{magn} for all cases with equal welding speed. The white arrows are drawn along sections of individual streamlines to represent the local directions of the fluid flow. The black line shows the contour for $f_s = 0.5$, which represents the weld-pool boundary. In order to facilitate the analysis, only the relevant section of the calculation area is depicted in the image. All cases were calculated on the whole mesh, as given in Figure 1.

For all parameters, after melting, at the front of the weld pool, the melt flow is parallel to the welding direction. As expected the velocities in front of the keyhole are low with a stagnation point at the symmetry line. The local velocity increases towards the side of the keyhole.

In case of $u_{weld} = 6$ m/min, the calculated length of the weld pool was with viscosity 7.6 mm and without viscosity 7.4 mm. Both agree well with the average length of the weld pool $l_{wp} = 8.2\ \mathrm{mm}^{+1.0\ \mathrm{mm}}_{-1.2\ \mathrm{mm}}$ which was measured in the videos presented in previous work [39].

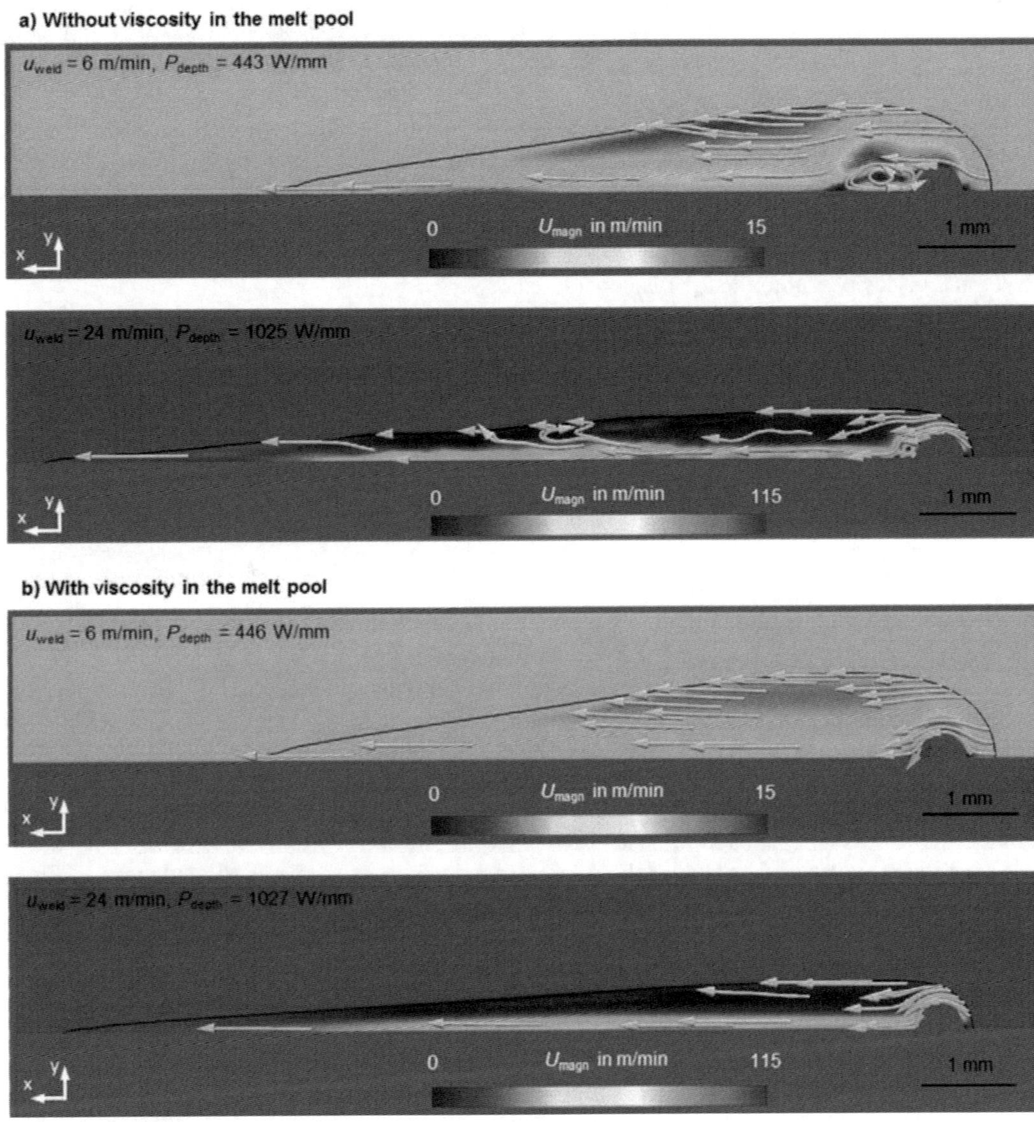

Figure 2. Fluid flow field a) without viscosity in the melt ($\nu = 0\,\mathrm{m^2/s}$). b) with viscosity in the melt ($\nu = 4.08 \cdot 10^{-7}\,\mathrm{m^2/s}$). Top: $u_{weld} = 6$ m/min. Bottom: $u_{weld} = 24$ m/min. Both cases were calculated with a *slip* boundary condition at the keyhole wall.

The flow field for $\nu = 0\,\mathrm{m^2/s}$ in Figure 2a shows different characteristics for both welding speeds. For $u_{weld} = 6$ m/min a counter-clockwise rotating eddy occurs behind the keyhole with fluid flow opposite to the welding direction and high velocities at the centre line and fluid flow in direction of the weld with high velocities at half of the weld-pool width. The diameter of the eddy is approximately twice the diameter of the keyhole. Behind the eddy, the flow re-attaches to the centreline and continues parallel towards the rear end of the weld pool. Next to the re-attached flow, towards the side of the weld pool, a wake region occurs with low velocities, as indicated by the blue area. The highest local velocities are about three times as high as the welding speed and occur at the side of the keyhole and in the region of the eddy. When the welding speed is increased to $u_{weld} = 24$ m/min, only a small instability forms after the keyhole. The flow re-attaches to the centreline directly behind this instability. The wake region at the side of the melt pool is more pronounced, compared to the case with $u_{weld} = 6$ m/min with velocities of 0 m/min. The highest local velocities are reached at the side of the keyhole and at the centreline and are approximately five times as high as the welding speed.

As shown in Figure 2b, when viscosity within the melt is considered, at both welding speeds the melt flows around the keyhole and re-attaches to the centreline directly behind it. In both cases this leads to the formation of a wake region towards the side of the melt pool, as indicated by the blue areas. Similarly to the case without viscosity, this wake region is more pronounced for $u_{weld} = 24$ m/min. Except for the absence of the small eddies behind the keyhole for $\nu = 4.08 \cdot 10^{-7}\ \mathrm{m^2/s}$, similar characteristics of the flow fields occur with and without viscosity in the melt. As expected, the maximum local velocities are lower when the viscosity within the melt is considered.

The results indicate that the consideration of the viscosity in the melt leads to a significant change of the characteristics of the fluid flow only in the region behind the keyhole in case of laser beam welding of aluminium. For both cases, a more distinct separation of the areas behind the keyhole is apparent at higher welding speeds with a small region of high local velocities at the centreline and a wide region of low local velocities towards the side. The absorbed power is similar with and without viscosity and increases from $P_{depth} \approx 440\ \mathrm{W/mm}$ at $u_{weld} = 6$ m/min to $P_{depth} \approx 1030\ \mathrm{W/mm}$ for $u_{weld} = 24$ m/min. The results show no significant influence of the viscosity in the melt on the resulting weld-pool geometry.

3.2. Influence of friction at the keyhole walls

In order to investigate the interaction between the wall of the keyhole and the melt flow, two cases with different boundary conditions were simulated. In the first case, a flow with friction on the keyhole wall was simulated. This was implemented by a *no slip* boundary condition, where the fluid has zero velocity relative to the keyhole. In the second case, a frictionless flow around the keyhole was modelled. This was implemented by a *slip* boundary condition which sets the radial component of the flow velocity u_{rad} at the wall of the keyhole to zero and keeps the tangential component u_{tan} unchanged.

The flow field resulting from the simulation with friction at the wall of the keyhole (*no slip*) is depicted in Figure 3a with $u_{weld} = 6$ m/min (top) and $u_{weld} = 24$ m/min (bottom). In order to facilitate the comparison of these two cases, the flow field without friction at the keyhole as shown above in Figure 2b is shown again in Figure 3b. Both cases were simulated with $\nu = 4.08 \cdot 10^{-7}\ \mathrm{m^2/s}$.

For both welding speeds, the flow with friction at the keyhole, as shown in Figure 3a, has low velocities where the solid material melts and where a stagnation point in front of the keyhole forms. This is similar to the results shown in Figure 2. Behind the keyhole the flow field differs significantly from the case without friction at the keyhole. Towards the side of the keyhole, the flow detaches and high local velocities are reached.

For $u_{weld} = 6$ m/min, the flow continues parallel to the centre line, as can be seen by the red area. This leads to the formation of a long eddy behind the keyhole. The width of the eddy is approximately equal to the keyhole's diameter. The local velocities in this eddy are low, as indicated by the dark blue area behind the keyhole. In contrast to the flow without friction at the keyhole, the flow does not attach to the centreline, which leads to higher local velocities in the region at the side of the melt pool.

For $u_{weld} = 24$ m/min, the flow re-attaches to the centreline. This leads to the formation of two eddies. The first is a counter clockwise rotating eddy behind the keyhole with smaller diameter, compared to the case with the lower welding speed. In the centre of the eddy, low local velocities down to 0 m/min occur. Towards the side of the eddy, these velocities are significantly increased. The second small eddy rotates clockwise and occurs towards the side of the melt pool, where the main flow is re-attached to the centreline. Inside of the wake, an area with zero velocity is formed.

The results indicate that friction at the keyhole walls pronounces the detachment of the flow at the side of the keyhole. This leads to significant changes in the flow field. The characteristics of the flow field change significantly when the welding speed is changed. The maximum local velocity as well as the geometry of the weld pool are not significantly affected by friction at the keyhole wall.

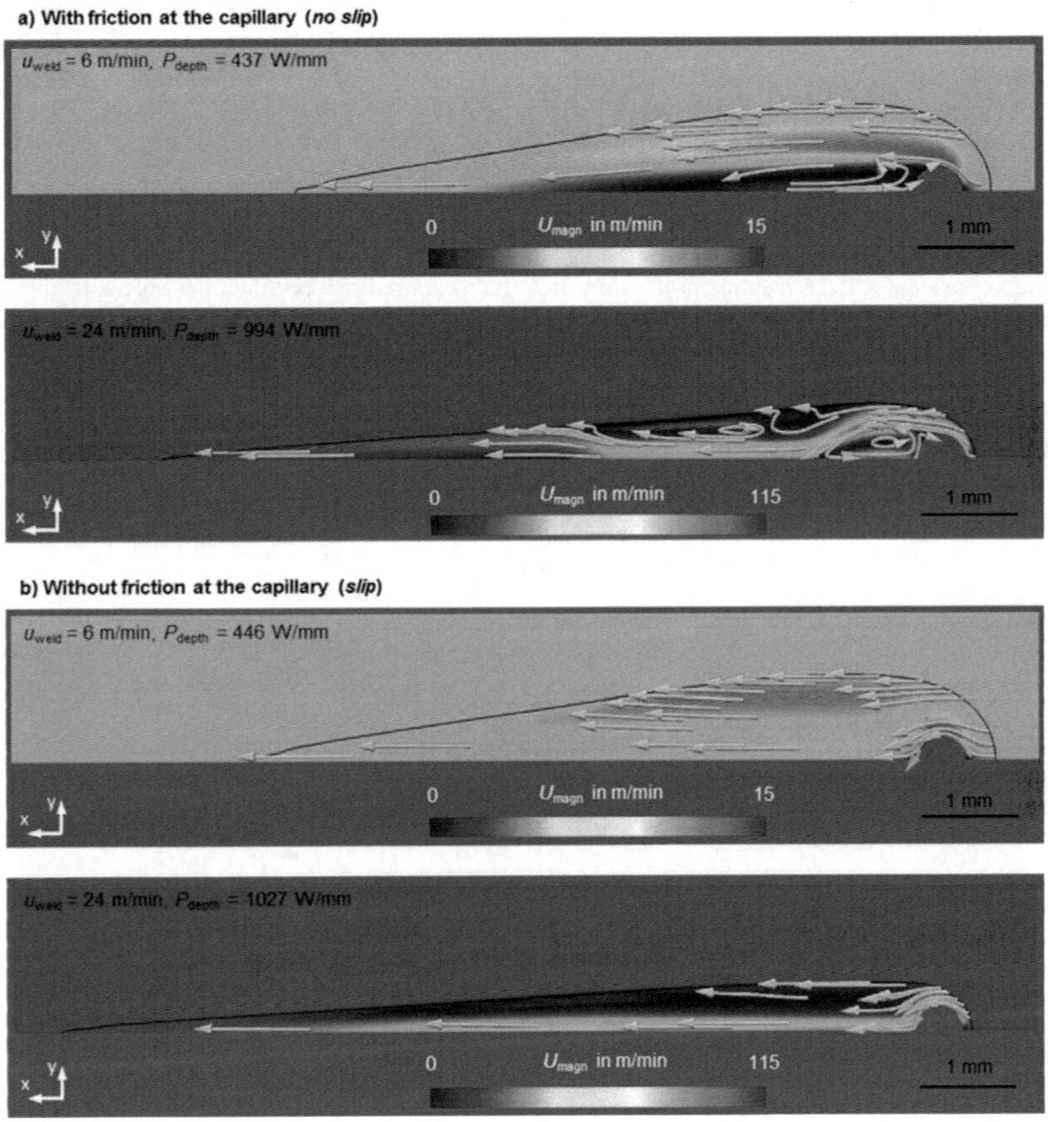

Figure 3. Fluid flow field. a) with friction at the keyhole wall (*no slip*) b) without friction at the keyhole wall (*slip*). Top: $u_{weld} = 6$ m/min. Bottom: $u_{weld} = 24$ m/min. Both cases were calculated with friction in the fluid ($\nu = 4.08 \cdot 10^{-7}$ m^2/s).

4. Conclusion

The proposed reduced numerical model allows for an analysis of the fluid flow during laser beam welding. Different aspects like viscosity in the melt and at the wall of the keyhole can be considered by changing the boundary conditions. This enables the separate analysis of the particular effects.

It was found that the consideration of viscosity in the melt leads to the formation of small eddies behind the keyhole. Besides this no additional effects on the characteristics of the fluid flow were found for laser beam welding of aluminium. A more pronounced effect may occur if materials with higher kinematic viscosity, e.g. steel are considered.

Under the assumption of friction at the wall of the keyhole, a significant influence was found. The flow detaches at the side of the keyhole, which leads to the formation of a counter-clockwise rotating eddy behind the keyhole. For u_{weld} = 6 m/min the flow continues parallel to the centre line with high local velocities whereas for u_{weld} = 24 m/min the fluid flow re-attaches at the centreline. This leads to higher velocities in the eddy behind the keyhole and at the centreline and subsequently to the formation of a second, clockwise rotating eddy towards the side of the melt pool.

If no friction at the wall of the keyhole is assumed, the overall characteristics of the flow are not changed when the welding speed is increased but the effects are more pronounced: higher local velocities at the centreline and lower velocities towards the side of the melt pool. Furthermore, a more distinct separation of these areas is apparent at higher welding speeds.

Due to the significant difference in the fluid flow the proposed model offers the potential to gain more insight into the interaction between the keyhole wall and the fluid by comparison with experimental results. It is planned to prove experimentally the presence of friction between keyhole and melt.

All results show no significant influence of these boundary conditions on the weld-pool geometry. A reason for this might be that due to the high heat conductivity of aluminium conductive heat transfer is more pronounced compared to convective heat transfer.

Future work will consider laser beam welding of further materials and the implementation of various geometries of vapour capillaries. Due to the short simulation times, the model offers the potential to design optimized keyhole shapes based on needed features in the fluid flow.

Acknowledgements
This work was funded by the Deutsche Forschungsgemeinschaft (DFG, German Research Foundation) – 266218804.

ORCID iDs
J. Wagner https://orcid.org/0000-0002-9906-3911
P. Berger https://orcid.org/0000-0003-4232-3732
P. He https://orcid.org/0000-0001-6054-2184
F. Fetzer –
R. Weber https://orcid.org/0000-0001-8779-2343
T. Graf https://orcid.org/0000-0002-8466-073X

References

[1] Rai R, Elmer J W, Palmer T A and DebRoy T 2007 *J. Phys. D: Appl. Phys.* **40** 5753–66
[2] Berger P, Hügel H, Hess A, Weber R and Graf T 2011 *Physics Procedia* **12** 232–40
[3] Collur M M, Paul A and DebRoy T 1987 *Metall and Materi Trans B* **18** 733–40
[4] Hugger F, Punzel E and Schmidt M 2019 Experimental results and modelling of element loss in continuous laser beam welding of aluminum alloys *LiM Lasers in Manufacturing Conference 2019 (Munich, June 24-27)* (Munich: WLT Wissenschaftliche Gesellschaft Lasertechnik e.V.)
[5] Mundra K, DebRoy T and Kelkar2 K M 1996 *Numerical Heat Transfer, Part A: Applications* **29** 115–29
[6] Ki H, Mazumder J and Mohanty P S 2002 *MTA* **33** 1817–30
[7] Vázquez R G, Koch H M and Otto A 2014 *Physics Procedia* **56** 1334–42
[8] Sonti N and Amateau M F 1989 *Numerical Heat Transfer, Part A: Applications* **16** 351–70
[9] Zhao H and DebRoy T 2003 *Journal of Applied Physics* **93** 10089–96

[10] Kaplan A 1994 *J. Phys. D: Appl. Phys.* **27** 1805–14
[11] Schuoecker D and Kaplan A F H 1994 Overview of modeling for laser applications *Laser Materials Processing: Industrial and Microelectronics Applications Europto High Power Lasers and Laser Applications V (Vienna, Austria, Sunday 3 April 1994) (SPIE Proceedings)* ed E Beyer *et al* (SPIE) p 236
[12] Tan W and Shin Y C 2014 *J. Phys. D: Appl. Phys.* **47** 345501
[13] Cho J-H and Na S-J 2006 *J. Phys. D: Appl. Phys.* **39** 5372–8
[14] Otto A and Schmidt M 2010 *Physics Procedia* **5** 35–46
[15] Gao X S, Wu C S, Goecke S F and Kügler H 2017 *Journal of Materials Processing Technology* **242** 147–59
[16] Otto A, Koch H and Vazquez R G 2012 *Physics Procedia* **39** 843–52
[17] Rodgers T M, Madison J D and Tikare V 2017 *Computational Materials Science* **135** 78–89
[18] Han R, Li Y and Lu S 2017 *International Journal of Heat and Mass Transfer* **106** 1345–55
[19] Wei H L, Elmer J W and DebRoy T 2017 *Acta Materialia* **126** 413–25
[20] Wei H L, Elmer J W and DebRoy T 2017 *Acta Materialia* **133** 10–20
[21] Draxler J, Edberg J, Andersson J and Lindgren L-E 2019 *Weld World* **63** 1489–502
[22] Bakir N, Üstündağ Ö, Gumenyuk A and Rethmeier M 2020 *Weld World* **64** 501–11
[23] Wei Y H, Dong Z B, Liu R P and Dong Z J 2005 *EWeld J (NY)* **13** 437–54
[24] Prieto C, Vaamonde E, Diego-Vallejo D, Jimenez J, Urbach B, Vidne Y and Shekel E 2020 *Procedia CIRP* **94** 596–600
[25] Jarwitz M, Lind J, Weber R, Graf T, Speker N and Haug P 2018 *International Congress on Applications of Lasers & Electro-Optics* **2018** 703
[26] Wang L, Mohammadpour M, Yang B, Gao X, Lavoie J-P, Kleine K, Kong F and Kovacevic R 2020 *Applied optics* **59** 1576–84
[27] Jasak H and Weller H G 2000 *Int. J. Numer. Meth. Engng* **48** 267–87
[28] Weller H G, Tabor G, Jasak H and Fureby C 1998 *Int. J. Numer. Meth. Engng* **12** 620
[29] Blomberg T 1996 *Heat conduction in two and three dimensions: Computer modelling of building physics applications (Report / TVBH* vol 1008*)* (Lund)
[30] Rösler F and Brüggemann D 2011 *Heat Mass Transfer* **47** 1027–33
[31] Fetzer F, Stritt P, Berger P, Weber R and Graf T 2017 *Journal of Laser Applications* **29** 22012
[32] Issa R 1986 *Journal of Computational Physics* **62** 40–65
[33] Moukalled F, Mangani L and M D 2016 *The finite volume method in computational fluid dynamics: An advanced introduction with OpenFOAM and Matlab* (Cham, Switzerland: Springer)
[34] Ferziger J H, Perić M and Street R L 2020 *Computational Methods for Fluid Dynamics* (Cham: Springer International Publishing)
[35] Voller V R and Prakash C 1987 *International Journal of Heat and Mass Transfer* **30** 1709–19
[36] Fetzer F, Hagenlocher C, Weber R and Graf T 2021 *Optics & Laser Technology* **133** 106562
[37] Kammer C 1998 *Aluminium-Taschenbuch* 15th edn (Düsseldorf: Aluminium-Verlag)
[38] Leitner M, Leitner T, Schmon A, Aziz K and Pottlacher G 2017 *MTA* **48** 3036–45
[39] Hagenlocher C, Weller D, Weber R and Graf T 2019 *Science and Technology of Welding and Joining* **24** 313–9

IOP Conf. Series: Materials Science and Engineering 1135 (2021) 012011 doi:10.1088/1757-899X/1135/1/012011

Root formation and mechanical properties in laser keyhole welding of 15 mm thick HSLA steel

Ivan Bunaziv[1,5], Cato Dørum[2], Steen Erik Nielsen[3], Pasi Suikkanen[4], Xiaobo Ren[1], Bård Nyhus[1], Magnus Eriksson[1], Odd Magne Akselsen[1]

[1]SINTEF Industry, P.O. Box 4760 Torgarden, NO-7465 Trondheim, Norway
[2]Norwegian Public Roads Administration, Bridge Section, Hamar, Norway
[3]FORCE Technology, Park Allé 345, DK-2605 Brøndby, Denmark
[4]SSAB Europe Oy, Rautaruukintie 155, FI-92100, Raahe, Finland

ivan.bunaziv@sintef.no

Abstract. Deep penetration laser welding is promising in joining thick (> 10 mm) steel sections. Focused laser beam by drilling vapour cavity, the keyhole, generates deep and narrow welds. Full penetration single-pass joining has a persistent problem with root quality where humping is one of the most frequent imperfection. This strongly hampers the use of high-power laser for thick plate welding. A 16 kW disk laser was used for single-pass welding of 15 mm thick plates in a butt joint configuration. Root humping occurred within a wide range of welding parameters. This provides narrow processing window. By adding an arc source to the laser beam process, the tendency of root humping increases. To achieve humping-free welds and consistent root quality over length, a delicate balance of process parameters is required. High heat input (> 0.50 kJ/mm) was positive to achieve a combination of low hardness (< 325 HV) and good Charpy toughness at −50 °C (> 50 J).

1. Introduction

Thick steel plates (> 10 mm) are frequently used for structural application in assembling bridges, ships, structural beams etc. Arc welding is the most widely used process for fusion joining of such components due to reliability and lower cost investments. Laser beam welding (LBW) and laser-arc hybrid welding (LAHW) can become a viable alternative [1]. Their ability in joining plates with thicknesses beyond 10 mm may be restricted due to several factors such as sensitivity to various imperfections and high hardness. When short-wavelength Yb:fibre/disk lasers (1030-1070 nm wavelength) are used, a very few investigations have been made for thick plates using single-pass welding technique. Single-pass welding is of high relevance due to simplicity and readiness for production without the need to turn the sections and where access is limited, e.g. very large plates or orbital pipeline welding. However, full penetration single-pass welding often causes imperfections such as root humping or sagging. Haug et al. [2] showed that 1070 nm wavelength disk laser had very narrow process window in case of 12 mm thick steel joining. By contrast, a 10 600 nm long wavelength (CO_2) laser provided much wider power process window. Short-wavelength lasers have higher absorption coefficient and subsequent melt overheat with unfavourable melt dynamics at the keyhole exit. Frostevarg [3] reported that LAHW often provides humping due to unfavourable melt dynamics in the weld pool and wider root width compared to LBW. There are still many unknown

IOP Conf. Series: Materials Science and Engineering 1135 (2021) 012011 doi:10.1088/1757-899X/1135/1/012011

factors in why and how to avoid root imperfection in welding thick plates. Application of electromagnetic backing can be a solution [4-9]. However, its viability on a full industrial scale (for welds longer than 5-10 m) is not proven. Therefore, further investigations are required on root formation.

Another challenge in LAHW/LBW is narrow and deep welds which tend to have excessive hardness due to the rapid cooling rates associated with lower heat input. In LBW of 20-25 mm carbon steel with 0.10 wt.% C, hardness up to 500 HV was reported [10]. With the use of lower carbon steels (< 0.08 wt.% C), the hardness can be up to 380 HV in partial penetration welds [11]. Such high hardness can be reduced by optimizing process parameters; an increase of laser beam power is effective to slow down the cooling rate [11], or alternatively preheating can be applied [12]. Up today, numerical simulations of root humping formation physics are not widely published. The main reason may be related to the complexity of the process since root formation requires an explicit model of the keyhole with considerable calculation time. Moreover, the precise mixture of filler wire and base metal must be identified in the molten state and during solidification, in addition to temperature dependent thermophysical properties such as surface tension, viscosity, density etc.

The present work is based on experimental observations using a wide range of parameter variations having heat input difference up to two times (0.48-0.82 kJ/mm). A 16 kW disk laser was used to weld 15 mm thick plates with a single-pass welding technique with the objective to obtain acceptable quality welds, level B according to ISO 12932 [13], and high toughness (min. 27 J at −50 °C).

2. Methodology

2.1. Materials

The steel plates were 15 mm in thickness and cut to dimensions of 500 mm in length and 150 mm in width. The base metal (BM) is thermomechanically rolled steel (TMCP) with a microstructure consisting of banded ferrite-pearlite providing min. 40 J impact toughness at −60 °C. A commercial 1.2 mm solid wire was selected. The steel and filler wire chemical compositions are shown in table 1. The BM has a carbon equivalent value of $CE_{IIW} = 0.38$ and $P_{cm} = 0.25$ representing good weldability. The base metal has 420 MPa min. yield strength ($R_{p0.2}$) and 520-680 MPa ultimate tensile strength (R_m). The solid filler wire provides yield and tensile strength of 420-460 MPa and 500-680 MPa, respectively.

Table 1. Chemical composition (wt.%) of base metal S420ML (mill certificate values, with small amounts of Nb, V and Ti < 0.05 wt.%) and wire (nominal values in manufacturer's datasheet).

Material	C	Si	Mn	P	S	Fe
Base metal	0.14	0.50	1.60	0.020	0.015	balance
Wire	0.10	1.00	1.70	0.025	0.025	balance

The Y-bevelling geometry is shown in figure 1a. Here, LBW or LAHW were used for the root with filling by the chosen arc welding method (GMAW). A standard (non-pulsed, spray) droplet transfer mode was chosen. Machined edges were used with fine surface quality ($R_a = 0.8$ μm). Run-in and run-out plates were tack welded to the plates.

2.2. Equipment and setup

A continuous wave multi-mode 16 kW disk laser (TruDisk16002 TRUMPF) was used in the experiments with 200 μm fiber core diameter, 8 mm·mrad beam parameter product, and 1030 nm wavelength. The laser beam had 600 mm focal length focused to a spot size of 300 μm in diameter.

The LAHW setup is shown in figure 1b. The laser beam had inclination angle was 7° from normal to eliminate high back reflections. A gas metal arc welding (GMAW) torch was tilted by 75° to the welding surface. The welds were deposited with an articulated robot. The filler wire stick-out was 20±2 mm, and the shielding gas composition was 80% Ar mixed with 20% CO_2 for LAHW and pure Ar for LBW with flow rate of 25 l/min.

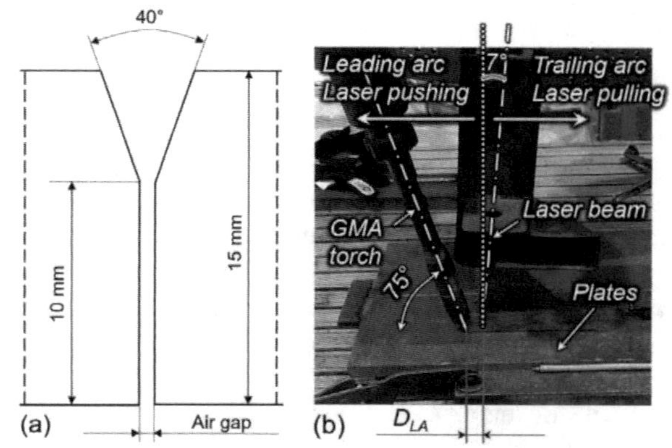

Figure 1. (a) Y-bevelling geometry and (b) welding setup.

2.3. Process variables

The process variables for LAHW and LBW are listed in table 2 and table 3, respectively. The main parameters are air gap, laser beam power (P_L, kW), welding speed (v_s, m/min), laser-arc distance (D_{LA}), wire feed rate (*WFR*, m/min), focal point position (*FPP*) and arc position (*AP*). Calculation formula of heat inputs which is measured in kJ/mm (Q_L for the laser, Q_A for the arc, Q_H for the LAHW) can be found in [14].

Table 2. LAHW parameters. *LA* – is leading arc, *TA* – is trailing arc. *a* – air gap (mm). The filling pass was made with arc using *WFR* of 9 m/min, travel speed of 0.5 m/min and heat input of 0.78 kJ/mm.

Weld	*AP*	*a*	P_L	v_s	*WFR*	D_{LA}	*FPP*	Filling pass	Q_L	Q_A	Q_H
1H	*LA*	0.5	10	1.0	4	15	0	–	0.42	0.14	0.56
2H	*LA*	0.5	10	1.0	2	5	0	–	0.42	0.06	0.48
3HA[a]	*LA*	0.5	10	1.0	2	15	0	+	0.42	0.06	0.48
3HB[a]	*LA*	0.5	10	1.0	2	15	0	+	0.42	0.06	0.48
4H	*LA*	0.5	12	1.0	2	5	0	–	0.50	0.06	0.56
5H	*LA*	0.5	12	1.0	4	15	0	–	0.50	0.14	0.64
6H	*LA*	0.5	12	1.0	2	15	+25	–	0.50	0.06	0.56
7H	*TA*	0.0	16	0.9	2	15	+25	–	0.75	0.07	0.82
8H	*TA*	0.0	16	1.2	2	15	+25	–	0.56	0.05	0.61
9H	*TA*	0.0	16	1.3	2	15	+25	–	0.52	0.05	0.57

[a] These two experiments had identical parameters and A/B is used to distinguish runs.

Table 3. LBW parameters. *LP* – is laser position. *PS* – is laser pushing, *PL* – is laser pulling. *a* – air gap (mm). Filling pass made with arc with *WFR* of 6-10 m/min, travel speed of 0.5 m/min and heat input of 0.41-0.95 kJ/mm.

Weld	*LP*	*a*	P_L	v_s	*FPP*	Filling pass	Q_L
1L	*PS*	0.0	9	1.0	0	+	0.38
2L	*PS*	0.0	10	1.0	0	+	0.42
3L	*PS*	0.5	9	1.0	0	+	0.38
4L	*PS*	0.5	10	1.0	0	+	0.42
5L	*PS*	0.5	11	1.0	0	–	0.46
6L	*PL*	0.0	16	1.2	+25	+	0.56
7L	*PS*	0.0	16	1.2	+25	+	0.56

2.4. Testing and characterization

The Charpy V-notch (CVN) impact toughness test was performed according to ISO 9016 [15] with standard specimen dimensions of 55×10×10 mm^3 and V-notch type geometry. Three test series were included; (*i*) the notch located at the weld metal (WM) centre in the arc zone, (*ii*) WM centre in the laser zone, and (*iii*) the fusion line (FL) of the laser zone (where the fusion line is straight and nearly perpendicular to the plate thickness direction). The samples were cut transverse to the welding direction (WD) with parallel fracture path during Charpy test. All tests were performed at −50 °C, using three specimens for each notch position.

Metallographic studies were performed according to the ISO 17639 [16] standard. Etching was carried out in a 2% nital solution for 5 s. Macrographs were examined with optical microscope for microstructure characterization. Measurements of Vickers microhardness ($HV_{0.5}$) were conducted with a 500 gf load according to the ISO 22826 [17] standard.

3. Results and discussion

3.1. Laser-arc hybrid welding

Initial experimental runs were conducted to achieve full penetration by using 0 mm *FPP* and leading arc position with low *WFR* (2-4 m/min). Welds No. 1H and 2H had incomplete penetration by using 10 kW laser power and 1.0 m/min weld speed at different D_{LA}. A reduction of *WFR* to 2 m/min with 15 mm D_{LA}, full penetration was achieved Weld No. 3HA, see figure 2. Thus, lower amount of added filler wire does not affect the effective thickness of the plates. However, full penetration was unexpected. The process shows very high sensitivity to D_{LA} and small changes in *WFR*. Moreover, both humping or dropout and melt ejection (melt expulsion or cutting) were present within the same weld. Such behaviour is very unusual in welding, and this implies that there is process instability issues or lack of melt flow control in the root. This observation may be related to the fact that higher laser power is required to achieve full penetration in welding of thick plates (> 12 mm). Hence, a larger weld pool volume and higher pressure inside the keyhole is reached due to high temperatures with excessive evaporation. When the keyhole penetrated the whole plate, turbulent melt flows were achieved with unstable behaviour. The experiment was repeated (Weld No. 3HB), and similar results were achieved with few large underfills. By increasing the laser power to 12 kW, the strong melt ejection mode (similarly to laser cutting) is achieved at different D_{LA}.

Different parameter set was applied including defocused laser at +25 mm *FPP* with increase of laser power, zero air gap, and trailing arc setup. The humping was generated more frequently (Welds No. 6H–8H, see figure 2) at any welding speed with inconsistent top weld quality. By increasing welding speed to 1.3 m/min (Weld No. 9H), the humping was significantly reduced but still variation of underfill and humping was present. This point may indicate that *FPP* of 0 mm with narrower keyhole at the bottom is more beneficial for the root quality. In addition, a small air gap (0.5 mm) provided better results since less BM was melted and the laser beam went partially through the plates. Thus, a lower pressure inside the keyhole was formed. It is challenging to achieve high quality of the root with the LAHW process due to the extra heat from the arc and unstable behaviour of the keyhole by adding filler material.

3.2. Laser beam welding

Autogenous LBW was tested with 0 mm *FPP*. Promising results can be achieved at zero air gap and 10 kW laser power (Weld No. 2L, figure 3). However, the weld root had inconsistent quality with humping and slight underfill at the same time. The root can be qualified to level B according to ISO 12932 [13]. However, the geometrical inconsistency can provide low fatigue properties and may not be suitable for application with dynamical loads. Such inconsistent quality of the root formation might be related to transversal laser beam oscillations which can be natural or forced (e.g. by equipment and/or optics, thermal lensing effect) with unequal amount melting of base metal, weld plume oscillations causing irregular laser power supply, and unequal laser power distribution due to multi-mode technology.

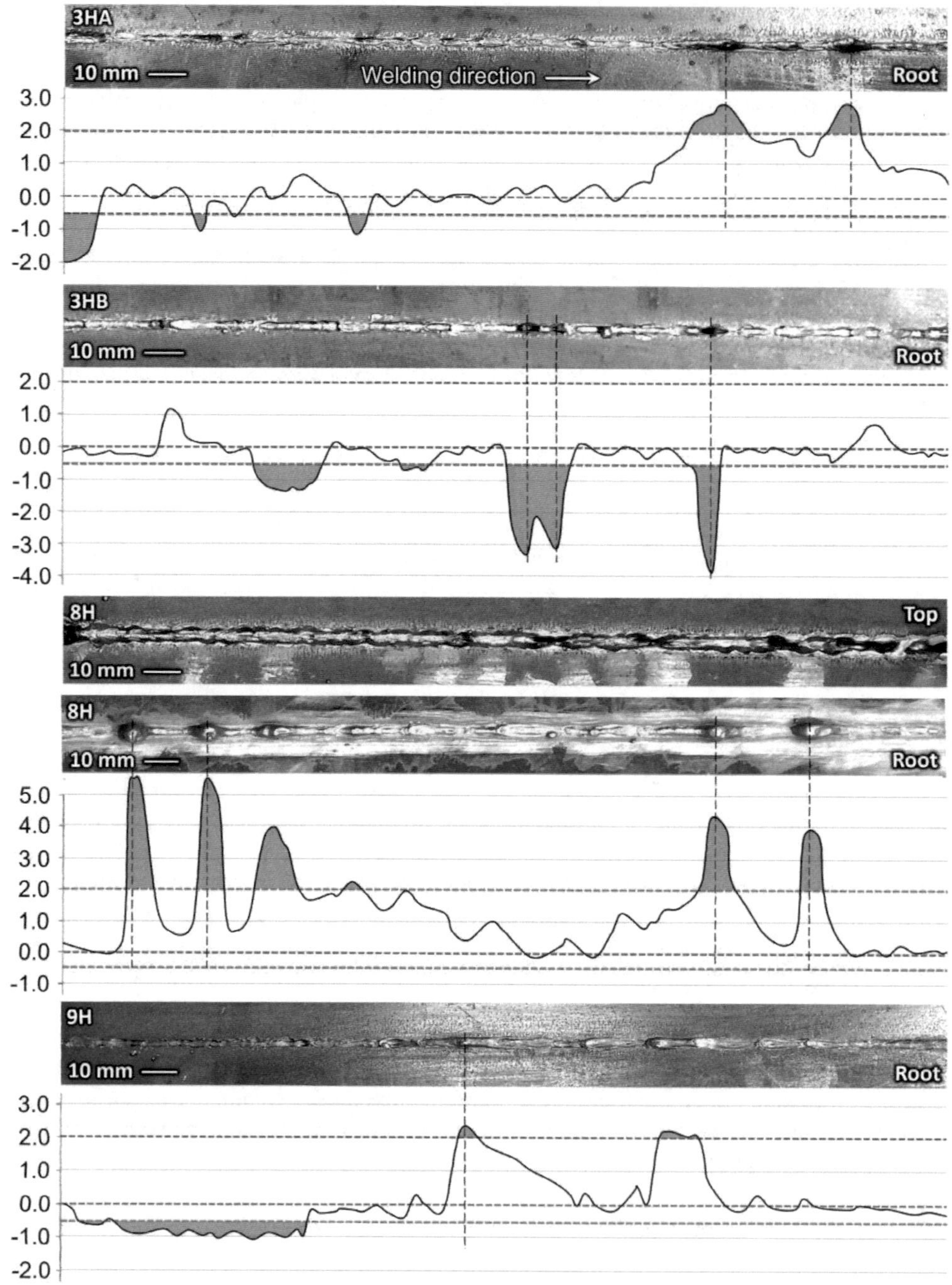

Figure 2. Weld seam appearance from root side of LAHW. Red dashed lines represent tolerance limits for imperfections, level B according to ISO 12932 [13]. *y*-axis shows deviation (in mm) of humping (upwards from 0.0) and underfill (downwards from 0.0).

By increasing the air gap to 0.5 mm, the quality was insufficient (Weld No. 3L). Further increase in laser power (by 10-20%), generated even more underfill. This implies that the process switched to a melt ejection mode. An increase in beam power only by 1 kW (< 10%) caused more underfilling through the whole weld length (Weld No. 4L). A further increase of laser power by 1 kW, generated more underfill (Weld No. 5L). This is another indication of the process sensitivity to a slight change in certain parameters (e.g. laser power).

By changing the *FPP* to +25 mm, combined with higher laser energy of 16 kW, zero air gap and welding speed of 1.2 m/min, the underfilling from the root was diminished (see Weld No. 6L). However, still some humping was present within the limits of the level B quality. A flatter and more consistent weld root quality was achieved (Weld No. 7L) by using laser pushing setup. In this case, the process is in melt ejection mode and showed good process repeatability (more than 3 m long welds were made in total). As a result, more stable process can be established with zero air gap using higher laser power, in melt ejection mode. In this case, the hanging humps were dropped out (cut out) from the root of the weld pool by generating higher pressure at the keyhole exit.

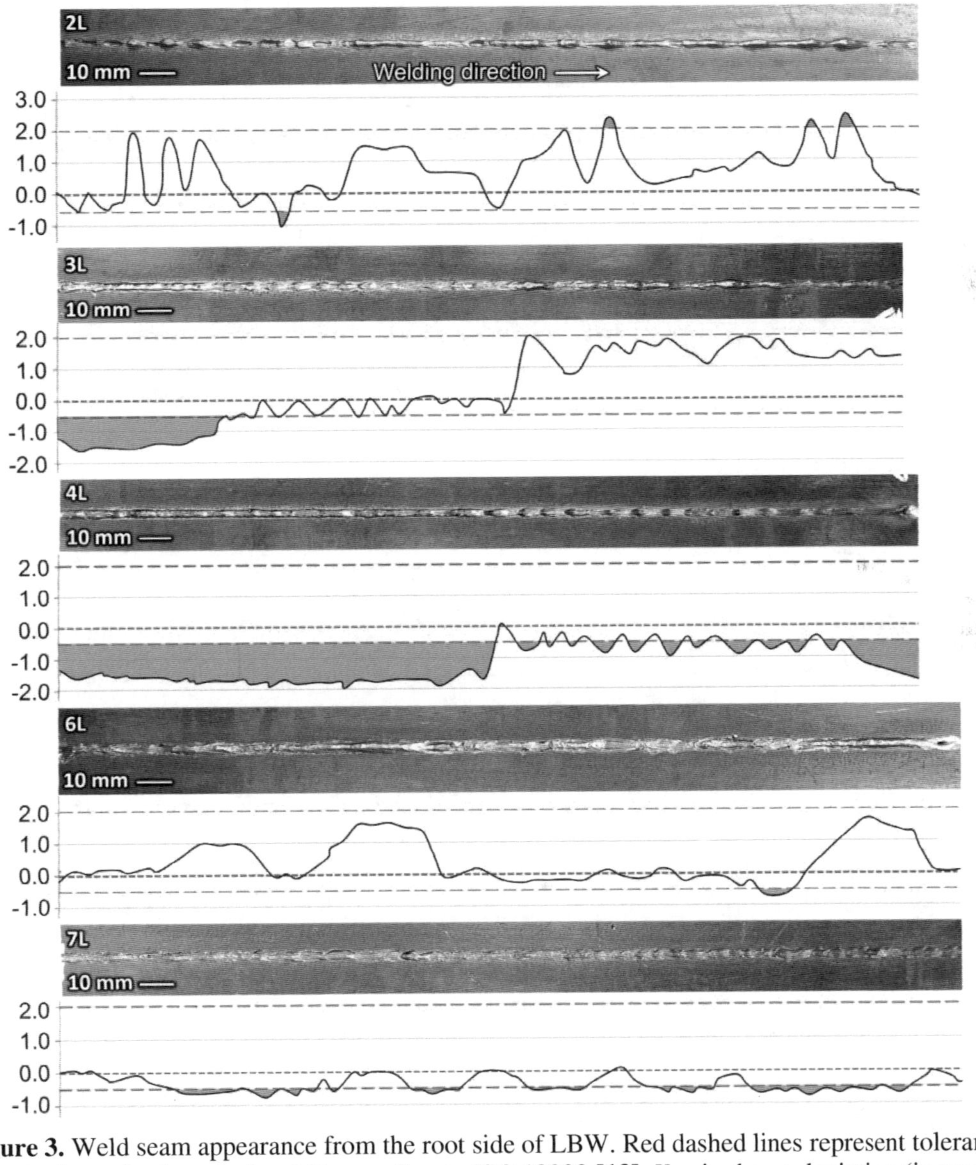

Figure 3. Weld seam appearance from the root side of LBW. Red dashed lines represent tolerance limits for imperfections for level B according to ISO 12932 **[13]**. *Y*-axis shows deviation (in mm) of humping (upwards from 0.0) and underfill (downwards from 0.0).

In figure 4, a comparison of the present study with published results is shown. The present case with carbon steel and 15 mm thickness, has a very narrow process window compared to thinner plates since higher heat inputs were used. Higher power density is required to achieve full penetration and leads to higher temperatures in keyhole, larger melt pool with faster melt flows at keyhole front (> 15 m/s [18]). As soon as keyhole penetrated the bottom surface, there is low control over melt dynamics, thus humping is produced. Here, the balance of forces is interrupted since surface tension is too low to sustain dropout and large weld pool with high melt forces are increased by gravity [3]. As a result, the balance is interrupted leading to high susceptibility to humping. A further increase in the laser beam power provides smooth root appearance with a slight reinforcement. Subsequent increase in laser power, generates melt ejection mode generating underfill/undercuts in the root area. More advanced studies are required in such case to understand the real reason of humping and to develop preventive methods. One of the possibilities to prevent humping is to use much smaller spot size at the same time lower laser power is needed. However, it will lead to high hardness increase. Chemical composition of the base metal and shielding gas may be important to reduce humping as shown by other researchers. The effect of filler wire also should be considered since it can influence on the physical properties of the molten metal during solidification with surface tension probably being the most influential during root formation [3, 19].

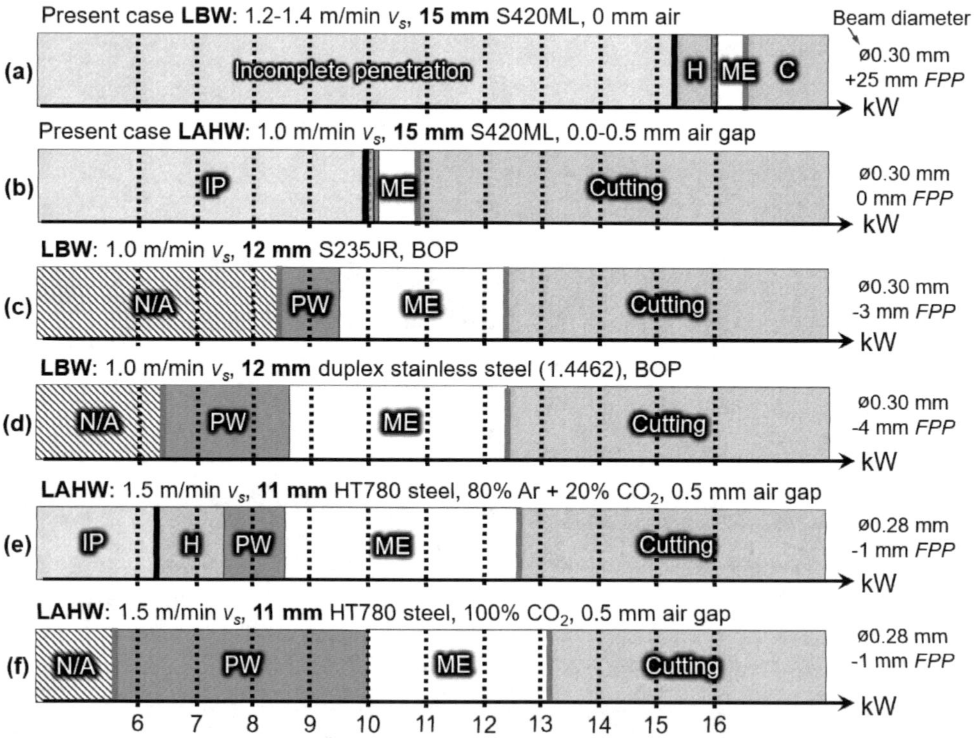

Figure 4. Comparison of (a, b) present case results with published results from (c, d) TRUMPF [2, 20] and (e, f) Osaka University [21]. *IP* is incomplete penetration, *ME* is melt ejection, *PW* is process window, *H* is humping, *C* is cutting, *N/A* is not reported. BOP is bead-on-plate experiments.

3.3. Microstructure and mechanical properties

In the upper part of the welds, a fine-grained acicular ferrite was developed (nearly the same heat input was used in almost all experiments, see table 2) and shown in figure 5. Therefore, high impact toughness is expected. In the laser part in WM, regardless of heat input, a mixture of upper bainite with some lath martensite (possibly with small fraction of retained austenite) was observed (figure 6).

As a result, lower impact toughness is expected. A mixture of martensite and bainite was found in the CGHAZ in all welds, see figure 7.

Hardness results are shown in figure 8a,b for low and high heat input. Hardness in the arc part in WM was low, 250 HV, similarly to the HAZ of the arc part, 300 HV. Higher hardness was found in the laser part. Low laser heat input (0.38 kJ/mm) provided 381 and 351 HV in the HAZ and WM, respectively. Higher heat input (0.56 kJ/mm) provided hardness of 346 HV and 333 HV, which satisfies the acceptance criterion (< 350 HV).

The Charpy test results are shown in figure 8c. As expected, the lowest toughness occurred for specimens extracted from the laser zone with the notch located at the weld centreline. The average impact toughness was 54 J. Such toughness is still much higher than the minimum acceptable value of 27 J at −50 °C. Slightly higher toughness was found for the arc zone with an average of 84 J. This observation agrees with the higher acicular ferrite volume fraction. The fusion line toughness results showed very high values with the average of 223 J. By inspection of the specimens after testing, there is a clear indication that the fracture deviated towards the fine-grained heat affected zone and the BM. Whereas in case of WM, the fracture was propagating within the weld metal. It is possible that the toughness at the fusion line may provide lower values if the crack did not propagate into other softer zones due to strength mismatch in the heterogeneous weld. Another reason of deviation may be due to narrow width (50-150 μm) of HAZ.

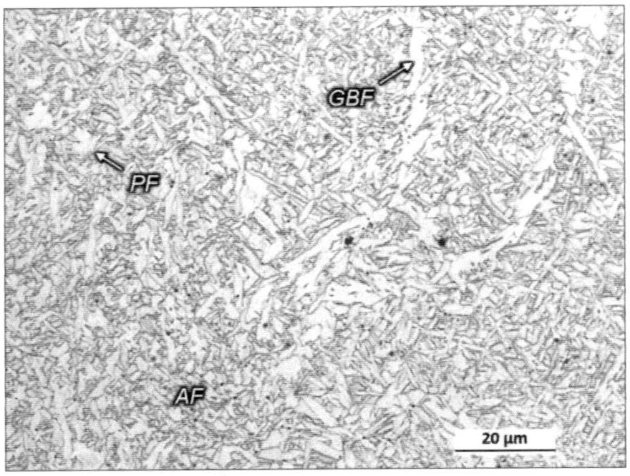

Figure 5. Weld metal microstructure in arc part (Weld No. 7L) representing large fraction of small grained acicular ferrite (AF) with some polygonal ferrite (PF) and grain boundary ferrite (GBF).

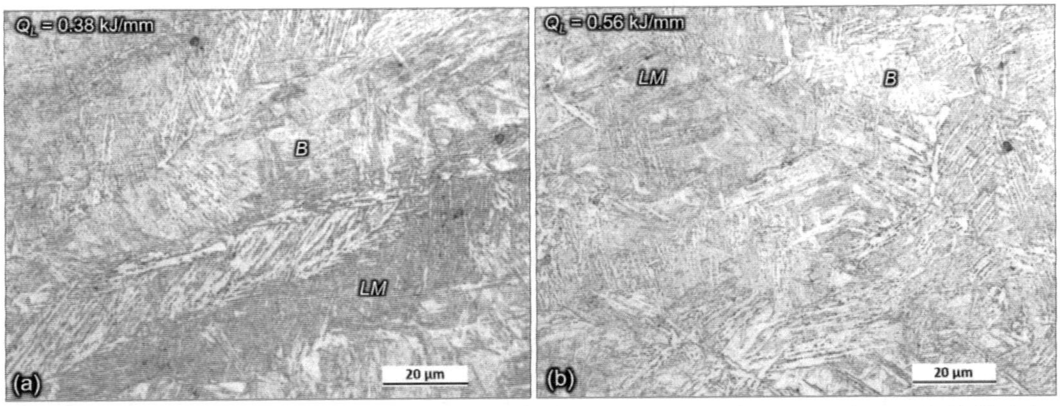

Figure 6. Weld metal microstructure in laser part in root for (a) low (Weld No. 3L) and (b) high heat input (Weld No. 7L) showing mixture of bainite (B) and lath martensite (LM).

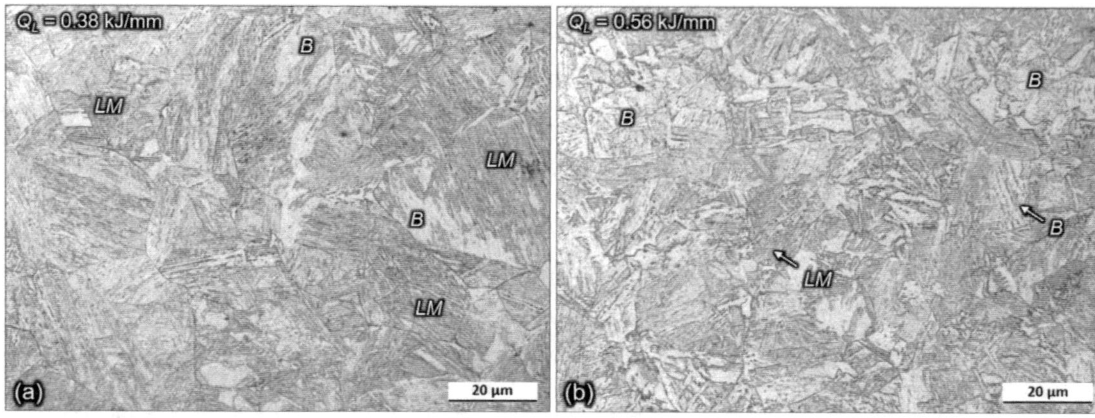

Figure 7. Microstructure in CGHAZ for (a) low heat input weld (Weld No. 3L) and (b) high heat input (Weld No. 7L) showing mixture of bainite (B) and lath martensite (LM).

Weld No. 7L did not qualify due to root underfill. However, based on X-ray radiographic testing, no cracks were detected, and only few pores were observed. Solving the issue of root humping is of primary concern since this has been found to represent the most frequent imperfection in single-pass welding according to this work and noted by other several researchers [2, 3, 22]. LBW/LAHW may gain increased acceptance in industry, even though some hardness spikes may occur. These spikes are easily prevented by using more modern TMCP steels with finer grain size and decreased carbon content (< 0.07% C).

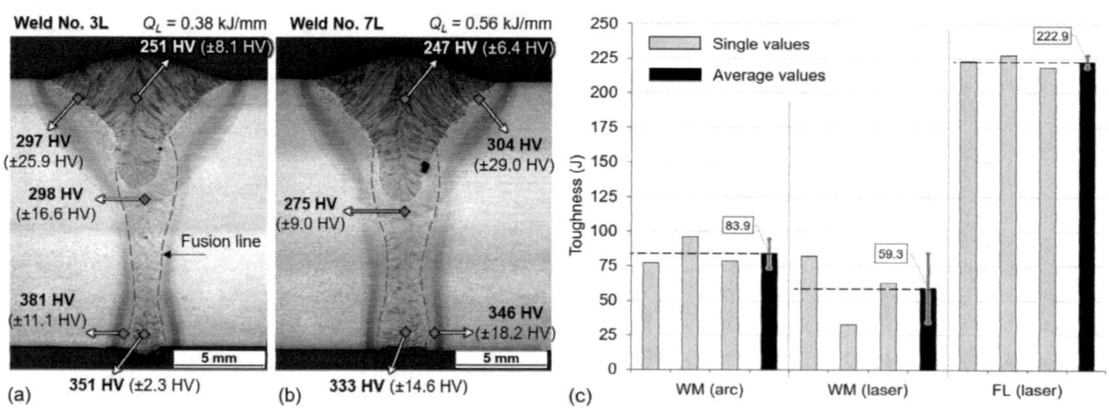

Figure 8. Hardness results of (a) low heat input weld and (b) high heat input weld. (c) Charpy V-notch toughness results of Weld No. 7L at −50 °C.

4. Conclusions

Laser beam and laser-arc hybrid welding has been conducted for a 15 mm thick 420 MPa structural steel. Based on the experimental results, the following conclusions can be drawn:

- LBW and LAHW very sensitive processes for the root formation using high powers (> 9 kW) and large thickness of plates causing high temperatures, pressure in the keyhole with unstable melt conditions.
- LBW showed better root control compared to LAHW due to absence of the secondary heat source. The added arc has an adverse effect on the root formation affecting melt flows. Therefore, LBW showed higher robustness due to repeatability compared to LAHW.

- The process is more robust within the melt-ejection mode due to narrow process window providing acceptable root quality. However, it may generate a slight root underfill and in some cases undercut.
- Appropriate hardness can be achieved (< 350 HV) by applying more than 0.50 kJ/mm heat input.
- LBW provided microstructure consisting of mixture of lath martensite and bainite. However, good toughness (> 27 J at 50 °C) was achieved showing promising results for industrial use.

Acknowledgements

The authors wish to acknowledge Statens vegvesen AS (The Norwegian Public Roads Administration) for funding the project. This publication has been partly funded by the SFI PhysMet (Centre for Research-based Innovation, 309584). The authors gratefully acknowledge the financial support from the Research Council of Norway and the partners of the SFI PhysMet. FORCE Technology (Odense, Denmark) and SSAB Europe Oy (Raahe, Finland) are appreciated for the cooperation.

ORCID IDs

I Bunaziv https://orcid.org/0000-0001-6925-7507
C Dørum https://orcid.org/0000-0002-0284-5002
X Ren https://orcid.org/0000-0003-1923-1733
M Eriksson https://orcid.org/0000-0003-2677-5077
O M Akselsen https://orcid.org/0000-0003-4894-1398

References

[1] Webster S, Kristensen J K and Petring D 2008 Joining of thick section steels using hybrid laser welding *Ironmak. Steelmak.* **35** 496-504

[2] Haug P, Rominger V, Speker N, Weber R, Graf T, Weigl M and Schmidt M 2013 Influence of Laser Wavelength on Melt Bath Dynamics and Resulting Seam Quality at Welding of Thick Plates *Phys. Procedia* **41** 49-58

[3] Frostevarg J 2018 Factors affecting weld root morphology in laser keyhole welding *Opt Lasers Eng* **101** 89-98

[4] Bachmann M, Avilov V, Gumenyuk A and Rethmeier M 2014 Experimental and numerical investigation of an electromagnetic weld pool support system for high power laser beam welding of austenitic stainless steel *J. Mater. Process. Technol* **214** 578-591

[5] Fritzsche A, Avilov V, Gumenyuk A, Hilgenberg K and Rethmeier M 2016 High Power Laser Beam Welding of Thick-walled Ferromagnetic Steels with Electromagnetic Weld Pool Support *Phys. Procedia* **83** 362-372

[6] Qi Y and Chen G 2018 Root defects in full penetration laser welding of thick plates using steady electromagnetic force *J. Mater. Process. Technol* **260** 97-103

[7] Qi Y, Chen G, Deng S and Zhou D 2019 Periodic root humps in thick-plate laser welding using steady electromagnetic force *J. Mater. Process. Technol* **273** 116247

[8] Üstündağ Ö, Gook S, Gumenyuk A and Rethmeier M 2019 Mechanical Properties of Single-pass Hybrid Laser Arc Welded 25 mm Thick-walled Structures Made of Fine-grained Structural Steel *Procedia Manuf.* **36** 112-120

[9] Avilov V, Fritzsche A, Bachmann M, Gumenyuk A and Rethmeier M 2016 Full penetration laser beam welding of thick duplex steel plates with electromagnetic weld pool support *J. Laser Appl.* **28** 022420

[10] Sokolov M, Salminen A, Kuznetsov M and Tsibulskiy I 2011 Laser welding and weld hardness analysis of thick section S355 structural steel *Mater. Des.* **32** 5127-5131

[11] Bunaziv I, Akselsen O M, Frostevarg J and Kaplan A F H 2018 Deep penetration fiber laser-arc hybrid welding of thick HSLA steel *J. Mater. Process. Technol* **256** 216-228

[12] Turichin G, Kuznetsov M, Pozdnyakov A, Gook S, Gumenyuk A and Rethmeier M 2018 Influence of heat input and preheating on the cooling rate, microstructure and mechanical properties at the hybrid laser-arc welding of API 5L X80 steel *Procedia CIRP* **74** 748-751

[13] ISO 12932 - Welding - Laser-arc hybrid welding of steels, nickel and nickel alloys - Quality levels for imperfections. International Organization for Standardization; 2013. p. 25.

[14] Bunaziv I, Frostevarg J, Akselsen O M and Kaplan A F H 2018 The penetration efficiency of thick plate laser-arc hybrid welding *Int. J. Adv. Manuf. Technol.* **97** 2907-2919

[15] ISO 9016 - Destructive tests on welds in metallic materials - Impact tests - Test specimen location, notch orientation and examination. International Organization for Standardization; 2012.

[16] ISO 17639 - Destructive tests on welds in metallic materials - Macroscopic and microscopic examination of welds. International Organization for Standardization; 2003. p. 9.

[17] ISO 22826 - Destructive tests on welds in metallic materials - Hardness testing of narrow joints welded by laser and electron beam (Vickers and Knoop hardness tests). International Organization for Standardization; 2005. p. 16.

[18] Eriksson I, Powell J and Kaplan A F H 2013 Melt behavior on the keyhole front during high speed laser welding *Opt Lasers Eng* **51** 735-740

[19] Petring D, Fuhrmann C, Wolf N and Poprawe R. Progress in Laser-MAG Hybrid Welding of High-Strength Steels Up to 30 mm Thickness. 26th International Congress on Applications of Lasers and Electro-Optics (ICALEO). Orlando, Florida, USA: Laser Institute of America; 2007. p. 300-307.

[20] Romminger V, Haug P, Speker N and Holzer M 2013 High-power Full Penetration Welding Behavior *Laser Tech. J.* **10** 36-40

[21] Pan Q, Mizutani M, Kawahito Y and Katayama S 2016 Effect of shielding gas on laser–MAG arc hybrid welding results of thick high-tensile-strength steel plates *Weld World* **60** 653-664

[22] Zhang M, Zhang Y, Mao C, Hu Y, Chen G and Bi Z 2019 Experiments on formation mechanism of root humping in high-power laser autogenous welding of thick plates with stainless steels *Opt Laser Technol* **111** 11-19

Hybrid additive manufacturing of hot working tool steel H13 with dissimilar base bodies using Laser-based Powder Bed Fusion

Lukas Langer[1], Matthias Schmitt[1], Jaime Cuesta Aguirre[1], Georg Schlick[1], Johannes Schilp[1,2]

[1]Fraunhofer Institute for Casting, Composite and Processing Technology IGCV, Am Technologiezentrum 10, 86159 Augsburg, Germany

[2]University Augsburg, Eichleitnerstr. 30, 86159 Augsburg, Germany

lukas.langer@igcv.fraunhofer.de

Abstract. Hybrid additive manufacturing (HAM) describes the combination of additively built structures onto a conventionally manufactured base body. The advantages of both manufacturing processes are combined in one process chain. As a result, new applications can be achieved with higher cost-effectiveness. With the Additive Manufacturing (AM) process a bonding zone is created that is comparable to a welded joint. In order to evaluate the quality and mechanical properties of the bonding zone, two steels (42CrMo4 and 25CrMo4) are investigated as base body materials with the hot working tool steel X40CrMoV5-1 (AISI H13) for the AM structure. Process parameters for Laser-based Powder Bed Fusion of X40CrMo4V5-1 are developed to achieve a crack and defect free structure as well as an optimized bonding zone in dependency of the base body material. Furthermore, the chemical and mechanical properties are examined in the as-built and heat-treated state. It is observed that a crack-free material bonding is possible and samples with relative densities above 99.5% are obtained. The size of the bonding zone depends on the material of the base body as well as post-process heat treatment. An average hardness of 600 HV1 can be achieved in the "as-built" state.

Keywords: dissimilar joint; hybrid manufacturing; hot working tool steel; bonding zone; AISI 4130; AISI 4140

1. Introduction

1.1. *Initial situation*

The motivation for Hybrid Additive Manufacturing (HAM) is the combined use of the advantages of Conventional Manufacturing (CM) and Additive Manufacturing (AM). Due to short lead-times and design freedom of prototypes AM and especially Laser-based Powder Bed Fusion of Metals (PBF-LB/M) is gaining relevance, not only in research applications [1]. The advantages of CM like turning and milling are the very well approved and common technologies, as well as fast and price-efficient manufacturing with high lot-sizes. By using HAM it is possible to manufacture economical and resource-efficient parts with challenging design or individualized structures on top of repeating and simple structures in the base body, as well as worn out parts [2–4].

By HAM 2D multi-material can be realized for even higher resource efficiency and ideal material properties of the final part. In Fig. 1 a schematic illustration of HAM is displayed. In comparison to regular PBF-LB/M the AM part will be not built upon a base plate and separated from it after the build job. The final, hybrid part consists of the base body and the AM part. By re-melting the base body and solidifying the powder layers of the AM part, a bonding zone is created. The joining of the two alloys is critical for the mechanical properties of the whole part. In this research the properties of the bonding zone will be investigated.

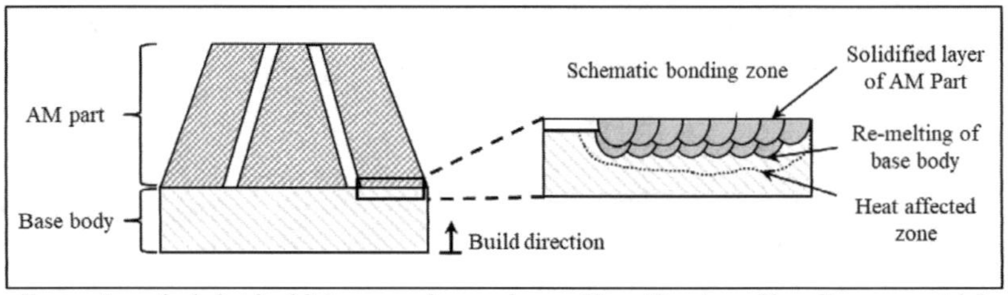

Fig. 1. Schematic illustration of a hybrid additive manufactured part. Magnification of bonding zone with heat affected and re-melting zone in base body with first, solidified layer in AM part. AM part with stylized melt pools in the bonding zone.

1.2. Material combination

Two alloys as base body material will be compared. Since 25CrMo4 and 42CrMo4 are only differentiated by their nominal carbon content, the influence of the carbon content in the base body for the properties and quality of the bonding zone will be directly compared. In Table 1 the nominal element composition of the alloys are displayed. The actual element composition of the powder feedstock (X40CrMoV5-1) was measured by Optical Emission Spectroscopy.

Table 1. Element composition of 25CrMo4 and 42CrMo4 according to DIN EN 10250 [5], X40CrMoV5-1 according to ASTM-A681 [6].

	Element composition				
Alloy	C	Cr	Mo	V	Fe
25CrMo4	0.22-0.29 wt.%	0.90-1.20 wt.%	0.15-0.30 wt.%	-	balance
42CrMo4	0.38-0.45 wt.%	0.90-1.20 wt.%	0.15-0.30 wt.%	-	balance
X40CrMoV5-1	0.32-0.45 wt.%	4.75-5.50 wt.%	1.10-1.75 wt.%	0.8-1.2 wt.%	balance

For better readability X40CrMoV5-1 will be named according to the ASTM standard AISI H13, while the base body alloys will be named according to their shortened European names 25CrMo4 and 42CrMo4. Since there are broad applications for hot working tool steels like H13 several research for AM has been conducted. The Carbon content of H13 is high, compared to alloys commonly processed in PBF-LB/M (e.g. 0.02 wt.% Carbon content in 316L), so weldability is reduced with tendency for cracks [7]. This can be compensated by higher platform heating or optimized process parameters [8].

In this research process parameters with 200 °C platform heating have been developed for producing dense and crack free parts of H13 and dissimilar joinings with 25CrMo4 and 42CrMo4.

1.3. Objective and approach

Hot working tool steels offer a broad range of industrial applications. Due to the design freedom especially in AM there is a rising demand for new materials [1]. Objective of this research is to develop process parameters for PBF-LB/M and to analyze the bonding zone of two material combinations in hybrid manufactured parts for gaining insight in 2D material combinations and hybrid manufactured parts in particular.

Initially the powder feedstock is characterized, a parameter study is conducted, hybrid parts are manufactured by PBF-LB/M and subsequently heat treated. Metallurgical and mechanical properties are systematically investigated on cross sections.

2. Experimental methods

Before specimens were produced, the powder feedstock of AISI H13 from Nanoval GmbH, Germany, was characterized. Gas atomized powder with a Particle Size Distribution (PSD) of +10/-45 µm was used for all the investigations. Samples were taken from virgin, unused powder, as well as after approx. 10 recycling stages used in PBF-LB/M process and sieved with 63 µm mesh size. Carbon content was determined with a LECO CS-200 in accordance to ASTM E1019-18 [9] and DIN EN ISO 15350-10 [10]. For qualitative inspection of powder morphology, the microstructure of etched cross-sections and local element composition of cross-sections a Hitachi TM3030Plus Scanning Electron Microscope (SEM) with Energy Dispersive X-ray Spectroscopy (EDS) was used. Heat treatment was performed with mono material, as well as HAM specimens in a chamber furnace from Nabertherm GmbH, Germany. The chamber was constantly flushed with 8 L/min Argon inert gas. For determining the influence of the PBF-LB/M process and the heat treatment on the element composition, Optical Emission Spectroscopy (OES) was carried out with a SpectroLab M10. The surfaces analyzed by OES were grinded before measurement. The qualitative metallographic analysis of cross-sections was performed with an Olympus BX53M upright light microscope. Relative density was measured by threshold analysis with a Keyence VK-9710 Laser microscope. Micro hardness was measured with a Zwick ZHU/Z2.5 hardness machine. Cross-sections are polished and etched with 5% Nital at room temperature for approx. 120 s.

All specimens were manufactured on a SLM 125 HL PBF-LB/M machine from SLM Solutions Group AG, Germany. It is equipped with an Ytterbium fiber laser with 1070 nm wave length and spot diameter of 65 µm. All specimens were manufactured with 200°C platform heating and a layer height of 30 µm.

For manufacturing dense and crack free mono material (MP) and hybrid parts (HP) a parameter study was performed. The PBF-LB/M process parameters laser power (P), scanning velocity (v) and hatch distance (h) are initially based on literature research [11–15].

The parameter study consists of 60 parameter sets within the following range:

- P: 120-400 W
- v: 350-1000 mm/s
- h: 0.09-0.11 mm

Three parameter sets with highest density were used to manufacture HP on two different base materials (42CrMo4 and 25CrMo4). The base materials were selected by means of carbon content (high and low), price of bulk material and by promising material properties for the application in injection parts of piston engines.

After investigating the micro hardness and microstructure of MP in as-built (AB) and two different quenched and tempered conditions (QT_I, QT_{II}) the bonding zone of HP was investigated by means of micro hardness, microstructure (qualitatively) and local element composition. In Table 2 the pre-selected heat treatment parameters are given. All heat treated specimens were austenitized and quenched according to Table 2, differing in the temperature during the tempering step.

Table 2. Parameters of heat treatment. QT_I corresponds to Austenitization + Quenching + Tempering I. QT_{II} differs in the temperature during tempering (610 °C).

Stage	Temperature	Time (holding)	Medium
Austenitization	1050 °C	15 min	Argon
Quenching	Room Temperature	-	Oil bath
Tempering I	550 °C	4 h	Argon
Tempering II	610 °C	4 h	Argon

The transition from the base body to the AM part in HP is described by the following terminology:

- Interface: The interface in HP is described as the physical boundary between the base body and the AM part. It is defined by the first solidified lower melt pool on the base body, (see Fig. 1 and Fig. 2). It is visualized in cross sections by etching with 5% Nital. The difference in microstructure causes a visible contrast in the interface. Re-melting, and heat affecting the base body lead to an expanded interface in build direction. Below the interface it is assumed to be pure base body material. In order to characterize the size more accurately the interface is used to define a zero point for micro-hardness profile and element composition.
- Bonding zone: The bonding zone is defined as the area, where due to re-melting and diffusion in particular a transition in metallurgical and mechanical properties is measured. Below the bonding zone it is assumed to be pure base body material and above the bonding zone in build direction it is pure AM part material. In the investigated two material combinations the base body has a lower hardness than the AM part. Therefor the end of the bonding zone is defined by an abrupt increase in hardness with a high gradient. It is followed by a region with a low gradient until the mono material properties are reached.

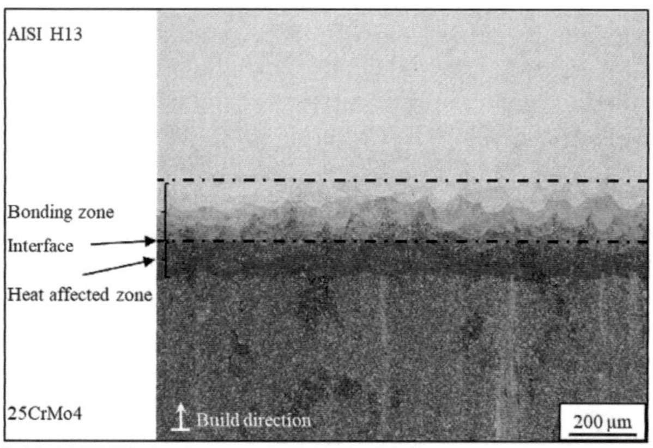

Fig. 2. Cross section of a 25CrMo4 | H13 hybrid part parallel to build direction. To visualize the terminology of interface and bonding zone the specimen is etched for 120 s with 5% Nital and relevant regions are marked.

3. Results and discussion

3.1. Powder characterization

Since in PBF-LB/M only a small fraction of the utilized powder is solidified, the majority is sieved after the buildjob and can be reused. The powder feedstock in virgin (unused) and recycled (approx. 10x sieved and reused powder) state is compared.

Compared to the virgin feedstock no significant loss in qualitative morphology, carbon content (C_{virgin} 0.386 wt.%, $C_{recycled}$ 0.381 wt.%) was measured. The Basic Flow Energy (BFE) of the Rheometer analysis is a parameter for comparing the cohesive force between the powder particles by applying a constant force. In Table 3 BFE, Stability Index (SI, tendency for de- or agglomeration) and Flow Rate Index (FRI, change in flowability by applying changing force) for virgin and recycled Powder is given. The reduced BFE is in the same scale as the standard deviation (SD), hence no significant change in flowability could be found.

Table 3. Results of Rheometer analysis of virgin and recycled AISI H13 powder. BFE: Basic Flow Energy; SI: Stability Index; FRI: Flow Rate Index; SD: Standard Deviation

Powder condition	BFE	SI	FRI
Virgin	601.7 mJ	0.95	1.18
SD	38.8 mJ	0.09	0.04
Recycled, 10x	575.0 mJ	0.91	1.12
SD	32.3 mJ	0.07	0.04

3.2. Mono material specimens

Initially a full parameter study with 60 parameter sets (see section 2) was conducted. Cubes with an outline of 10 x 10 x 10 mm³ were manufactured directly on the build plate (material: X2CrNiMo17-12-2). The relative density was measured by Archimedes method. The highest densities achieved were between 98.0 and 98.3% (with 7.9 g/cm³ as reference). Due to the quite low densities, the four best parameter sets were used to manufacture MP specimens again. These specimens were cut parallel to the build direction for metallographic analysis. Relative density was measured by threshold analysis and resulted in densities over 99.9%. In Table 4 three parameter sets are given with the corresponding densities by Archimedes and laser microscopy. A detailed comparison of both methods is given by Spierings et al. with a similar deviation in densities using microscopy compared to Archimedes method [16]. Since the size, morphology and distribution of defects were taken into account, densities were calculated by microscopy for selecting the best process parameter. Images were taken at 50x magnification at three random points of the cross section.

Two parameter sets were selected by their highest density (E1.2 and A1.6). Since the scanning velocity is effecting the process productivity, one set with highest scanning velocity v (C1.8) was selected for comparison. These three parameters were used for manufacturing hybrid parts.

Table 4. 3 of 60 parameter sets for further analysis, chosen by means of max. relative density and scanning velocity

Parameter set	h	P	v	Archimedes density	Microscopy density
E1.2	90 μm	245 W	650 mm/s	98.36%	99.96%
A1.6	100 μm	306 W	700 mm/s	97.95%	99.93%
C1.8	100 μm	369 W	1000 mm/s	97.87%	no data

No cracks could be determined in MP cross-sections. EDS analysis shows an evenly distributed element composition. In Fig. 3 two polished cross sections are displayed. There are partially non-spherical defects, as well as spherical pores visible. No cracks in the volume are visible. In the etched cross section (Fig. 3b) the melt pools are visible.

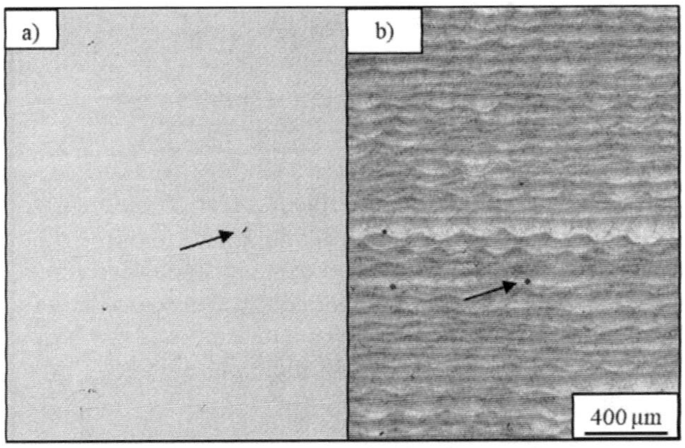

Fig. 3. Polished cross sections at 50x magnification. Parameter E1.2. a) Arrow is marking a non-spherical shaped defect. b) Etched with 5% Nital. Arrow is marking a spherical pore.

The main alloying elements (Iron, Chromium, Molybdenum and Vanadium) are measured by OES comply with the requirements of ASTM A681-08 (see Table 2). The comparison between as-built and heat treated conditions shows no significant influence of quenching and tempering in terms of element composition. The inspected surface of each specimen was grinded before measurement, therefor the values given in Table 4 are mean values close to the surface. A mean carbon loss of 2.5% is measured, comparing the powder feedstock before the build job and the mean OES values of the manufactured parts. Carbon and other alloying elements loss after solidification can be expected due to the high energy density and the process temperature. In this case just a small amount of carbon is lost due to vaporization in the PBF-LB/M process.

Table 5. Element composition in wt.% measured by OES. As-built and two heat treatments compared with reference according to ASTM A681-08 [6].

Element	As-built	QT_I	QT_{II}	Reference
Fe	Balance	Balance	Balance	Balance
C	0.372	0.369	0.38	0.32-0.45
Cr	5.13	5.14	5.09	4.75-5.50
Mo	1.34	1.35	1.34	1.10-1.75
V	0.97	0.98	0.97	0.80-1.20

Micro hardness was measured across the whole cross section to obtain a mean value. In as-built condition hardness over 600 HV1 is reached. This can be established due to the rapid cooling rates in the PBF-LB/M process. In QT_I condition hardness is similar to ASTM reference of wrought material. By SEM of the etched micro structures the dissolving cellular fine grain in as-built condition is observed with raising tempering temperature and longer holding time. Though QT_{II} results in a hardness below the reference a high toughness is expected. Therefor both heat treatments were carried out at HP, too. Higher deviations in As-built condition were reduced by QT_I, without a huge loss in hardness.

Table 6. Mean hardness and standard deviation (SD) of as-built and two heat treatments were measured at bottom, middle and close to the top in build direction. ASTM reference (QT_I) is given as reference.

Hardness		
Heat treatment condition	Mean	Standard deviation
As-built	683.3 HV1	23.1 HV1
QT_I	570.2 HV1	4.4 HV1
QT_{II}	436.7 HV1	4.5 HV1
ASTM reference QT_I	525 HV1 (min. 52 HRC)	

3.3. Bonding zone in hybrid manufactured parts

For characterizing the HP the defects and cracks are measured by microscopy in the full length of the bonding zone (diameter of specimens 24 mm). Parameter E1.2 and A1.6 result in almost the same mean and total length of defects. Although using E1.2 a slightly higher total defect length of 340 µm is measured, higher density in the AM volume was achieved (see Table 4). Parameter set C1.8 with the fastest scanning speed resulted in far longer defects in total (> 1000 µm). It is assumed, that the melt pool morphology in the bonding zone is negatively affected by too high scanning velocity. For further research a combination of E1.2 in the bonding zone and C1.8 in the AM volume will be investigated for optimal bonding and high productivity.

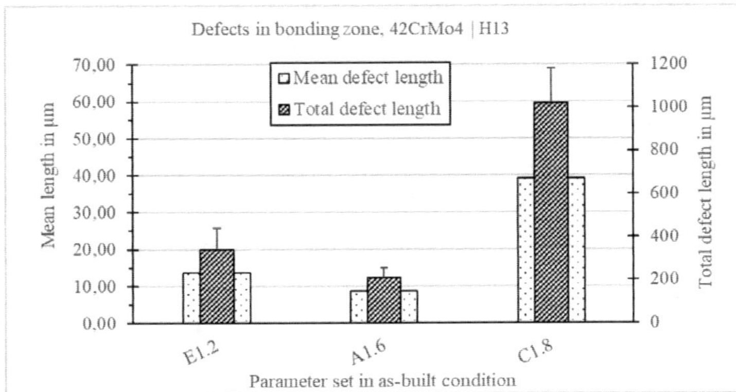

Fig. 4. Measured defect length in the bonding zone of 42CrMo4 | H13 related to parameter sets. Mean length in µm displayed with standard deviation.

In Fig. 5 the bonding zones of parameter sets E1.2 and C1.8 are displayed. On the lower half the base material (42CrMo4) is identified by the regular pores and defects in build direction. This is due to the rolling direction of the wrought material while production and was already determined in cross sections of the base body without PBF-LB/M. In the first 5-10 layers of the AM part (150-300 µm) accumulated cracks and pores were investigated in all HP. Starting at the interface (see Fig. 5) the bonding zone is formed (see micro hardness profile of HP). Due to their small size (mean defect length < 20 µm), using parameter E1.2 and their presence in the base body before PBF-LB/M it is assumed to be not critical for realizing hybrid parts. This will be validated e.g. by tensile tests in future research.

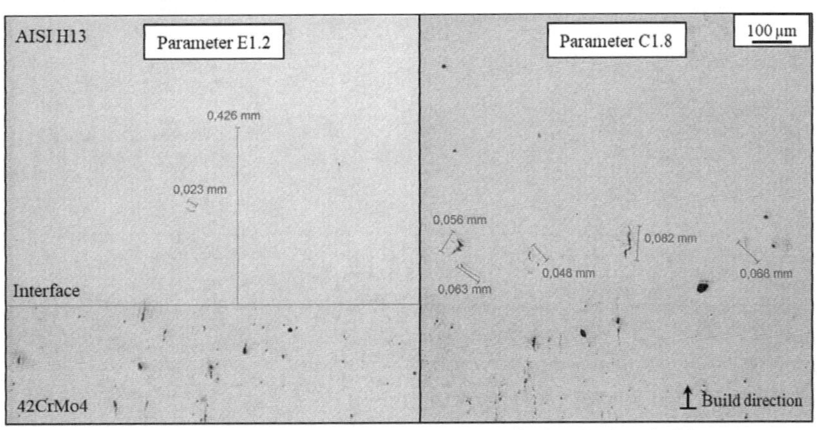

Fig. 5. Cracks in bonding zone of parameter set E1.2 (highest density) and C1.8 (highest build rate).

In Table 7 the results of defect analysis in HP with 25CrMo4 as base body are displayed. The base body with less carbon content resulted in comparable mean length of defects but almost twice the total length of defects, compared with parameter E1.2 and 42CrMo4. It cannot be explicitly explained by the lower carbon content, since lower carbon content should lead to less susceptibility for cracks. Bischof et al. investigated the crack affinity of hot working tool steel H11 with and without elevated carbon content in detail [8]. It is assumed, that the height of the first powder layer in HAM is crucial for crack propagation, since Bischof et al. are linking melt pool stability and geometry to the layer height, too. Cracks were reduce by adjusting process parameters in the bonding zone, though further research demand concerning the influence of layer height is present. Heat treatment, especially quenching does not lead to propagation of cracks in the bonding zone and have no negative effect on the microstructure in the bonding zone.

Table 7. Defects in HP with 25CrMo4 as base body. It is compared as-built and two heat treatment conditions

Sample	Total number of defects	Mean length	Total length of defects
As-built (E1.2)	52	12.42 µm	646 µm
QT_I	64	8.85 µm	566 µm
QT_{II}	59	9.55 µm	564 µm

For quantifying the bonding zone, micro hardness profiles parallel to the build direction were conducted. In Fig. 6 the hardness measurement in a HP with 42CrMo4 is displayed. The starting point of the measurement is marked by the interface. The heat treatments (QT_I and QT_{II}) did not influence the size of the bonding zone, though a slightly lower gradient was determined compared to the as-built condition. The elevated tempering temperature of QT_{II} should lead to faster diffusion and therefor to a change in hardness profile, which could not be stated in this research. The hardness in the base body was slightly lower with 25CrMo4 (270-370 HV0.5, compared to 300-400 HV0.5) due to the lower carbon content. The influence of the carbon content on the bonding zone was one scope of this research. There is no significant effect on the size of the bonding zone, though on the resulting hardness. The as-built hardness in the AM part is slightly lower in HP (600 HV0.5) than in MP (683 HV1). This was assumed to be correlated to the 3x higher thermal conductivity of the base body materials compared to the build plate material of the MP build jobs. It was stated, that as long as a sufficient, solid bonding is manufactured in a hybrid manufactured part, after 180-200 µm of build height similar mono material conditions can be reached.

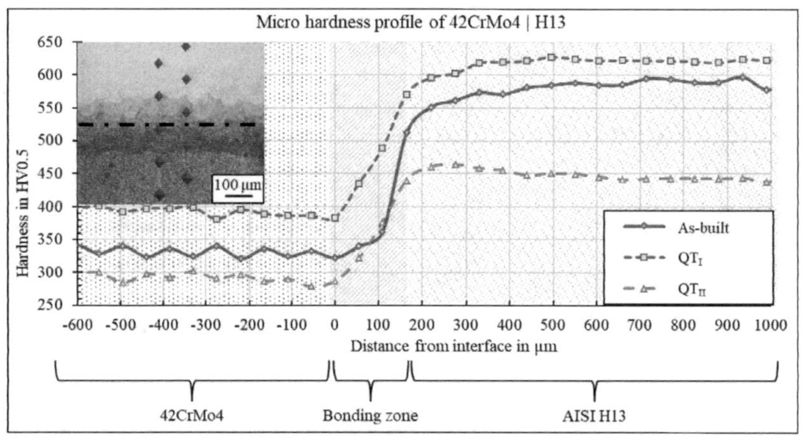

Fig. 6. Micro hardness measured in 42CrMo4 | H13 part parallel to the build direction.

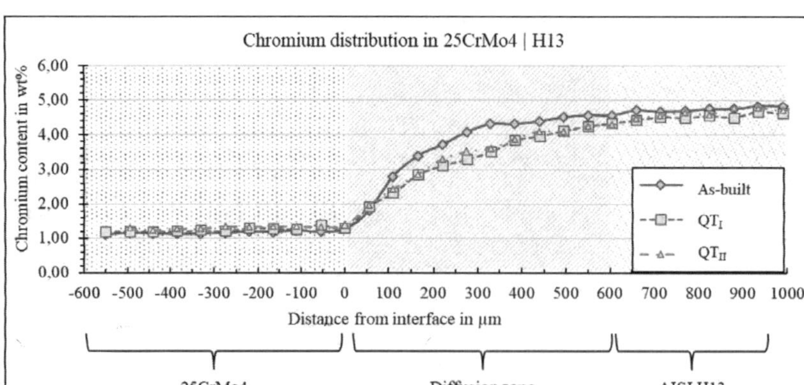

Fig. 7. Chromium distribution by EDS in 25CrMo4 | H13 specimens.

Additionally EDS measurements parallel to the build direction were performed (Fig. 7). Comparing the Chromium distribution, after 600-1000 µm of build height the mono material conditions were reached. This means, in terms of e.g. corrosion resistance the bigger diffusion zone of the elements needs to be considered in hybrid manufactured parts.

This research was finalized with demonstrator geometries (see Fig. 8) for measuring surface and down-skin properties. Further optimization is necessary for reducing post-process machining. For testing the hybrid material combination with bigger volumes a hybrid demonstrator with cooling channels was manufactured on 25CrMo4 (Fig. 8 c). Similar microstructure with low porosity and no cracks in the AM part was observed. The bonding zone was similar to the HP of the previous study.

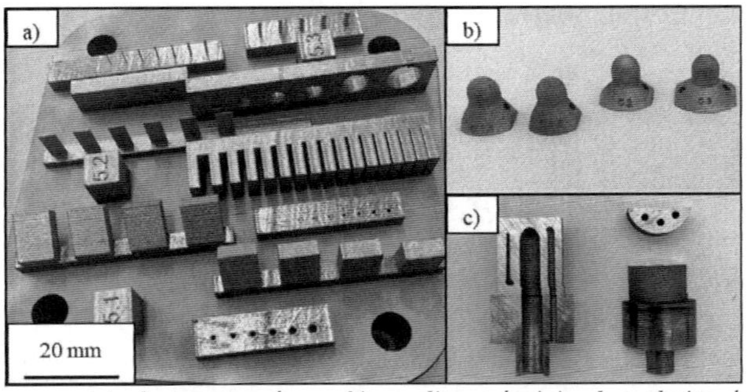

Fig. 8. a) Test geometries for geometrical accuracy, down-skin quality and minimal resolution. b) Mono material demonstrator for piston application. c) Hybrid demonstrator with cooling channels on top of base body.

4. Conclusion and outlook

The conducted parameter study resulted in mono material parts with density > 99.5%. It was shown, platform heating with 200°C is sufficient for manufacturing crack-free parts, although only small sized geometries (diameter 20 mm, build height 35 mm) were inspected in cross sections.

The influences of base body materials, process parameters, and post-process heat treatment on hybrid manufactured parts were investigated. For evaluating a hybrid part, characterizing the bonding zone is crucial for in HAM. The gained knowledge for manufacturing hybrid parts with dissimilar base bodies can be summarized in:

- Process parameters for manufacturing mono material parts were utilized without further adjustment necessary for manufacturing hybrid parts in two material combinations 42CrMo4 | H13 and 25CrMo4 | H13. No defects in the bonding zone were observed and the density is comparable with mono material parts.
- For evaluating the interface between base body and AM part a bonding zone was defined. It is quantified and characterized by micro hardness profile, element composition and optical inspection of etched cross sections. This procedure can be transferred to other material combinations and LPBF-LB/M machines.
- Depending on the heat treatment and the base body material, a bonding zone of 200-500 µm is determined. Element diffusion, re-melting and heat affected zone are affecting the size and hardness in the bonding zone and need to be considered in hybrid parts according the application and the heat treatment.
- Inside the bonding zone no cracks greater than 50 µm, in total 350 µm after optimization and a gradient in hardness is observed. Outside the bonding zone the mono material properties are present (hardness >600 HV1). Outside the bonding zone the element composition in quenched and tempered AISI H13 is not affected by the material combination and no cracks in the AM part are observed.

For a transfer to industrial applications the following topics need to be investigated in further research:

- In piston engine application not only hardness, but also toughness are relevant. Static tensile tests at room and elevated temperature need to be performed at hybrid manufactured specimens.
- The poor surface roughness (Ra=12.3 µm, Rz=128 µm at vertical walls) makes post-process machining crucial for industrial applications. To improve surface roughness and reduce machining costs, advanced contour and down-skin parameters have to be developed.
- The manual positioning of the AM part to the base body resulted in off-sets > 0.1 mm. An enhanced positioning method supported by software is necessary for accuracy < 0.1 mm of the AM part to the base body.

ORCID iDs

L. Langer https://orcid.org/0000-0002-5170-5428
M. Schmitt https://orcid.org/0000-0002-2632-5052
G. Schlick https://orcid.org/0000-0001-5448-0482

References

[1] Wohlers T, Campbell R I, Huff R, Diegel O and Kowen J 2019 *Wohlers Report 2019: 3D printing and additive manufacturing: state of the industry* (Fort Collins, Colo: Wohlers Associates)
[2] Jacob A, Steimer S, Stricker N, Häfner B and Lanza G 2019 Integrating product function design, production technology optimization and process equipment planning on the example of hybrid additive manufacturing *Procedia CIRP* **86** 222–7
[3] Strong D, Sirichakwal I, Manogharan G P and Wakefield T 2017 Current state and potential of additive – hybrid manufacturing for metal parts *Rapid Prototyping Journal* **23** 577–88
[4] Zhang X, Cui W, Li W and Liou F 2019 A Hybrid Process Integrating Reverse Engineering, Pre-Repair Processing, Additive Manufacturing, and Material Testing for Component Remanufacturing *Materials (Basel, Switzerland)* **12**
[5] 2020 DIN EN 10250-3, Open die steel forgings for general engineering purposes - Part 3: Alloy special steels (Berlin: Beuth Verlag GmbH)
[6] 2008 A 681 – 08 Standard Specification for Tool Steels Alloy (West Conshohocken, PA: ASTM International)
[7] Wenz T, Kirchner A, Klöden B, Weißgärber T and Jurisch M 2020 Processing of high carbon steel by Selective Electron Beam Melting (SEBM) *steel research int.*
[8] Bischof C, Scheitler C and Kneisel L 2016 Influence of preheating Temperature and Carbon Content on crack formation during Laser Beam Melting of AISI H13 Tool steel *iCat 2016*
[9] 2018 ASTM E1019-18, Standard Test Methods for Determination of Carbon, Sulfur, Nitrogen, and Oxygen in Steel, Iron, Nickel, and Cobalt Alloys by Various Combustion and Inert Gas Fusion Techniques (West Conshohocken, PA: ASTM International)
[10] 2010 DIN EN ISO 15350, Steel and iron - Determination of total carbon and sulfur content - Infrared absorption method after combustion in an induction furnace (Berlin: Beuth Verlag GmbH)
[11] Åsberg M, Fredriksson G, Hatami S, Fredriksson W and Krakhmalev P 2019 Influence of post treatment on microstructure, porosity and mechanical properties of additive manufactured H13 tool steel *Materials Science and Engineering: A* **742** 584–9
[12] Šafka J, Ackermann M and Voleský L 2016 Structural properties of H13 tool steel parts produced with use of selective laser melting technology *J. Phys.: Conf. Ser.* **709** 12004
[13] Hengsbach, F., Koppa, P., Holzweissig, M.J. et al. 2018 Inline additively manufactured functionally graded multi-materials: microstructural and mechanical characterization of 316L parts with H13 layers *Prog Addit Manuf* **3** 221–31

[14] Laakso P, Riipinen T, Laukkanen A, Andersson T, Jokinen A, Revuelta A and Ruusuvuori K 2016 Optimization and Simulation of SLM Process for High Density H13 Tool Steel Parts *Physics Procedia* **83** 26–35
[15] Ren, B., Lu, D., Zhou, R. et al. 2019 Preparation and mechanical properties of selective laser melted H13 steel *Journal of Materials Research* **34** 1415–25
[16] Spierings A B, Schneider M and Eggenberger R 2011 Comparison of density measurement techniques for additive manufactured metallic parts *Rapid Prototyping Journal*

IOP Conf. Series: Materials Science and Engineering 1135 (2021) 012013 doi:10.1088/1757-899X/1135/1/012013

Determination of the geometry of laser-cutting fronts with high spatial and temporal resolution

Michael Sawannia[1], Peter Berger[1], Rudolf Weber[1] and Thomas Graf[1]

[1]Institut für Strahlwerkzeuge, University of Stuttgart, Pfaffenwaldring 43, 70569 Stuttgart, Germany

michael.sawannia@ifsw.uni-stuttgart.de

Abstract. The melt flow velocity and the local surface angles of the cutting front during laser fusion cutting of 10 mm AISI 304 were determined for a laser power of 8 kW and a feed rate of 2 m/min. The cut front was recorded with a polarization goniometer, which uses the polarization of the process emission to determine the local surface angles, allowing to calculate the orientation of the normal vector of the surface. The records in this work were carried out with a frame rate of 75 kHz and a spatial resolution of about 30 µm. This allowed to identify big and small structures moving down the cutting front and to determine their velocities. The approximate velocity of the small structures was 9.1 m/s and for the big structures approx. 2.5 m/s. The information of a usual high-speed video was compared with the additionally obtained geometry information.

1. Introduction

The quality of laser fusion cuts decreases for higher cut depths through dross formation and striations on the cut flanks [1]. These effects also result from the melt flow in the kerf, which in turn depends on the locally absorbed laser power at the cutting front and the kerf flanks [2]. In order to analyze the melt flow, mainly high-speed recordings have been used so far. This technique is used not only in laser beam cutting, but also in laser beam welding. The resulting images typically show bright spots on a dark background moving downward the cutting front or the capillary front during laser beam welding. They are usually interpreted as geometric structures called steps [3], shoulders [4], shelves [5], bumps [6] or humps [7, 8]. For their interpretation a distinction should be made between observation methods that illuminate externally or that record the direct emission from the process. In the case of the external illumination, the brightest spots are created by direct reflections of the illumination source via the front onto the camera. For the case of the direct process emission, the brightness of the spots depends not only on the observation angle but also on the local temperature. The temperature in turn is higher for lower angles of incidence between the surface of the structures and the absorbed laser beam. For both cases the velocity of the bright spots can be determined [1, 6, 8], but exact statements about the structure size and surface inclination of the cutting front are not possible. To determine the real geometry of the cutting front, a new measuring device (polarization goniometer) was used. It uses the polarization information of the thermal process emission to determine the local surface angles, allowing to calculate the orientation of the normal vector of the surface, which is called in the following the surface orientation of the cutting front. With the surface orientation the 3D-surface can be determined, which was already shown in [9, 10]. The primary goal of this work is to determine and describe the actual structures and their resulting velocity.

2. Measurement method

The polarization goniometer records four single images of hot metal surfaces, e.g. cutting fronts, simultaneously for one time step. The single images contain the intensity of the thermal process emission after four linear polarizers. The four polarizers are oriented under 0°, 45° 90° and 135°. With these recorded intensities the surface angles for each image pixel of each time step can be calculated. Two angles are needed to describe all possible surface orientations of a flat surface element, in our case the angle of inclination φ and the angle of rotation α are used, see Figure 1.

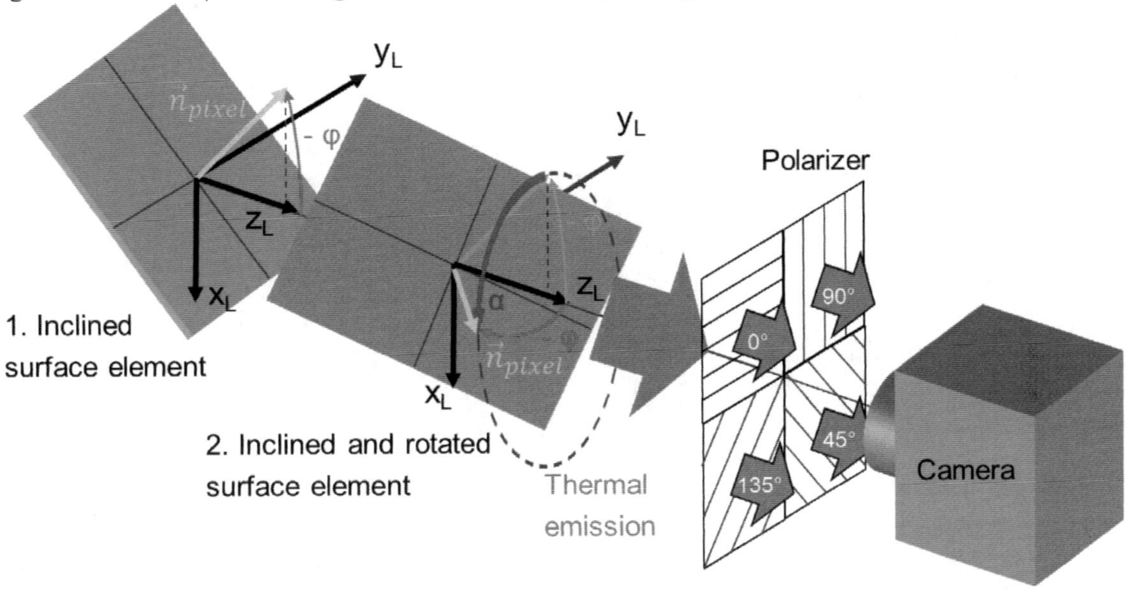

Figure 1. Schematic sketch of a camera taking four single images through four different linear polarizers from a thermally emitting surface (the real device is more complex). To describe the orientation of the surface in three-dimensional space, two angles are needed, which can be calculated by the four single images. The first one is the angle of inclination φ that inclines the surface around y_L. The second is the angle of rotation α, which rotates the inclined surface around z_L. By rotating around both angles, the spatial orientation of the surface element is now fully described.

To describe the angles α and φ, first consider a surface element rotated only about the y_L axis relative to the projection of a surface element on the x_L y_L plane. The tilt angle φ is then the angle between the normal vector of this inclined surface element (see Figure 1) and the camera axis z_L. This is a positive angle if the rotation was counter clockwise about the positive y_L axis. The angle of rotation α describes the following rotation of the normal vector around z_L. For the calculation of the angles only the refractive index n and the extinction coefficient k for the emitting material are needed. The spectral range of the recorded emission is limited by a bandpass filter in front of the camera, to allow the assumption of a constant refractive index over the observed spectral range. A more detailed explanation is given in [10]. For the angle calculation, quotients are calculated from the polarized emission parts. The use of quotients has the further advantage that the temperature influence on the refractive index and on the extinction coefficient is negligible, which is shown in [11]. The two angles already tell a lot over the surface. To detect rising structures, the angle of inclination φ directly indicates steep regions, whereas the angle α indicates in which direction the structure is inclined. The measurement method was applied to laser beam fusion cutting [9, 10].

3. Experimental Setup

A schematic sketch of the used experimental setup is shown in Figure 2. For the investigations, 10 mm thick stainless steel (AISI 304) was cut with an 8 kW thin-disk laser (Trumpf, TruDisk 8001). The cutting head was a ProCutterET 100/150 from Precitec. A 100 μm fiber connected the laser with the head. The EdgeTec module of the cutting head was used, which creates a ring shaped intensity distribution of the focused laser beam. With $M^2 = 51$, a beam diameter of 824 μm and a Raleigh length of 9.9 mm was calculated.

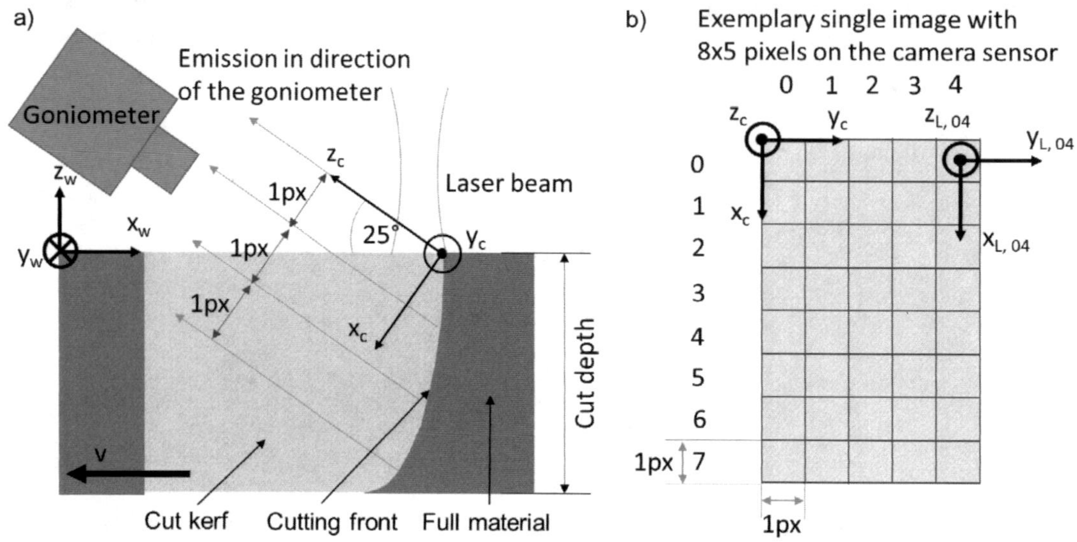

Figure 2. Schematic sketch of the experimental setup in a). Schematic single image of a goniometer recording in b) to define the coordinate systems and pixel position.

A standard cutting nozzle was used with an inner diameter of 2.5 mm. The standoff distance was 1 mm. The nozzle was abraded on the outer contour in Goniometer direction to allow a better view on the cutting front. Nitrogen was used as process gas with a pressure of 12 bar. The sample had a dimension of 150 mm x 6 mm with a thickness of 10 mm. A cut of 40 mm was carried out by moving the sample in feed direction v, see Figure 2. The used laser power was 8 kW with a feed rate of 2 m/min.
For the experiments, the goniometer was set at 25° to the workpiece surface, as shown in Figure 2. The spectral range of the bandpass filter was 857 nm ± 15 nm. The high-speed camera was a Photron FASTCAM SA5, with a recording frame rate of 75000 frames per second. The spatial resolution was around 30 μm, like for the experiments shown in [9].

4. Results obtained by the goniometer

In the following section, calculation results are shown and explained for the local inclination angle φ and rotation angle α. Furthermore the resulting structures are compared with classic high-speed recordings. Thereafter, corresponding velocities of the structures moving down the front are given.

4.1. Polarization goniometer - measurement and calculation

Figure 3 a) shows the four single images of a time step. Despite different polarizer orientations, they look quite similar. The small differences in the intensity, which are caused by different surface orientations are hardly noticeable, just that the 45° and 135°images are darker, which can be attributed to losses in the corresponding two beam paths that are compensated for the calculation. Before the calculation of the angles, the four single images were filtered with a Gaussian blur filter [12] to reduce the noise impact of the images. For the smoothing the program ImageJ (v1.53i) was used with sigma 1 for the filter. The resulting average image of the four intensities (AVG) and the angles φ and α are given in Figure 3 b). The average image is identical to the result of a classic high-speed record. The speed of the structures on the front is later evaluated along the three the drawn lines. These lines are also used for the φ and α images. All images are oriented to the camera coordinate system (x_c, y_c and

z_c). The angles φ are all positive and are mostly in the range of 15° to 40°. In the φ image round spots can be discovered, some are marked in Figure 3 c) with purple rectangles. They mostly coincide with the bright spots in the AVG image and indicate that there are significant changes in the surface geometry. The phi and the alpha angle can be used to estimate elevations or dents on the surface, thus obtaining geometry information about the front. The angles α are in the range of -90° to 90°. In the α image almost all α's are positive on the left side and negative on the right sight of the cutting front. This would also be the alpha distribution for a hollow half cylinder, which can be thought of as a simplified cutting front. For the middle of the cutting front, all α values vary around α = 0°, which is in full agreement with a geometry similar to a hollow half cylinder. In the middle, the angle φ corresponds to the inclination of the half cylinder, whereby in our case it must still be taken into account that we are looking at the front at an observation angle of 25°, if one wants to make statements about the inclination of the front in the workpiece coordinate system.

In the upper fifth and the lower third of the cut front, α changes its sign from plus to minus on the left side, which can be a result of an undercut or indicates bigger structures on the front.

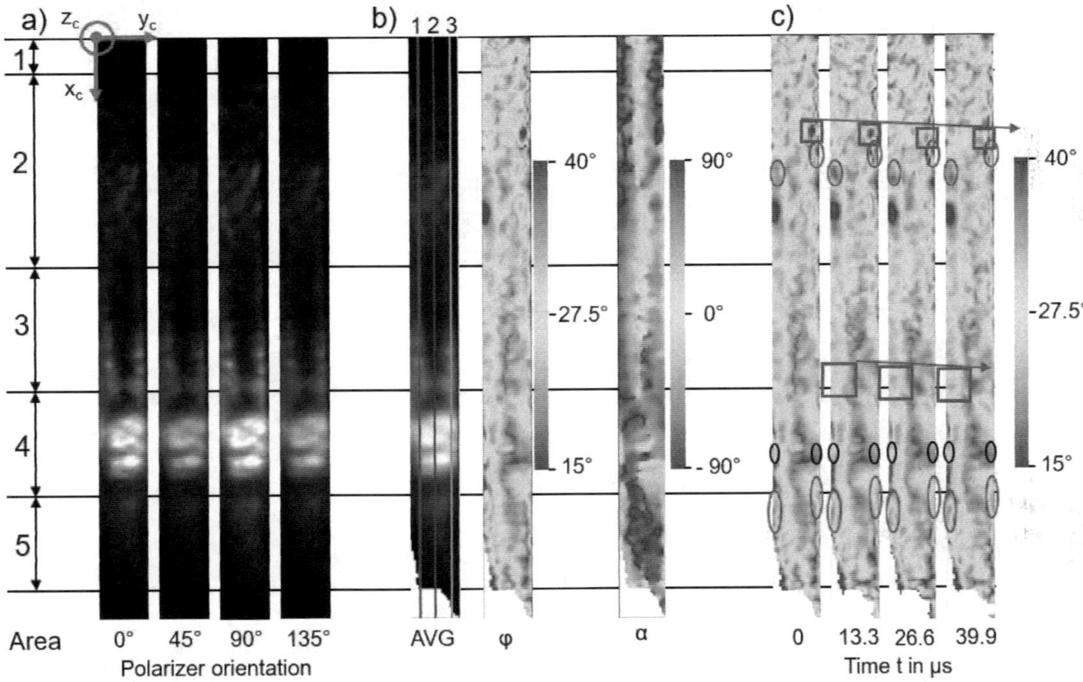

Figure 3. Recorded single images corresponding to the four polarizer orientations in a). In b) the average image (AVG) and the images of the angle of inclination φ and of the angle of rotation α, calculated from the four images in a), are presented. In c) four consecutive φ images are shown with a time difference of 13.3 µs to each other. The purple squares mark red spots moving down the front, the purple arrows indicate the downward movements over time. The circles mark positions where the spots are more or less at rest. The front is divided into five sections, see labels on the left.

In Figure 3 c) the angle φ is shown in four images of consecutive time steps. The time steps are 13.3 µs, which corresponds to the recording frame rate of 75000 Hz. The red spots can be followed, while they are moving down the front. Two examples are marked with purple squares. The moving spots are probably structural elevations moving down the front due to the cutting gas. The purple arrows show the downward movement of the spots on the front over time. There are also spots that increase or decrease in size over the time steps, which allows the structural evolution to be studied in detail, e.g. the marked spot with the purple rectangles in the upper area. The circles mark positions where the spots do not appear to move for a period much longer than depicted with the four frames in Figure 3. In addition, the black circles mark a special feature. Above these points, the bright spots from the camera image seem to decrease in their velocity and accumulate. This causes the melt to flow not only downwards, but also over the sides of the kerf.

4.2. Moving structures at the front

To show the dynamics at the cutting front, streak images along a line are a suitable option. The streak images were created for the AVG, φ and α images. The streak images shown in Figure 4 were created along the line on the right side of the cutting front (line 3 in the AVG image in Figure 3). 500 single frames were used, resulting in a displayed time interval of 6.7 ms (time steps of 13.3 µs between each image corresponding to the recording frame rate of 75000 Hz). In the AVG image of Figure 4, orange lines mark small fast moving spots, which occur periodically. The φ and α images also clearly show changes at the same locations. They mainly occur in the marked areas from area 2 to 3. Starting from area 4, most of the structures become larger and slower, which can be seen from the smaller slope over time t and the distance of the structure moved along x_C in the φ-image. In area 1 and 5 the signal is very low, so no statements are made for these areas. Big slow moving structures also occur in the streak images in the upper part (area 2). Three structures are marked in the streak images. The yellow one encloses a big bright spot in the AVG image (in area 2), whereas the red and black one enclose very steep structures in the φ image (in the areas 2 and 4). The markings of the structures were transferred to the respective other image and to the α image. The red marked area in the AVG image is always bright except for the tip at the beginning. In the φ image there is already a steep angle, which suggests, that the structure forms before it becomes really bright (hot). This is a considerable gain of information. The reason for the formation of the slow structures is not yet clear. One presumption for the large structures in the upper area is a pressure rise along the x_c axis in the upper area, whereby the melt is slowed down.

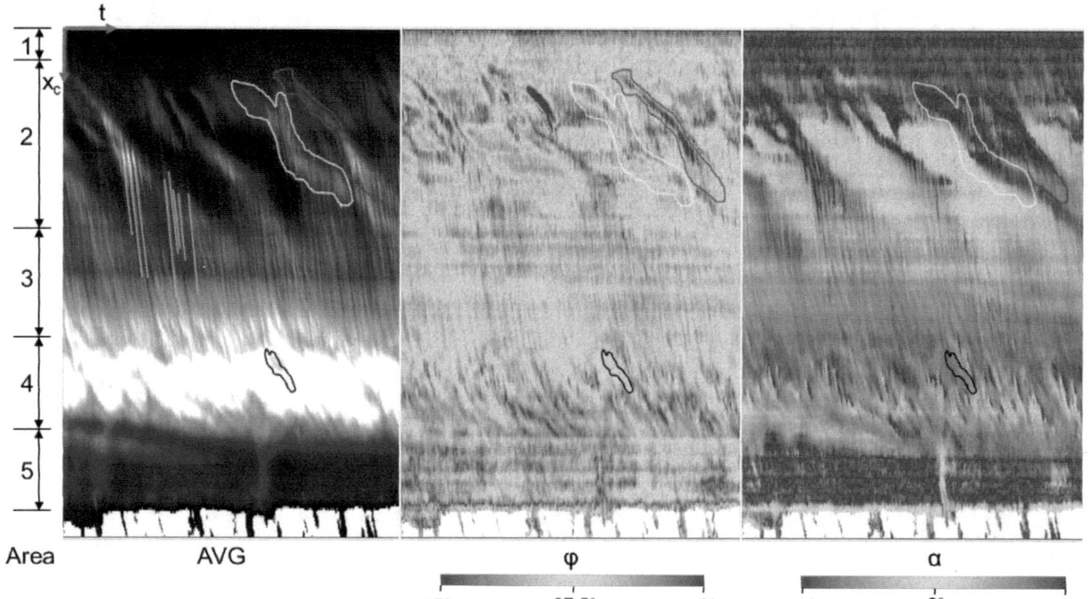

Figure 4. Streak images over 500 single images along line 3 shown in the AVG image in Figure 3. The streak images were carried out for the AVG, φ and α images. The time interval between each time step is 13.3 µs. In the AVG streak image small (orange) and bigger spots (yellow, red, and black) are marked. The larger structures (yellow, red, and black) were also inserted into the φ and α images. The front is divided into five sections, see labels on the left. The contrast of the AVG image is increased, the original image was not overexposed.

IOP Conf. Series: Materials Science and Engineering 1135 (2021) 012013 doi:10.1088/1757-899X/1135/1/012013

Thereby the angle of the surface changes and more laser power can be absorbed. In the following the structure heats up and thus becomes brighter. This applies, for example, to the area marked in yellow. In the lower area, a separation of the gas flow from the front is suspected as the cause. This creates a low pressure, which sucks in ambient air in the lower region of the front and slows down the melt flow (transition from region 4 to 5). This presumption is supported by strong oxidation in the lower region of the cut front. To make the brightness variations in the red and yellow spot easier to see, the contrast for the AVG image was increased. For this reason, it looks like the image was overexposed in area 4, which was not the case. In comparison, the φ image has a very even contrast, which is another advantage compared to classic high speed recordings. The angle α implies the rotation of the local surface element accordingly to Figure 1. It varies between - 90° and 90° for the structures in area 4 as well as for the two marked in structures area 2, i.e. the structures do not move directly downwards in x_C, but also have a lateral movement in y_C direction.

4.3. Velocities of the structures at the front

As mentioned before, the slope of the structures in the streak image correspond to their velocity. The velocity of the small and big structures were evaluated for the left side, the middle and the right side of the cutting front, line 1 - 3 in Figure 3. Ten structures were measured for each area, the error bar is the standard deviation. The results for the laser power of 8 kW and a feed rate of 2 m/min are plotted in Figure 5. For the fast structures, the average velocity in the upper part (area 2) of the front is around 10.6 m/s and in the middle part of the front (area 3) about 10.2 m/s. In the lower part the velocity is about 6.5 m/s. In addition, it is noticeable that the velocity on the centre line is always slightly higher than on the left or right side. The reason for this could be that the melt on the side not only flows directly downwards, but also has a directional component along the cut flank, which means that the observed velocity is lower from the point of view of the goniometer. The slow structures only occurred in the upper and lower part, therefore there are no velocities in the middle (area 3). In the upper part, the average velocity is around 1.8 m/s. In the lower part the structures are nearly twice as fast with 3.2 m/s, but as said before, the structures are also smaller. For the determination of the velocities in Figure 5, the slopes in the respective φ images were used. For comparison, the velocities were also determined from the average images. The deviation is not very large with a maximum of 0.72 m/s, which is within the range of the measurement accuracy.

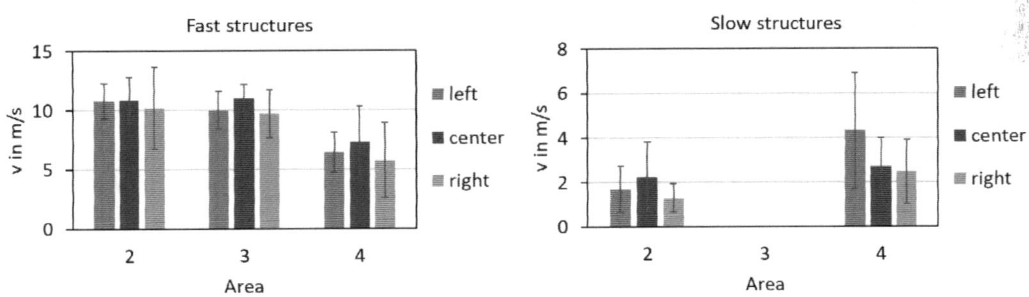

Figure 5. Measured velocities for the fast and slow structures in the area 2, 3, and 4. The velocities were measured on the left side, the center and the right side of the front. The error bars correspond to the standard deviation.

5. Conclusion and outlook

This paper presents the actual spatial and temporal resolved surface orientation of the cutting front, calculated by the polarization of the thermal process emission. With these, the bright spots in usual high-speed records get a geometrical orientation, which allows to estimate the shape on the front. Here it is remarkable that the inclination of the structures is already noticeable shortly before the appearance of bright spots. Small and big moving structures can be recognized. The small structures have an average

velocity of 9.1 m/s and the big structures a velocity of 2.5 m/s. The velocity determination over streak images works the same for usual camera images as for the determined angles.
The next steps in the future are to determine the actual structure sizes and to look more closely at the creation of the structures, with the aim of determining an optimal cutting front and melt flow that will produce high quality cuts.

Acknowledgement
The presented work was funded by the German Research Foundation (DFG) and experiments performed in the context of the project "ELS PoGo" (GR 3172/20-1) and the DFG project "FastShape" (426328417). The responsibility for this paper is taken by the authors.

ORCID iDs
M. Sawannia: https://orcid.org/0000-0003-0972-5975
P. Berger: https://orcid.org/0000-0003-4232-3732
R. Weber: https://orcid.org/0000-0001-8779-2343
T. Graf: https://orcid.org/0000-0002-8466-073X

References
[1] Arntz D, Petring D, Schneider F and Poprawe R 2018 *Proceeding of ICALEO*
[2] Bocksrocker O, Berger P, Fetzer F, Rominger V and Graf T 2018 *Lasers Manuf. Mater. Process.* **24** 52006
[3] Berger P, Schuster R, Zvyagolskaya M, Hügel H and Schäfer P 2011 *Schweißen und Schneiden* 20–8
[4] Berger P, Schuster R, Hügel H and Graf T 2010 *Proceeding of ICALEO*
[5] Golubev V S Laser welding and cutting: recent insights into fluid-dynamics mechanisms *Laser Processing of Advanced Materials* p 1
[6] Pocorni J, Petring D, Powell J, Deichsel E and Kaplan A FH 2015 *Physics Procedia* **78** 99–109
[7] Matsunawa A and Semak V 1997 *J. Phys. D: Appl. Phys.* **30** 798–809
[8] Hirano K and Fabbro R 2011 *J. Phys. D: Appl. Phys.* **44** 105502
[9] Michael Sawannia, Peter Berger, Michael Jarwitz, Rudolf Weber, Thomas Graf 2019 *Lasers in Manufacturing Conference*
[10] Sawannia M, Berger P, Jarwitz M, Weber R and Graf T 2018 *Proceeding of ICALEO*
[11] Weberpals J, Hermann T, Berger P and Singpiel H 2011 *Physics Procedia* **12** 704–11
[12] 2019 *ImageJ - Gaussian Blur filter: https://imagejdocu.tudor.lu/gui/process/filters*

On multi-sensor monitoring of fiber laser fusion cutting

Nikita Levichev[1], Joost R. Duflou[1]

[1]Department of Mechanical Engineering, KU Leuven, Celestijnenlaan 300, 3001 Heverlee, Belgium, Member of Flanders Make

nikita.levichev@kuleuven.be

Abstract. Laser cutting is a well-established industrial process for sheet metal applications. However, cutting thick plates is still accompanied by problems because of the characteristic limited process parameter window. Since cutting by means of fiber lasers has become dominant, tailored solutions are required in such systems for industrial applications. The development of a robust real-time monitoring system, which adapts the process parameters according to a specific quality requirement, implies a significant step forward towards automated laser cutting and increases the process robustness and performance. In this work, a coaxial multi-sensor monitoring system is tested for fiber laser cutting of stainless steel thick plates. A high-speed camera and a photodiode sensor have been selected for this investigation. Experiments at different cutting speeds, representing primary cut quality cases, have been conducted and various features of the obtained process zone signals have been examined. Finally, the feasibility of industrial application of the developed setup for high-power fiber laser cutting is discussed, followed by several implementation recommendations.

1. Introduction

Laser cutting of metals is a highly automated industrial process that offers excellent flexibility and productivity comparing to other technologies. Although the laser cutting process has unique advantages, it is susceptible to inner and outer influences, such as variability of plate thickness and chemical composition, changes of process parameters caused by the wearing of machine parts or any mistakes of the operator. Therefore, real-time quality monitoring techniques have to be applied to ensure a high-quality cut edge and cost-effective use of laser cutting machines in workshops.

Monitoring systems have been developed and applied for in situ quality inspection of laser processes [1]. However, laser cutting monitoring has several distinct difficulties. First, a confined process zone with a nozzle, which needs to be positioned close to the cutting front, limits the observation from the top. Second, a stream of molten material is ejected from the process zone, creating an aggressive environment that obstructs the observation from the bottom side of the plate. Third, the process is highly dynamic and typically accompanied by fast changes in cutting direction that complicates the off-axis observation. As a result, the natural solution to overcome these issues is to use a coaxial monitoring system that is proven to be able to provide essential information about cut quality [2].

One of the major challenges in laser cutting monitoring is identifying sensing parameters that correlate well with specific cut quality characteristics. In the ideal case, an appropriate sensing parameter based on a single sensor measurement should be found for all quality features. Several single sensors have been already tested and presented in literature. For instance, real-time monitoring of cutting with a fiber laser has been investigated applying a photodiode [3, 4] and a camera [5, 6].

In industrial practice, the combination of different sensors can offer increased robustness due to the redundancy between different information streams and the resulted reduced sensitivity for noise factors. The combination of parameters from different sensor types is better known as sensor fusion [7]. Sensor fusion is a promising technique because it can include complete information about the process and the corresponding cut quality by using the advantages of several optimally combined sensors.

Sensor fusion systems have been already investigated for laser material processing. For instance, a multi-sensor system consisting of two photodiodes and two high-speed cameras for different spectral ranges has been introduced for high-power disk laser welding [8]. Another multi-sensor data fusion system for laser welding monitoring with two photodiodes and a pyrometer has been developed to analyze different in-process defects [9]. An optical process monitoring system based on infrared, visual and polarization sensors has been designed for laser powder bed fusion [10].

The first example of sensor fusion in laser cutting can be represented by the combination of different information sources that is used by experienced operators responsible for process optimization. By interpreting visible and audible signals, they can typically obtain a general impression of the process status. As a result of such observations, high-power CO_2 laser cutting has been monitored by means of an acoustic microphone and photodiodes [11]. However, it has been concluded that the acoustic signal does not deliver any additional information compared to the photodiode signals and requires advanced signal processing that makes a microphone less suitable for industrial applications.

A few studies reported that the combination of different spectral ranges might be beneficial for laser cutting monitoring. Cross-correlation of InGaAs and Si photodiodes has been used for cut interruption detection of laser fusion cutting [12]. Heat accumulation during laser flame cutting has been detected by applying both visual and near-infrared photodiodes, which has helped to distinguish this issue from other possible process defects [13].

In this paper, an example of a multi-sensor approach based on a photodiode and a high-speed camera is provided for fiber laser fusion cutting. The experiments are performed with different cutting speeds in order to cause several levels of cut quality deterioration. The robustness of the sensor fusion is analyzed compared to the performance of a single sensor monitoring.

2. Experimental description

2.1. Laser cutting

The experiments were performed on a commercial 2D flat-bed laser cutting machine with a 4 kW IPG Ytterbium multimode fiber laser source emitting at 1.07 µm wavelength with a top-hat intensity distribution. The laser beam was delivered from a 100 µm core optical fiber into a Precitec ProCutter laser head equipped with a 200 mm focusing lens and a 100 mm collimation lens. The beam parameter product and the beam waist diameter were measured using an Ophir BeamWatch device as 2.84 mm mrad and 224 µm, respectively.

The cutting samples were obtained from stainless steel 316L plates of 12 mm thickness. As commonly used in industrial practice and obtained from the machine supplier, the following process parameters were applied for the cutting experiments for this material-thickness combination. Nitrogen of 99.999 % purity with 18.0 bar pressure was used as an assist gas together with a 3.5 mm diameter nozzle and a stand-off distance of 0.4 mm. The focus position of the laser beam was set to 10.0 mm below the top surface of the plate. The nominal cutting was performed with a speed of 700 mm/min, further referred to as 100 %, a laser power of 4 kW and a duty cycle of 100 %.

A Keyence Digital Microscope VHX-6000 was used for high-resolution imaging of cut edge profiles. The surface roughness of cut edges (Rz) was measured by means of a Mitutoyo Formtracer CS-3200 using a Gaussian filter with a sample length of 2.5 mm and a cut-off of 0.008 mm. The measurements of each specimen were performed for five lines with an evaluation length of 12.5 mm. The measuring lines were spaced by 2.5 mm, starting at 1 mm from the top surface. The average roughness value was calculated as a mean value of these five measurements. The dross area was calculated by image processing of a corresponding cut edge picture for an evaluation length of 50 mm [6].

2.2. *Monitoring system*

A coaxial monitoring setup, as depicted in Figure 1, was used with a dichroic mirror that transmits the radiation with wavelengths shorter than 800 nm and longer than 1200 nm towards the sensors. Next to the mirror, a diaphragm with a diameter of 1.5 mm was situated to decrease the optical aberrations after the laser head optics designed for the laser beam wavelength. After the diaphragm, a notch filter with an optical density of 6 and a bandwidth of 1064 ± 36 nm was placed to remove the scattered and reflected laser radiation. A non-polarizing beamsplitter cube that transmits half of the radiation and reflects the other half was used to ensure light propagation to both sensors.

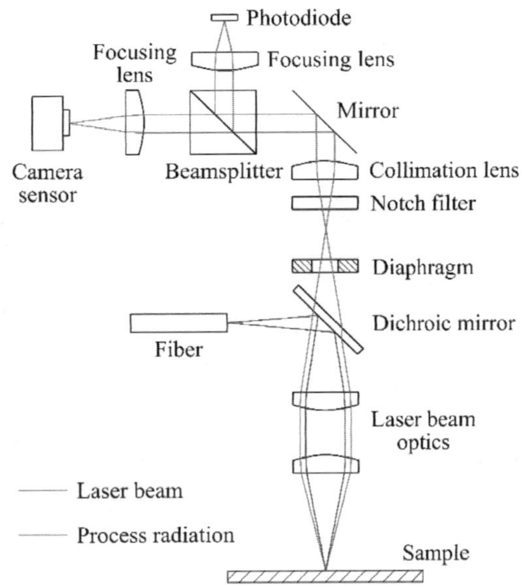

Figure 1. Optical scheme of the coaxial multi-sensor monitoring system.

Depending on the considered spectral interval, different phenomena can be observed by a monitoring system. At the same time, laser cutting is a highly dynamic process that brings additional restrictions on the choice of the sensors. The photodiode selection was made assuming that its spectral range should cover the maximum of thermal radiation from the process zone to use minimum amplification. However, the camera sensors typically have limitations in achievable frame rate, resolution and pixel size. Consequently, the selection of the camera should be based on its capabilities to acquire maximum geometrical information from the process zone with acceptable time resolution.

Since the radiation peak of the process temperatures is around 1500 ± 100 nm, an InGaAs photodiode (FGA21) responsive to the near-infrared radiation with wavelengths from 800 to 1700 nm was selected. The photodiode signals were recorded by means of a data acquisition card with amplification and low pass filtering, allowing sample rates of up to 50 kHz. The signal was processed using a moving average filter with a window length of 1 ms to reduce the noise.

In order to be able to capture dynamic changes of the cutting front geometry, a Mikrotron EoSens MC1362 camera with a Si-based sensor and a pixel size of 14 µm has been used. Considering the best performance of the camera, images with a size of 320 × 250 pixels were acquired at a frame rate of 5000 fps with an exposure time of 100 µs.

2.3. *Video processing*

The analysis of the videos has been performed after experiments using MATLAB. A video frame is represented by a matrix M with I columns and J rows where the matrix grayscale value $g_{i,j} \in [0, 255]$ with $i \in [0, I]$ and $j \in [0, J]$ is the pixel intensity.

In order to compare the results of the two sensors, the overall image intensity (I_{img}) is calculated as:

$$I_{img} = \sum_{i=1}^{I}\sum_{j=1}^{J} g_{i,j} \tag{1}$$

Then, in order to extract geometrical features, image transformations are performed as in detail described in [6]. An example of binarization with a threshold value of 50 and logarithmic transformation with an enhancement coefficient of 2.5 is shown in Figure 2.

In this paper, the dimensionless factor *l/w* defined as a ratio between the length and width of the process zone is used instead of several geometrical parameters since it can be considered more general. The length and width of the process zone are computed based on the centroid position identification of the blob obtained after image binarization as the blob dimension parallel and perpendicular to the cutting direction, respectively [6].

In order to reduce the noise, the calculated parameters were processed using a moving average filter with a window length of 5 ms.

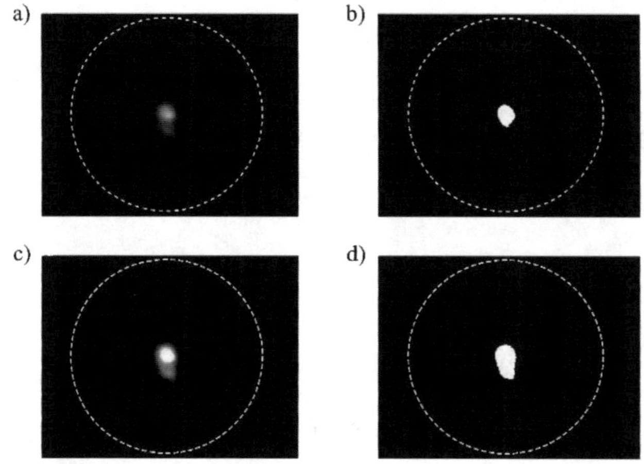

Figure 2. An example of video frame processing: a) – raw image with pseudo colors representation, b) – raw image after binarization, c) – raw image after logarithmic transformation with pseudo colors representation, d) – logarithmically transformed image after binarization. The dashed circle indicates the nozzle border.

As a variability feature, the standard deviation of the process zone length to width ratio is computed as:

$$\sigma_{l/w(n)} = \sqrt{\frac{\sum_{i=m}^{n}(l/w(i) - \overline{l/w})^2}{\tau - 1}} \tag{2}$$

where τ is the number of frames in a preceding time interval of 5 ms, $\overline{l/w}$ is the mean value on the time interval being used, n is a frame number for which the standard deviation is calculated and $m = n - \tau + 1$. For standard deviation values, a bandpass filter with a lower cut-off frequency of 250 Hz and an upper cut-off frequency of 750 Hz was applied.

3. Results and discussion

Experiments at various speeds, slower and faster than the nominal one, have been performed by stepwise speed changes, leading to significantly different quality issues. Due to several factors that are hard to control in an industrial environment, the sets of optimal parameters recommended by machine tool builders usually have a certain safety margin. It allows obtaining a good quality cut even when a material that is different from the reference one, for instance in surface quality, chemical composition or heat treatment, is being processed. As a result, a certain material can be cut not at the fastest speed or without optimal quality.

The cut edge profiles with the corresponding quality parameters are shown in Figure 3. In this case, cutting at a speed of 120 % leads to the best quality with the lowest amount of dross. However, a further increase in speed results in an occurrence of the irregular striation pattern that is characterized by unacceptably high roughness and increased risk of a cut failure [14]. Cutting at speeds higher than 140 % led to a loss of full penetration. Cutting at low speeds, such as 60 % and 80 %, causes a lot of dross which is unsatisfactory for industrial practice.

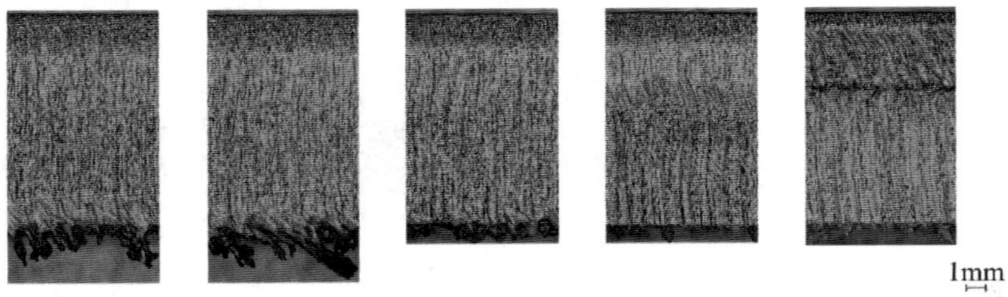

Cutting speed [%]	60	80	100	120	140
Dross area [mm^2]	50.9	48.5	27.6	14.6	17.6
Roughness [µm]	61.4	57.8	54.7	60.9	110.1

Figure 3. Cut edge profiles with corresponding quality obtained during cutting at different speeds.

The photodiode signal while cutting at different speeds is depicted in Figure 4. The signal level monotonously rises with an increase in speed. Similar behavior of the photodiode signal has been noticed before for laser flame cutting of mild steel since the cutting front inclination changes, leading to the elongation of the process zone and increase of the absorptivity [3]. However, a significant difference with the results of flame cutting might be observed when cutting too fast. The signal is considerably higher at a speed of 140 % which corresponds to the poor quality of the cut edge. It should be noted that relatively large nozzles are typically used for fusion cutting of thick plates in order to ensure appropriate gas flow, creating fewer obstacles to observe the process zone, especially in the cutting direction.

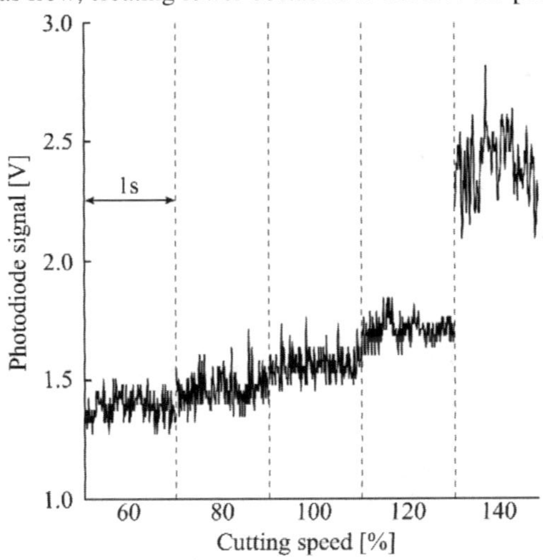

Figure 4. Photodiode signal during cutting at different speeds that were changed stepwise.

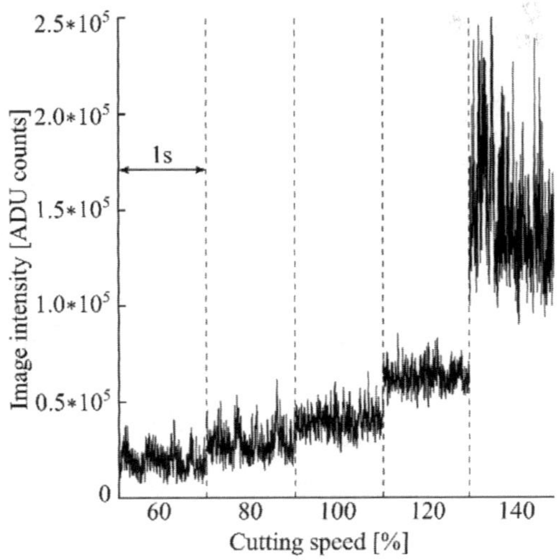

Figure 5. Overall image intensity during cutting at different speeds that were changed stepwise.

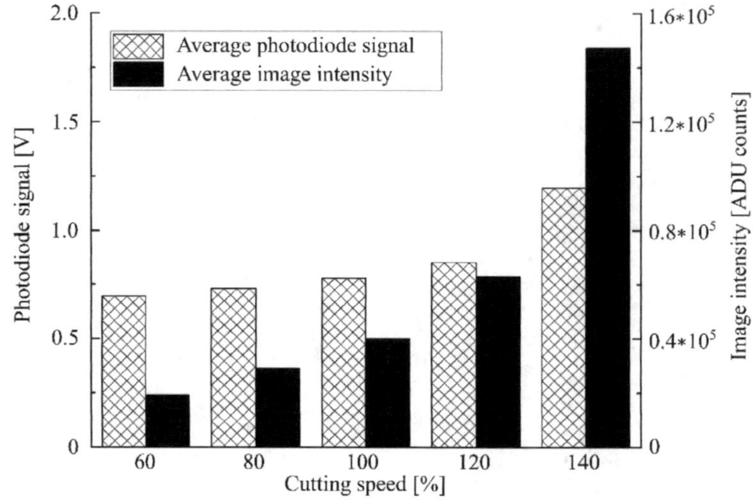

Figure 6. Comparison of the average photodiode signal and average camera image intensity.

The overall image intensity of the camera is shown in Figure 5. Similar to the photodiode signal, the image intensity is higher at faster speeds and a significant leap can be observed in the transition from 120 % to 140 %. Figure 6 depicts the comparison of the average photodiodes signal and camera image intensity. The image intensity has been calculated so that it represents the overall amount of light coming to the camera sensor, imitating the principle of the photodiode light acquisition.

As can be noticed in Figure 6, the difference between 120 % and 140 % speed for the camera parameter is more pronounced. It might be explained by the fact that the camera and the photodiode acquire radiation of different spectral ranges. Laser fusion cutting at elevated speeds is characterized by the phenomenon of local vaporizations accompanied by intense flashes. During evaporation, the temperature becomes much higher and iron atomic emission lines (Fe I) can be observed in the spectrum emitted from the process zone [15]. Therefore, taking into account Planck's law, the intensity of the emission in the visual spectral range is significantly higher, explaining a slightly different behavior of the camera image intensity.

Another essential issue of the local vaporizations occurrence is their duration. An example of three consecutive frames when a flash appears is presented in Figure 7. The flashes may be observed for a very short time which is mostly less than 0.4 ms. Therefore, a sensor with an acquisition rate higher than several kfps is required to detect the flash formation without missing any of them. In theory, state-of-the-art high-speed visual cameras can provide process observations with an acceptable resolution with a frame rate of up to 100 kfps. However, using such scientific cameras for real-time applications in industrial practice seems unrealistic due to their costs and the enormous amount of created data that needs to be transferred and processed.

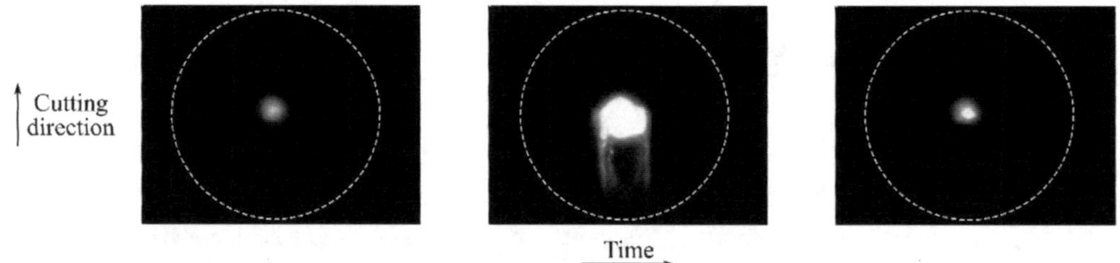

Figure 7. An example of the temporal evolution of the process zone at 140 % speed. The time interval between consecutive frames is 0.2 ms. The dashed circle indicates the nozzle border.

Since information about process zone intensity is sufficient for detecting local vaporizations on the cutting front, using relatively low-cost photodiodes that are available with ultra-high bandwidths might be relevant for possible industrial implementation. Moreover, bandpass filters designed for certain emission lines of Fe I can be coupled with photodiodes to monitor only the spectral range of interest. However, as can be seen in Figure 4 and Figure 5, the detection of flashes does not guarantee cutting with the best edge quality and performance since the increase of the process zone intensity occurs stepwise at low and moderate speeds before the start of local vaporizations appearance. Therefore, in order to select the optimal set of cutting parameters, the geometrical features of the process zone have to be evaluated.

The evolution of the length to width ratio of the process zone for different cutting speeds is shown in Figure 8. Cutting faster causes an increase in the process zone length since the cutting front inclination becomes less steep. As a result, even though the process zone width slightly increases, the process zone has minor ellipticity. However, the dross attachment has the main influence on the length to width ratio since it leads to a sudden increase in the process zone length. Consequently, as can be observed in Figure 8, the amplitude of the length to width ratio is high for slow and nominal speeds even taking into account that a moving average filter has been used. At the same time, cutting at 140 % speed does not lead to significant ratio changes since both length and width of the process zone typically increase during an intense flash (Figure 7).

Since the absolute value of the length to width ratio is affected by the stability of the process zone shape, its standard deviation shown in Figure 9 has been calculated in order to measure the variability of the process zone geometry. The consistency of the process zone shape directly influences the amount of generated dross. The lowest standard deviation level for 120 % speed corresponds to the best cut edge quality, while the highest values of standard deviation for 60 % speed correlate with the largest amount of dross (Figure 3).

An example of consecutive frames during cutting at 100 % speed is shown in Figure 10. Due to the fluctuations of the melt ejection stream, some of the molten material can attach to the kerf walls on the bottom edge. As a result, a relatively low-intensity tail of the process zone can be observed that is a typical indication of possible dross formation. As can be seen in Figure 10, the process zone can arbitrarily increase in length, keeping the width relatively constant. It can be concluded that the detection of the absolute value for such an elongation provides less information about the process status than statistical analysis of how frequently that happens.

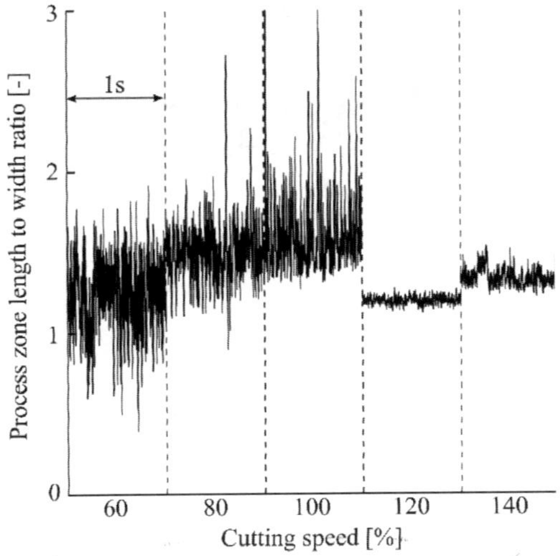

Figure 8. Process zone length to width ratio evolution during cutting.

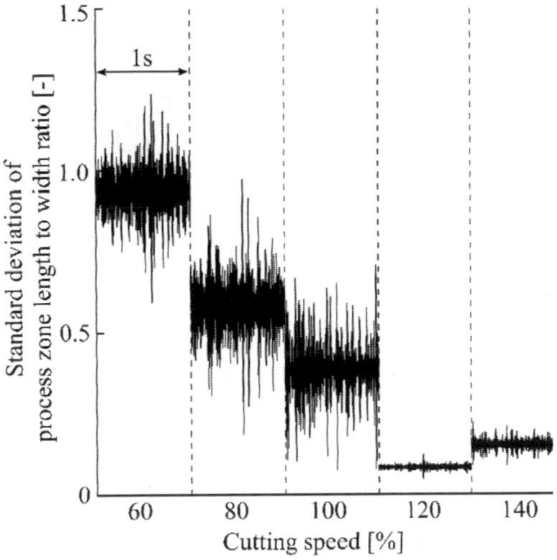

Figure 9. Standard deviation of process zone length to width ratio evolution during cutting.

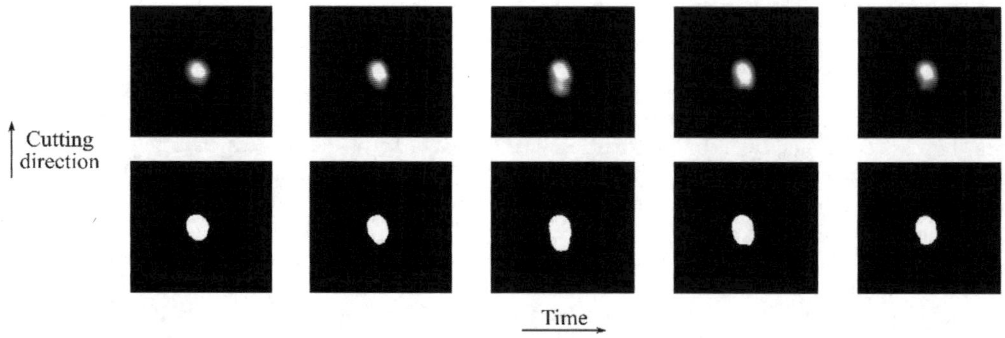

Figure 10. An example of the temporal evolution of the process zone at 100 % speed. The time interval between consecutive frames is 0.2 ms. The top row images underwent the logarithmic transformation. The bottom row pictures represent the top row after binarization.

The formation of dross during fusion cutting occurs if the lateral melt flow adheres when flowing backwards from the cutting direction along the lower edge [16]. As a consequence, the material on the bottom edge can still be hot and emit radiation even a few millimetres away from the cutting front. This typically results in the appearance of several blobs on the processed camera image as shown in Figure 11. In general, the detection of a couple of blobs can also be used as a monitoring signal for dross recognition. For instance, the number of such frames during cutting at 120 % speed is neglectable while dozens of such images have been observed per 1 mm during cutting at slow speeds. However, high-intensity flashes of local vaporizations typically are dominant and do not allow using this approach for high speeds since a flash irradiates the kerf walls creating additional blobs, which are not related to the dross attachment.

On the contrary to laser fusion cutting, laser flame cutting is characterized by higher process zone temperatures with noticeable gradients in the thickness direction [17] which might be used as a monitoring signal in a multi-sensor system if dross attachment cannot be observed.

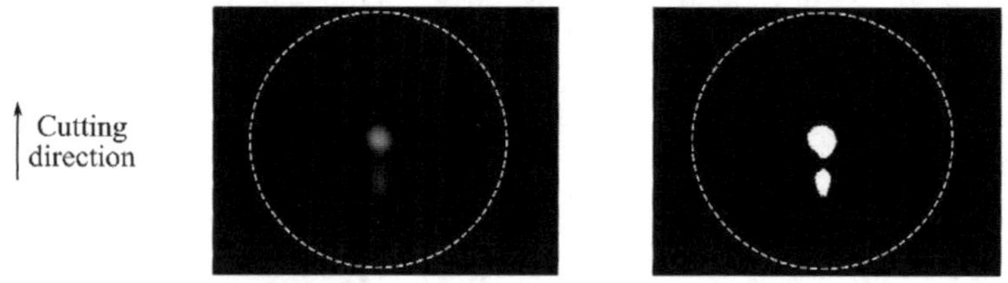

Figure 11. An example of a video frame where several blobs can be detected: raw image with pseudo colors representation on the left and its processed version on the right. The dashed circle indicates the nozzle border.

4. Conclusion

Laser fusion cutting is a rather complicated process that involves several phenomena. In order to efficiently distinguish different problems occurring during cutting, a multi-sensor approach that accumulates advantages of different sensors can be used. Sensor fusion in a coaxial optical configuration offers specific advantages, which are mainly relevant for potential industrial implementations. A fast and low-cost photodiode signal can be used to detect local vaporizations on the cutting front since it only requires the high-speed acquisition of process zone intensity data and provides great flexibility for the spectral range being used. A significant increase in the intensity of both sensors has been observed when cutting too fast that can be used as an indication of a cut failure risk. However, defining the set of process parameters that allows cutting with the lowest dross area demands geometrical information about the process zone, especially about its dimension in the cutting direction. As it was found, the lowest values of the standard deviation of the process zone length to width ratio correlate well with the lowest amount of dross. Therefore, a camera is preferred for real-time optimization of cut edge quality in terms of dross amount. In this case, ultra-fast cameras and computationally expensive data processing are not necessary, significantly reducing the costs and complexity of a monitoring system.

Future work will be dedicated to further exploring the synergy of using multi-sensing for laser flame cutting where the process zone information is limited due to the obstruction by the nozzle. Moreover, other sensors that provide spectral information might be considered for monitoring both fusion and flame cutting processes.

ORCID IDs
N Levichev https://orcid.org/0000-0001-8929-9843
J R Duflou https://orcid.org/0000-0002-7265-9686

References

[1] Purtonen T, Kalliosaari A and Salminen A 2014 Monitoring and adaptive control of laser processes *Phys. Proc.* **56** pp 1218–31.
[2] Fallahi Sichani E, De Keuster J, Kruth J-P and Duflou J R 2010 Monitoring and adaptive control of CO_2 laser flame cutting *Phys. Procedia* **5** pp 483–92.
[3] Levichev N, Costa Rodrigues G and Duflou J R 2020 Real-time monitoring of fiber laser cutting of thick plates by means of photodiodes *Proc. CIRP* **94** pp 499–504.
[4] Schleier M, Adelmann B, Neumeier B and Hellmann R 2017 Burr formation detector for fiber laser cutting based on a photodiode sensor system *Opt. Laser Technol.* **96** pp 13–17.
[5] Pacher M, Franceschetti L, Strada S C, Tanelli M, Savaresi S M and Previtali B 2020 Real-time continuous estimation of dross attachment in the laser cutting process based on process emission images *J. Laser Appl.* **32** 042016.
[6] Levichev N, Costa Rodrigues G, Vorkov V and Duflou J R 2021 Coaxial camera-based monitoring of fiber laser cutting of thick plates *Opt. Laser Technol.* **136** 106743.
[7] Mitchell H B 2007 *Multi-sensor data fusion: An Introduction* (Springer-Verlag Berlin Heidelberg) p 282.
[8] Gao X, Sun Y, You D, Xiao Z and Chen X 2016 Multi-sensor information fusion for monitoring disk laser welding *Int. J. Adv. Manuf. Technol.* **85** pp 1167-75.
[9] Deng F *et al.* 2020 A multi-sensor data fusion system for laser welding process monitoring *IEEE Access* **8** pp 147349-57.
[10] Peng X, Kong L, Chen Y, Shan Z and Qi L 2020 Design of a multi-sensor monitoring system for additive manufacturing process *Nanomanuf. Metrol.* **3** pp 142-50.
[11] De Keuster J, Duflou J R and Kruth J-P 2007 Monitoring of high-power CO_2 laser cutting by means of an acoustic microphone and photodiodes *Int. J. Adv. Manuf. Technol.* **35** pp 115–26.
[12] Schleier M, Adelmann B, Esen C and Hellmann R 2018 Cross-correlation-based algorithm for monitoring laser cutting with high-power fiber lasers *IEEE Sens. J.* **18** pp 1585–90.
[13] Levichev N, Costa Rodrigues G, Dewil R and Duflou J R 2020 Anticipating heat accumulation in laser oxygen cutting of thick metal plates *J. Laser Appl.* **32** 022018.
[14] Wandera C, Salminen A and Kujanpaa V 2009 Inert gas cutting of thick-section stainless steel and medium-section aluminum using a high power fiber laser *J. Laser Appl.* **21** pp 154-61.
[15] Bocksrocker O, Berger P, Kessler S, Hesse T, Rominger V and Graf T 2020 Local vaporization at the cut front at high laser cutting speeds *Lasers Manuf. Mater. Process.* **7** pp 190-206.
[16] Stoyanov S, Petring D, Arntz-Schroeder D, Günder M, Gillner A and Poprawe R 2020 Investigation on the melt ejection and burr formation during laser fusion cutting of stainless steel *J. Laser Appl.* **32** 022068.
[17] Levichev N, Staudt T, Schmidt M and Duflou J R 2021 Hypespectral imaging and trim-cut visualization of laser cutting *CIRP Ann. – Manuf. Technol.* **70** pp 207-10.

IOP Conf. Series: Materials Science and Engineering 1135 (2021) 012015 doi:10.1088/1757-899X/1135/1/012015

Visualization of laser back-reflection distribution during laser welding

Petr Horník[1], Hana Šebestová[1], Jan Novotný[1], Libor Mrňa[1]

[1]Institute of Scientific Instruments of the Czech Academy of Sciences, Královopolská 147, 612 64 Brno, Czech Republic
hornik@isibrno.cz

Abstract. There are several approaches to weld quality monitoring during laser welding. Reflected laser radiation carries partial information about the welding process. Fibre lasers has usually a built-in diode to detect excessive back-reflected laser radiation to protect the laser source from damage. Reflected laser radiation measured in the laser source is compared with reflected laser radiation measured in the welding head. Moreover, coaxial high-speed imaging with a narrow bandpass filter on laser wavelength is used to visualize the reflected laser radiation. The advantage of this solution is that no additional illumination is needed and the reflected laser intensity and spatial distribution can be obtained from the image. Keyhole inlet dimensions are measured and related to the laser power. The transition between laser welding modes is studied.

1. Introduction

Part of laser radiation is reflected during laser welding. High levels of reflections can lead to dangerous situations and damage. However lower levels of reflections may carry information about the condition of the welding process. Laser back reflection (BR) is a sum of reflections on optical elements of a processing head, reflection on the weld pool surface and keyhole walls, and scattering on metal droplets and microparticles above the weld pool.

Molten metal has greater reflectivity than non-molten metal and behaves like a mirror that reflects the laser beam. Movements of the free surface of the weld pool can focus, defocus, or deflect the beam, while the keyhole predominantly absorbs the beam. This article intends to investigate which areas of the weld pool reflect the most laser radiation during different welding modes. Such conclusions will help to interpret the effects of the individual components on one-dimensional photodiode measurements.

Photodiode-based monitoring systems are commonly used in industry due to the advantage of higher possible frame rates and robustness, but on the other hand, with such a simple tool you cannot obtain a comprehensive description of the process. Norman et al. found out that the signal is governed mainly by the weld pool size and liquid pool emissions [1]. The emissivity of the weld pool surface is strongly influenced by the oxidation state of the weld pool surface. You et al. measured emitted laser and visible radiation and compared them with the keyhole and plasma plume area, concluding that BR is sensitive to laser power and keyhole size [2]. The intensity of reflected laser radiation is also affected by focus position [3]. Eriksson et al. found out that variance of photodiode signals tends to be larger for bad welds while the mean value changes only a little [4]. This may cause problems with setting thresholds for weld quality classification and shows the challenges in finding principles for a universal monitoring system.

Stritt et al. investigated reflected laser radiation to distinguish between heat-conduction and deep-penetration mode during disc laser welding with modulated laser power [5]. The authors also observed higher and more variable BR during the heat-conduction mode compared with deep-penetration mode and noticed that the BR is higher during the keyhole formation than the BR corresponding to keyhole closure. However, no analyses have been made to determine whether the major source of BR is weld pool or keyhole.

Images of laser welding process are usually recorded by a side camera with additional illumination. Zhao et al. proposed a setup with coaxially mounted camera with a side lighting source [6]. Kaierle et al. tried an imaging setup with a coaxially mounted NIR camera and standard CMOS camera with additional coaxial illumination and bandpass filter [7]. Near-infrared images were too saturated and thermal radiation was visible. In contrast to previous studies, replacing the illumination laser with the processing laser suggest since it is also a powerful source suitable for illumination.

The welding process is a source of a wide range of emitted radiation. The highest level of intensity is reached at the wavelength of the processing laser. A narrow bandpass filter tuned to the wavelength of the laser is necessary to ensure that only desired wavelength reaches detectors. Nevertheless, Both the weld pool and laser emit in NIR spectrum. The contribution of laser and thermal radiation cannot be separated from the measured values. Fortunately, the weld pool radiation is an order of magnitude weaker than laser reflection and can be suppressed under the detection limit by neutral density filters. Such a solution should suppress the effect of thermal radiation and images carry an intensity value of BR in each pixel and therefore can be used for further measurement.

2. Experimental setup

The laser source IPG YLS-2000 with welding head Precitec YW30 with 100 mm focusing and 200 mm collimating lens were used. The feeding core diameter is 200 μm which theoretically results in a 400 μm beam waist diameter. The welding head is equipped with a beam splitter allowing coaxial observation of the process zone. Custom optics with InGaAs photodiode with amplifier and AOS Promon U750 camera were attached to the welding head.

Several bead-on-plate welds with different processing parameters were performed. Welding speed was 20 mm·s^{-1} and laser focus position 1 mm under the surface of the 10 mm thick S355J2 carbon steel sheet. Different weld dimensions were obtained by laser power variation from 100 to 2000 W with an increment of 100 W. The beam waist diameter was checked by contactless profilometer BeamWatch. The beam parameters vary depending on the laser power, therefore measurements for two laser power values were made. The average diameter was 391 ± 26 μm for 1000 W and 446 ± 10 μm for laser power 2000 W. The diameter of the laser spot on the sheet surface, thus 1 mm above the focus position, for both laser power values was 410 μm for 1000 W, respectively 460 μm for 2000 W. To ensure the correct focal position, the sheet standoff distance was measured by a triangulation sensor prior the welding. The welds were shielded with an argon protective atmosphere.

2.1. Imaging optics

Back-reflected laser radiation is split inside the welding head. A part of the laser beam continues through the optical fibre to the built-in photodiode in the laser while a part of the beam is directed outwards. Standard coaxial camera optics were replaced with a modular system, figure1. The reflected laser beam passes through the gas nozzle of inner diameter 5.5 mm, welding head focusing lens and beam splitter, simple optics, bandpass filter on laser wavelength 1070±10 nm and neutral density filter with optical density OD = 3 to suppress the intensity. An aperture with a 4 mm diameter is mounted before a 50/50 beam splitter. The beam splitter divides the beam in half and the lens focuses the beam on both InGaAs photodiode and camera chip. The lenses have an AR coating to eliminate reflections and clear aperture of 25.4 mm. The optical path was covered during the welding.

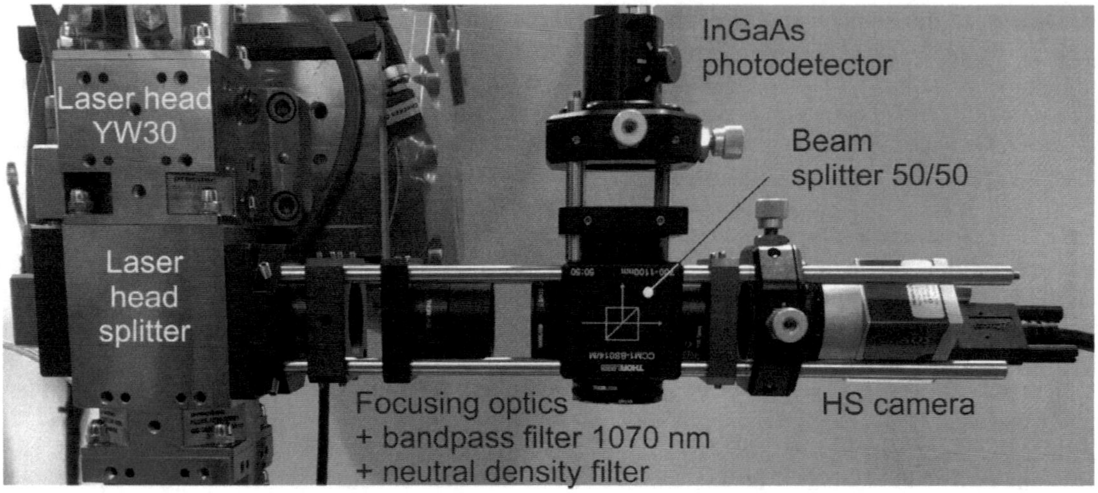

Figure 1. Optics for measuring the intensity and spatial distribution of reflected laser beam.

3. Results

Both photodiodes built into the laser and the head were sampled at 100 kHz. The camera framerate was 3003 fps which is the fastest possible at a resolution of 144 x 140 pixels. The exposure time was 59 µs which is the limit of the camera to ensure a sharp image. The camera can be treated as a single photodiode by summing the intensities of all pixels. Figure 2 shows the reflected laser intensity detected of all three detectors the photodiode built into the laser source, photodiode built into the welding head and sum of all pixels of coaxially mounted camera at 600 W laser power. The figure shows that all detectors have a similar pattern which indicates that the same phenomenon is measured. While photodiodes are suitable for industrial practice due to their simplicity, the camera allows for a more detailed spatial analysis of the back-reflected laser radiation.

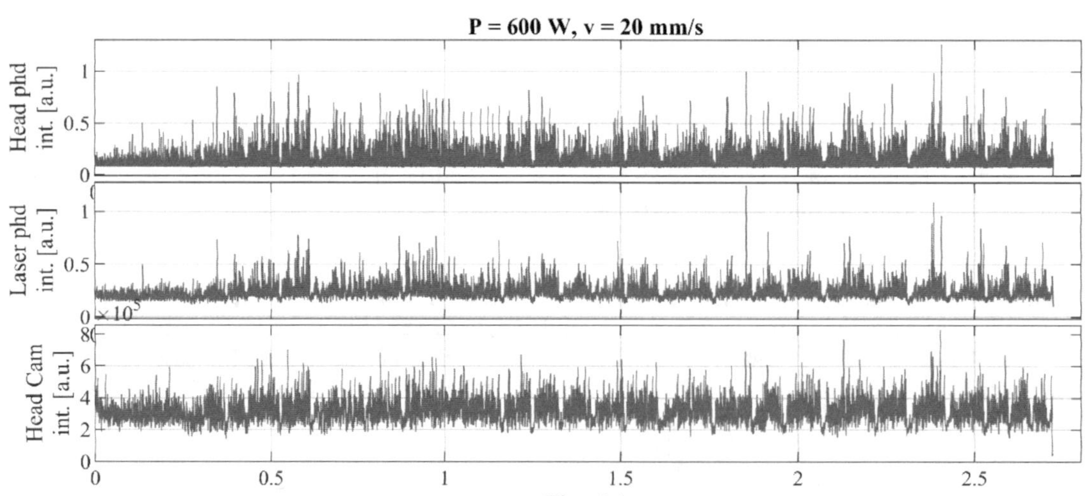

Figure 2. The intensity of reflected laser radiation measured with the built-in photodiode (phd) in laser, and coaxial photodiode and camera (cam) in welding head.

Figure 3 shows the intensity distribution of BR of consecutive images at three different laser power values. The laser is reflected only from a small area and most of the image is black. Correct intensity measurements are disrupted by small, saturated areas. Those can be eliminated by using a filter with higher optical density but at the cost of reducing lower intensity reflections. Nevertheless, they are not an issue for the subsequent evaluation. The recorded events are faster than the frame rate of the camera because the changes between frames are too rapid and the frames do not show the gradual changes. Thus, a framerate of 3003 fps is not enough to fully capture the dynamics of the laser welding process.

Welding with laser power of 300 W melts the sheet without creating a keyhole. The unabsorbed laser beam is shaped and partially reflected back. On the other hand, the laser power of 600 W begins to form a keyhole. The keyhole is present from time to time, but not fully developed. The small circular grey area is an early stage of the keyhole. The saturated high-intensity reflection still occurs. The last pictures of figure 3 show that the laser power of 1500 W is enough to form a stable keyhole with part of the weld pool visible. the keyhole changes the shape and size of the inlet opening. It is possible to look inside the keyhole but the details are not visible and the depth cannot be estimated.

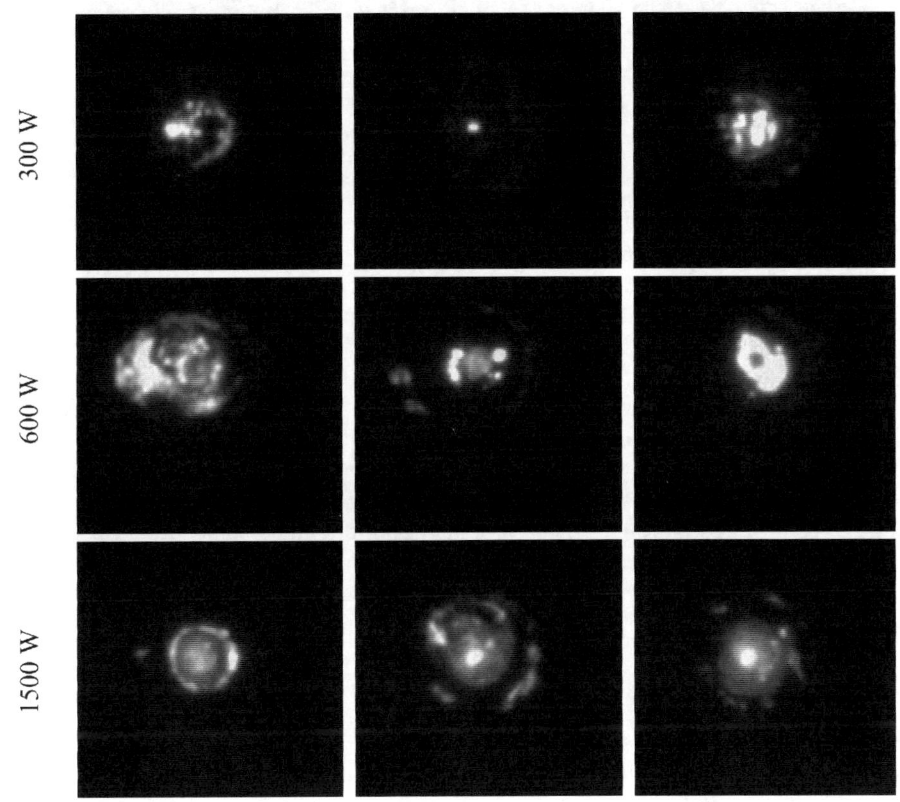

Figure 3. Consecutive images of reflected laser radiation for laser power 300, 600 and 1500 W.

It is quite challenging to determine what part of the welding process is present in the camera pictures. Laser irradiates a relatively small area, molten metal has smooth shapes and occasionally spatter occurs. This makes the images blurry and the surface of the weld pool does not always reflect the light upwards, making it difficult to determine the dimensions programmatically. To describe the process in a more statistical manner than a single measurement, averaged image from over 8000 images for each weld was calculated in figure 4.

The heat-conduction mode has the sharpest weld contour with a relatively uniform average intensity distribution except for the intermediate ring with lower intensity. On the other hand, the deep-penetration mode with a fully developed keyhole has the brightest spot in the centre and unsharp edges. Laser is absorbed inside deep and narrow keyhole by multiple reflections that is the reason why the keyhole is not much brighter compared to reflections from the free surface [8]. The bright area does not represent the weld pool, but the keyhole in this case. The transition mode is a combination of the previous two.

The images were calibrated to perform length measurements as well by capturing the scale. One pixel corresponds to 12.5 μm in the camera focal plane. By modifying the imaging system, better resolution can be achieved but in this case the frame rate was more important at the sacrifice of spatial resolution. Since laser spot diameter is 0.4 mm values are reasonable and will be discussed in more detail below.

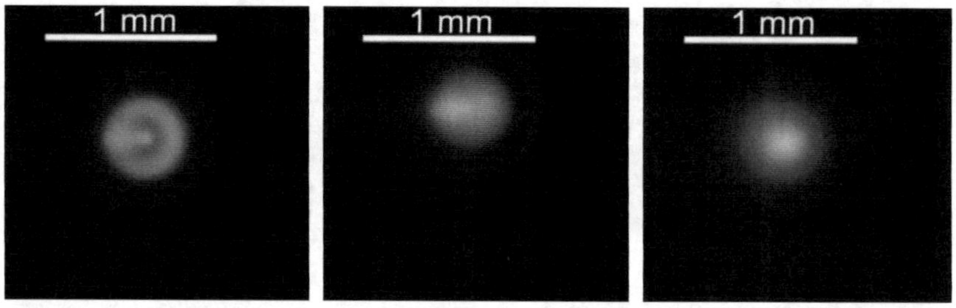

Figure 4. Average distribution of reflected laser intensity during laser welding with power values from the left 300, 600 and 1500 W.

Weld fusion zone dimensions of metallographic cross-sections were also measured and plotted in figure 5. The excessive width deviations for higher laser power values are caused by humping and the fact that only one cross-section was made from each weld. The penetration depth is only needed to assess the welding mode as depth cannot be directly measured by the proposed setup. The transition between both welding modes is around the point where measured weld width and depth intersects. In this case, it is at the laser power of 650 W. Figure 6 shows estimated bright spot dimensions from averaged camera images. The threshold was set fixed for all welds based on the visible contours of the weld pool in the heat-conduction mode. The dimensions of the area are approximately 0.44 - 0.53 mm for laser power under 500 W. While for laser power higher than 1000 W it is between 0.4 - 0.44 mm. The zone between them has a prolonged shape and the estimated length and width differ significantly. On this basis, the transition mode can be detected. During the heat-conduction mode, the bright spot represents the weld pool and on the other hand, during deep-penetration welding keyhole is observed.

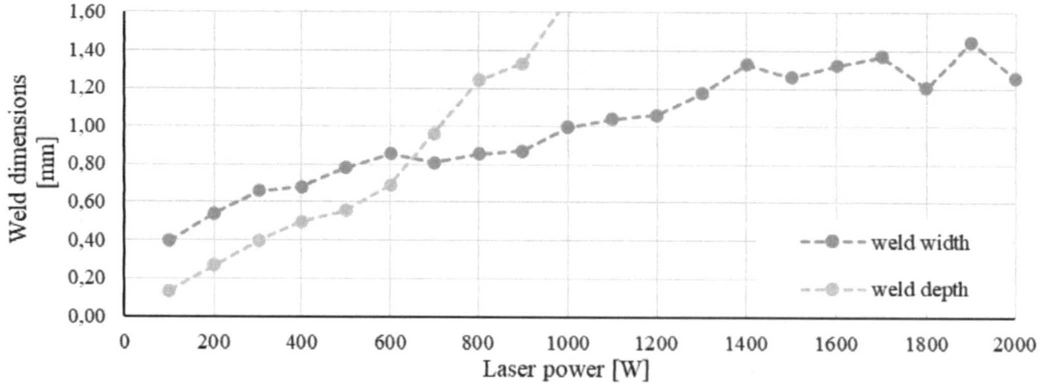

Figure 5. Weld dimensions measured on metallographic cross-sections.

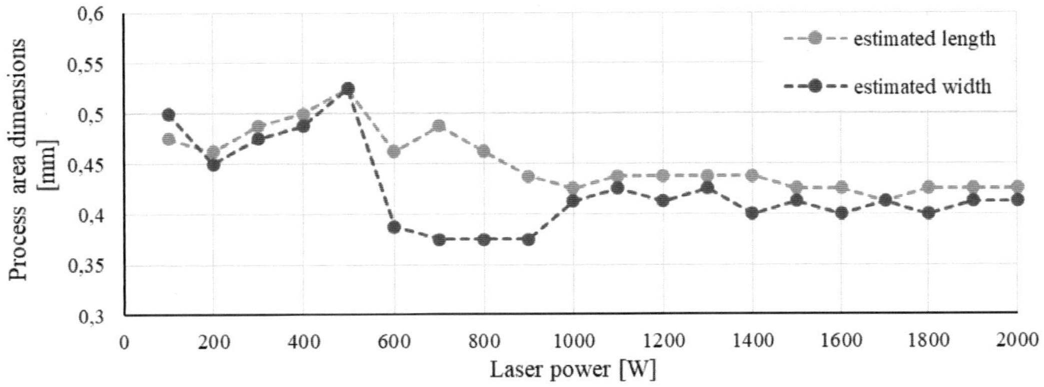

Figure 6. Process area dimensions estimated from averaged camera images.

4. Discussion

The intensity of reflected laser in figure 2 detected by the photodiode and signal calculated from the camera seems to be very similar on a macro scale. Yet on closer look, there are small differences that may be caused by different chip size, sensitivity, and alignment. Also, light detected by the built-in photodiode passes through the fibre while part of the beam detected in the welding head is deflected before the fibre.

The laser beam has the highest intensity within its nominal diameter of 0.4 mm. The beam measurement made at 1000 and 2000 W shows that with increasing laser power the beam diameter is also slightly increased. Therefore, the largest irradiated area detectable by the camera depends on the laser power. But it is difficult to verify this statement since the metal sheet is molten at higher laser power values. The irradiated area for laser power under 500 W is roughly 0.86 mm which is greater than the weld pool size and the solid surface of the sheet is visible.

The transition between welding modes is detectable by both photodiodes. The figures are very similar, hence only intensity measured in the welding head is presented. Figure 7 shows the mean values and standard deviations of signal averaged over the length of the weld. The three welding modes are distinguishable based on a variation of BR signal. The zone below 500 W with the greatest variability corresponds with the heat-conduction mode. The zone above 1000 W with the lowest variability is deep-penetration laser welding and the zone between them with medium variability is transition mode. The results are consistent with Stritt et al. Excessive variability in low power laser welding is caused by the data not following a normal distribution. Two or more effects are mixed together. The camera captures only the keyhole area without the weld pool at higher laser power values. Therefore, BR signals follow the normal distribution so that assumptions for correct standard deviation estimation are met. Image segmentation to distinguish the contributions of keyhole and weld pool should be done and is crucial for verifying the hypothesis of different contributions. One possibility is differentiating the regions based on a variation of intensity. The region with lower variance should be the keyhole and solid metal sheet. On the other hand, the weld bath should exhibit greater intensity variability as Stritt et al. noticed.

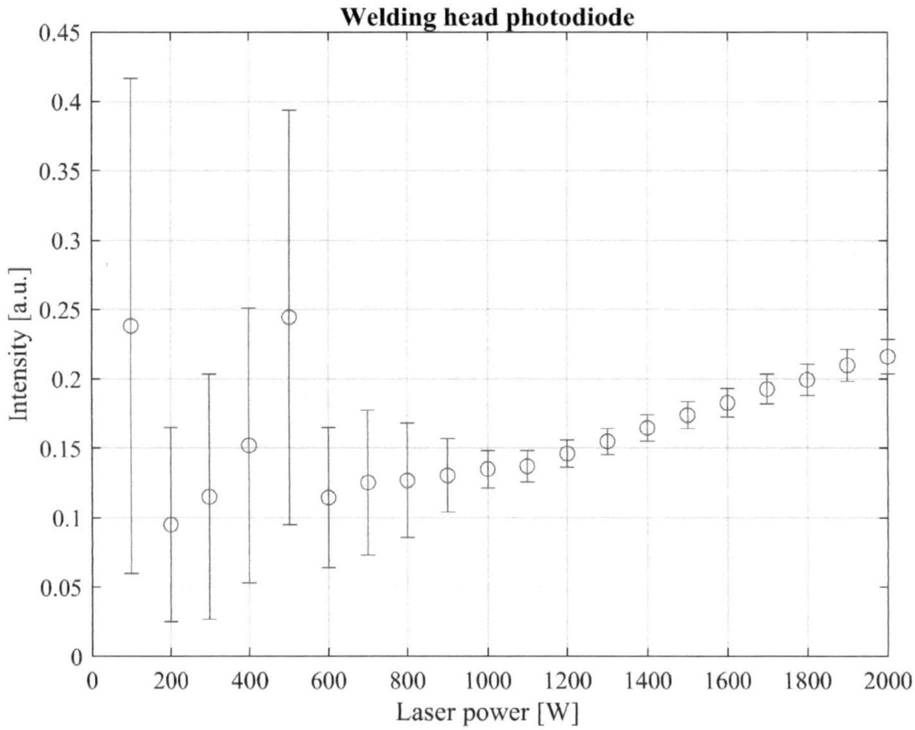

Figure 7. Reflected laser intensity measured by the photodiode in the welding head.

The common dividing line in defining the transition between laser welding modes is based on metallographic cross-sections. Asibu claims that heat-conduction welding mode typically results in a depth-to-width ratio of about 3:1 compared to about 10:1 or more for deep-penetration welding [9]. That is much more than in our case and thereby unusable. Another dividing line is the depth-to-width ratio lower than one for heat-conduction mode and greater than one for deep-penetration mode. Now compare at which laser power values the changes corresponding to the welding modes vary for different methods. The welding head photodiode indicates transition mode between 600 and 900 W and the observation of keyhole presence on images suggests the same values. A slight drawback is that detection of keyhole does not work on individual images at this stage but estimates the presence of a keyhole based on averaged values. The depth-to-width ratio equals one at 650 W according to the weld dimensions plot. The Depth-to-width ratio in figure 8 proposes that transition mode is shifted to lower laser powers between 400 and 700 W. Thus, all methods give similar results, but for a detailed comparison more measurements with finer power division would be needed. This was a simple verification that the coaxial camera imaging method does not give results that conflict with the other methods. Moreover, others [5] have shown, that capillary formation and closure is correlated with reflected laser radiation. There is the potential to monitor the keyhole presence during the welding process.

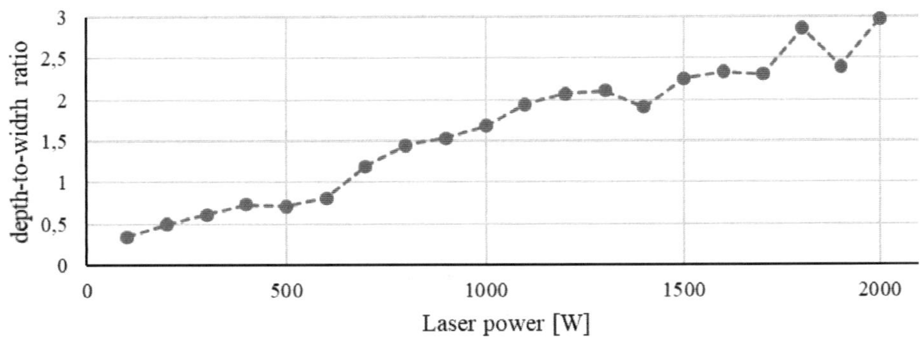

Figure 8. The effect of laser power on depth-to-width ratio.

Further development should be directed towards the development of algorithms for segmenting the images capable of distinguishing between weld pool and keyhole. The biggest challenge here is that in heat-conduction mode only the size of the weld pool is visible, whereas in deep-penetration mode only the keyhole is visible and the weld pool remains dark. This existence of multiple welding modes especially transition mode complicates implementation of universal recognition algorithm and interpretation of results. However, it may be sufficient to focus only on the study of the one welding mode, namely the deep-penetration mode.

5. Conclusion

For general monitoring purposes, it should not matter if reflected laser intensity is measured by a built-in photodiode which is primarily used for laser protection or photodiode solution implemented in welding head. It was proved that the BR signal from both photodiodes is the same except for minor differences. The sum of pixel intensities from all the pixels of the coaxially mounted camera shows similar behaviour as the photodiodes. Therefore, the camera can be used instead of a photodiode. The greatest advantage is the ability to visualize detailed spatial distribution of reflected laser radiation during the laser welding process. It is possible to coaxially observe the reflections from the weld pool surface and interior of the keyhole during laser welding. The laser beam is predominantly focused by the convex shape of the weld pool during heat-conduction mode. On the other hand, a major part of the laser beam incident on the weld pool surface during deep-penetration mode is absorbed in the keyhole or deflected. The image is not negatively affected by the plasma plume.

No external lighting source synchronized with a camera is needed. The advantage of this solution is that the images can be used for intensity measurement. On the other hand, the disadvantage of this method is that only a small area approximately the size of the laser beam is visible.

Three welding modes can be distinguished based on back-reflected laser intensity variation. Although distinguishing the modes can be based on intensity alone, so photodiode appears sufficient, the camera allows to check the presence of keyhole. The coaxial images of reflected laser radiation have made it possible to understand from which regions the reflections originate. The images were averaged for an objective assessment. It is important to be aware that the method of averaging images emphasizes different objects – weld pool for heat-conduction mode and keyhole for deep-penetration welding mode. The thresholding method is not suitable because some areas of the process are not always visible and the reflections from the weld bath have a random character. Averaging the images solves the problems with randomness but average values differ only slightly, and it is problematic to find the correct threshold for transition and deep-penetration welding mode. More advanced algorithms should be used to take full advantage of this imaging technique.

It has been verified that the intensity of reflected laser radiation for heat-conduction mode is proportional to the weld pool size. The greater the weld pool size estimated from images, the higher the overall intensity detected by the camera or both photodiodes. The size of the measurable weld pool is limited by the size of the illuminated area at the laser spot size scale.

Acknowledgements
This research was supported by the Technology Agency of the Czech Republic (projects TH04010366 and FW01010293) and with the institutional support of the Czech Academy of Sciences (RVO:68081731).

ORCID iDs
P Horník https://orcid.org/0000-0002-9320-9561
H Šebestová https://orcid.org/0000-0003-4445-1324
J Novotný https://orcid.org/0000-0001-6175-869X
L Mrňa https://orcid.org/0000-0002-3928-5619

References

[1] Norman P, Engström H and Kaplan A F 2008 Theoretical analysis of photodiode monitoring of laser welding defects by imaging combined with modelling *Journal of Physics D: Applied Physics* **41** 195502

[2] You D, Gao X and Katayama S 2013 Multiple-optics sensing of high-brightness disk laser welding process *NDT & E International* **60** 32–9

[3] Horník P, Šarbort M, Šebestová H and Mrňa L 2019 Study of the influence of focal position on back-reflected radiation during deep penetration laser welding and its simulation *Optics and Measurement International Conference 2019*

[4] Eriksson E A I, Norman P and Kaplan A F H 2009 Basic study of photodiode signals from laser welding emissions *Proc. 12th Nordic Laser Materials Processing Conference* (Denmark, Copenhagen: Kgs. Lyngby ATV-SEMAPP)

[5] Stritt P, Weber R, Graf T, Müller S and Ebert C 2011 Utilizing Laser Power Modulation to Investigate the Transition from Heat-Conduction to Deep-Penetration Welding *Physics Procedia* **12** 224–31

[6] Zhao P, Lu R, Wu J and Zhang Y 2020 Monitoring method of weld penetration in laser keyhole welding irradiated by laser auxiliary illuminant *Journal of Laser Applications* **32** 022037

[7] Kaierle S, Ungers M, Franz C, Mann S and Abels P 2010 Understanding the Laser Process *Laser Technik Journal* **7** 49–52

[8] Jin X, Cheng Y, Zeng L, Zou Y and Zhang H 2012 Multiple Reflections and Fresnel Absorption of Gaussian Laser Beam in an Actual 3D Keyhole during Deep-Penetration Laser Welding *International Journal of Optics* **2012** 1–8

[9] Kannatey-Asibu E 2009 *Principles of laser materials processing* (Hoboken, N.J, New Jersey: Wiley) pp 516–28

IOP Conf. Series: Materials Science and Engineering 1135 (2021) 012016 doi:10.1088/1757-899X/1135/1/012016

Gas flow study for development of a novel shielding gas nozzle for directed energy deposition processes using computational fluid dynamic simulations

Yogesh Nalam[1], Tobias Hauser[1,2], Philipp Peter Breese[1], Tobias Kamps[1], Yves Küsters[1], Rebar Hama-Saleh Abdullah[3], Joerg Volpp[2], Alexander F H Kaplan[2]

[1] Technology, Siemens AG, D-81739 Munich, Germany

[2] Department of Engineering Sciences and Mathematics, Luleå University of Technology, S-971 87 Luleå, Sweden

[3] Fraunhofer Institute for Laser Technology, Steinbachstrasse 15, 52074 Aachen, Germany

tobias.hauser@siemens.com

Abstract. Directed energy deposition (DED) enables the additive manufacturing of several materials such as molybdenum alloys that are very difficult to process by conventional methods. Some of these materials are highly reactive to gases in ambient atmosphere such as oxygen, and nitrogen. Oxidation during additive manufacturing significantly influences the mechanical properties of a part. In some cases, the shielding gas coverage of standard powder nozzles is not sufficient, and oxidation still takes place. A functional prototype of a compound multi flow path annular nozzle was developed using computational fluid dynamics simulations. Simulations were performed using multi-component miscible gas model. Prototypes were manufactured for several design iterations to test their functionality in cold flow conditions. In the end, an Inconel based prototype was built, using laser powder bed fusion. The volume of shielding gas cover over the substrate improved with the proposed design and the radial extent of 80 ppm oxygen concentration increased from 8 mm to 25 mm. Finally, Mo-Si-B alloy was deposited on a 1000 °C pre-heated substrate without significant oxidation or cracks.

1. Introduction

Additive manufacturing (AM) plays an important role in digitalization of manufacturing industries. Laser metal deposition (LMD) is a DED technique in which material powder is melted by laser beam and deposited layer by layer. The technique of LMD provides the aspects of flexibility of manufacturing in terms of materials and shapes [1]. One of the major limitations with this process is maintaining the quality of the build. Oxidation is a major factor determining the final mechanical properties of the build which occurs at different temperatures for different materials. Oxidation can be restricted by reducing the concentration of oxygen in the zone of metal deposition by displacing the oxygen with inert gases [2]. A popularly used method is to fill a physical chamber with inert gases and perform metal deposition inside the chamber [3]. This is economically and temporally a demanding process. Besides the use of a physical chamber, a localized inert atmosphere of shielding gases can be used. This local atmosphere can be achieved by using nozzle systems to effectively deliver the inert gas and shield the metal from oxidation. Mostly nozzles are designed to deliver material powder coaxially with laser beam. Material powder is transported with a carrier gas which is also inert in nature [3]. A lot of research has been performed on effects of shielding gas on metal deposition techniques and the parameters that effect the shielding phenomena such as gas flow rate, and the type of gas [4, 5]. For the LMD setup used in this work, a commercially available coaxial nozzle from Precitec is used. However, such commercial nozzles showed limitations in shielding the process zone. Especially, strong oxidation occurred when processing materials with high affinity for oxygen such as molybdenum alloys. Therefore, in this work, the influence of shielding gas flows from a conventional nozzle and a specially designed additional nozzle was investigated using computational fluid dynamic (CFD) simulations.

2. Material and Methods

2.1. Material

The material used in this study is an alloy of molybdenum, silicon and boron. Mo-Si-B is an alloy which can be used for parts operating in ultra-high temperature zones due to its melting temperature over 2000°C [6]. The substrate used in this study was a titanium-zirconium-molybdenum cylinder with a radius of 20 mm. The powder used for deposition was a Mo-Si-B alloy with the chemical composition listed in Table 1.

Table 1. Chemical composition of Mo-Si-B used in this study.

Component	Molybdenum	Silicon	Boron
Atomic %	76	16.5	7.5

2.2. Experimental set-up

The experimental set-up with the additional shielding gas nozzle is shown in Figure 1. The equipment, which is shown in Figure 1, is a specifically developed test setup from Siemens. This test setup consists of a motion control SINUMERIK 840D, a TRUMPF laser TRUDisk 8002, a medium-frequency generator MFG70 for preheating, and a vibration powder feeder from MediCoat for the powder supply. The SINUMERIK 840D controls the laser, the processing system, and the vibration powder feeder. The process chamber consists of a Precitec laser processing head, the developed additional shielding gas nozzle, a three-axis motion system, and an inductor with a pyrometer for control of the preheating temperature. The track height of the deposited material was monitored by a distance sensor of Prectiec, which measures the distance of the LMD nozzle to the melting zone to guarantee a uniform distance across all layers in the z-direction. The diameter of the laser beam was 800 μm and the focal length was 300 mm. The feeding fibre core diameter was 400 μm and the collimation was 150 mm. Argon has been used as a shielding gas as well as carrier gas for the powder. In the experiments, a cube with an edge length of 10 mm was built. The process strategy was based on an outer frame which was filled by a meander strategy. The process parameters used for the experiments are shown in Table 2.

Table 2. Process parameters used for the experiments.

Parameter	Total powder mass flow	Travel speed	Laser power	Laser wavelength	Hatching space	Track height	Preheating
Value	1.60 g/min	0.4 m/min	240 W	1030 nm	0.4 mm	0.4 mm	1000°C

Figure 1. Experimental setup with the additional shielding gas nozzle used for the deposition of Mo-Si-B alloy.

2.3. *Experimental measurement method*
The main metric measured in this study is the concentration of oxygen. Experimentally, a residual oxygen meter ORBmax from Orbitalum was used which gives the concentration in terms of parts per million (ppm). The device uses optical sensor technology and works on the principle of fluorescence extinction. It works with an accuracy of 10% of the value displayed. The concentration of oxygen was measured at the centre of the process zone, where the laser and powder focus are located, by drawing the gas through a small hole in a plate located below the nozzle.

2.4. *Computational simulation method*
In this study, the finite volume method approach was used to simulate the flow of multiple miscible gases and predict their concentrations. The computational domain is divided into smaller units called as control volumes. Several partial differential equations are discretised for these control volumes and solved for the required quantities in those equations. The main concept behind this approach is to conserve the mass, momentum and energy of a given flow in each control volume and the whole computational domain [7].

$$\frac{\partial}{\partial t}\int_V \rho Y_i \mathrm{d}V + \oint_A [\rho Y_i(\boldsymbol{v} + \boldsymbol{M}_i)] \cdot \mathrm{d}\boldsymbol{a} = \oint_A \left[J_i + \frac{\mu_t}{\sigma_t}\nabla Y_i\right] \cdot \mathrm{d}\boldsymbol{a} + \int_V S_{Y_i} \mathrm{d}V \quad (1)$$

In addition to flow conservation equations of mass, momentum and energy, one more equation known as species transport equation from Siemens Star-CCM+ was solved to determine the concentrations of the miscible gases in the domain as shown in equation (1). The parameters used are: V = volume, Y = mass fraction, i = component index, $\boldsymbol{a}$ = area vector, $\boldsymbol{M}$ = migration term in electric field, J = laminar (molecular) diffusive flux, σ_t = turbulent Schmidt number, S = source term, and μ_t = turbulent viscosity. Concentration of the required species was measured in terms of ppm which is derived from mole fraction computed from mass fraction in equation (1).

2.5. *Simulation setup*
The digital models for the nozzle were designed using Siemens NX. These models were later used for development of the additional nozzle. Geometry preparations, meshing and CFD simulations for this study were performed in Siemens Star-CCM+. Investigations of the designs were performed by reducing the computational domain to 2D axis-symmetrical models with a polygonal mesh [8]. Detailed analysis of the final design was performed by simulating 3D models with polyhedral mesh. Two layers of prismatic cells were generated along the wall to capture the wall effects inside the nozzle [9, 10]. Mesh was refined till the extent where an iso-line representing 1 ppm concentration of air was not affected by

IOP Conf. Series: Materials Science and Engineering 1135 (2021) 012016 doi:10.1088/1757-899X/1135/1/012016

changing the mesh size further. A base mesh size of 1 mm was applied on the whole computational domain with a minimum cell size of 0.1 mm to maintain at least 4 cells between two adjacent walls.

Simulations were performed in steady state where the flow is assumed to be not changing with time [7]. The fluids in this study were argon and air. Multi-component gas model was used for the simulations. It considers the fluids as miscible and non-reacting. Ideal gas model was chosen to perform the simulations as there were no huge density changes due to Mach number of the flow. Temperatures were not considered, and iso-thermal model was applied to the simulations. Realizable two-layer all y+ K-Epsilon turbulence model was used to model the flow in the viscous regime. Segregated solver from Siemens Star-CCM+ was used to perform the simulations.

2.6. Conventional nozzle

The existing conventional nozzle is a coaxial compound nozzle. Figure 2 shows the different flow paths of the nozzles. The main gas flow is used as shielding gas of the process and the side gas flow is used as shielding gas and as carrier gas of the metal powder which is necessary for an accurate powder focus. The baseline boundary conditions used for the CFD simulations are shown in Table 3. 'Outlet' is the outer boundary of the computational domain around the nozzle. For the 'Main Flow' and 'Side Flow', the flow rates which are mentioned in terms of liters per minute (lpm) are converted into velocities based on the cross-sectional areas of the nozzle inlets.

Table 3. Boundary conditions used for the CFD simulations of the standard LMD nozzle.

Boundary	Type	Value
Main Flow	Velocity Inlet	18 lpm
Side Flow	Velocity Inlet	12 lpm
Outlet	Pressure Outlet	0 Pa

2.7. Additional nozzle

Taking into consideration the analysis of flow behavior from the conventional nozzles, an additional nozzle was designed which could shield the entire substrate region. This nozzle was designed with two separate flow paths which could be mounted upon the conventional nozzle system. Figure 3 depicts the cross section of the computational domain with existing nozzle flow paths 'A' and 'B', additional nozzle flow paths 'C' and 'D', substrate, and workbench.

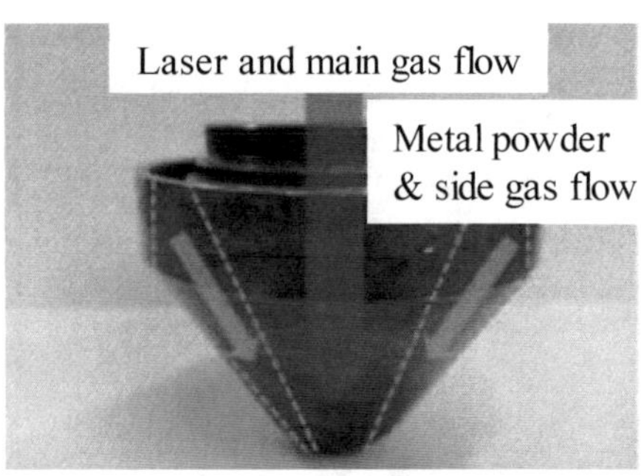

Figure 2. Standard LMD nozzle with a main gas flow and side gas flow path through which the powder is delivered.

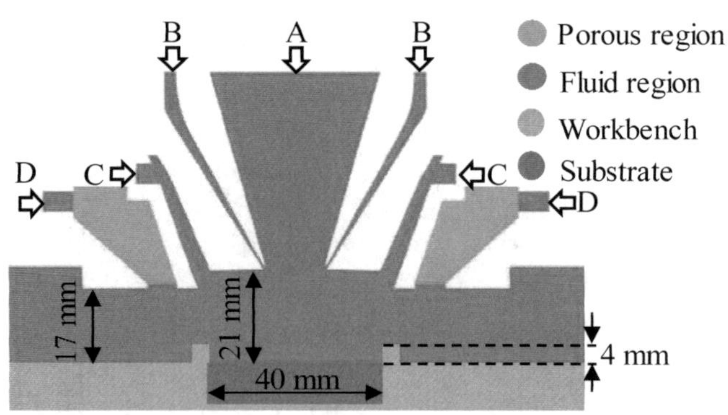

Figure 3. Cross-sectional view of computational domain of additional nozzle.

A porous region was incorporated into the outer flow path 'D'. The additional nozzle boundary conditions used in the CFD simulation are shown in Table 4. Nitrogen was used through the flow path 'D' as it was abundantly available at higher pressures. The porous region was designed as lattice structure which is feasible to manufacture in laser powder bed fusion (LPBF). The structural and porous properties of the lattice structure are shown in Table 5. The lattice structure was converted into a porous region to be used in Siemens Star-CCM+ for flow simulations. The porous properties were calculated by performing simulations on the lattice structure and using Darcy-Forchheimer formulation [11, 12].

Table 4. Boundary conditions used to simulate the additional nozzle.

Boundary	Type	Fluid	Value
A (Main Flow)	Velocity Inlet	Argon	14 lpm
B (Side Flow)	Velocity Inlet	Argon	6 lpm
C	Velocity Inlet	Argon	14 lpm
D	Velocity Inlet	Nitrogen	60 lpm

Table 5. Physical properties and calculated porous coefficients of the lattice structure.

Lattice Type	Pore edge (mm)	Porosity	Porous inertial coefficient (kg/m^4)	Porous viscous coefficient (kg/m^3-sec)
Tri-diametral	0.7	54%	12,763	5,828

3. Results

3.1. Conventional nozzle

To understand the flow behaviour with the conventional nozzle, simulations were performed and compared with recordings from experimental tests. Figure 4 shows the air concentration in a 2D computational domain for the base flow conditions of the conventional shielding gas nozzle. It can be imagined as an axis-symmetric slice of the nozzle.

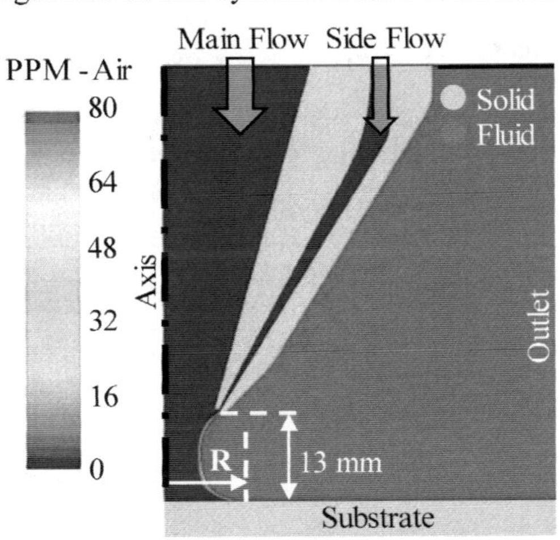

Figure 4. Scalar plot of concentration of air from simulation of conventional nozzle.

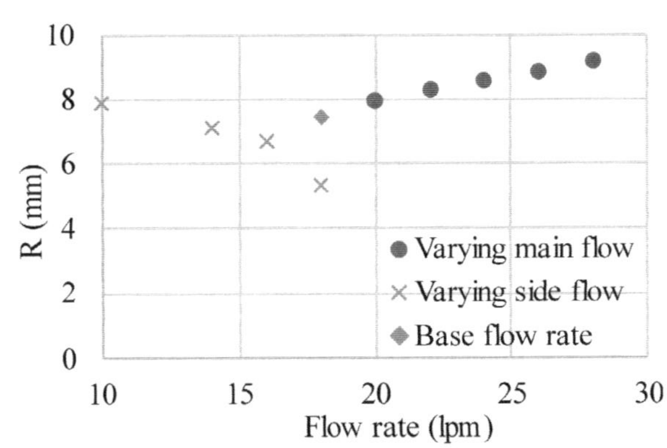

Figure 5. Flow rate sensitivity of conventional nozzle

It shows the different boundaries used in the simulation. '*R*' depicts the radial distance from the axis of the nozzle to the extent of 80 ppm of air on the substrate level which was found to be 8 mm. Another analysis was performed to check if the used gas flow rates from Table 3 were suitable and if the spread of shielding gas was sensitive to the flow rates. It was observed that radial distance 'R' depends directly on the main flow and inversely on the side flow. Figure 5 shows the plot of results from the simulation. The case of 'Base flow rate' uses the conditions mentioned in Table 3. Similar trend was observed from experimental recordings.

3.2. Additional shielding gas nozzle

Figure 6 shows the scalar plot of concentration of air for different nozzle-to-work distances when the additional shielding gas nozzle is implemented. It was observed that the entire substrate region of 20 mm radius was covered with shielding gas by maintaining the concentration of oxygen below 80 ppm, also for a nozzle-to-work distance of 21 mm. Therefore, the shielding is sufficient to build parts with a height of at least 10 mm.

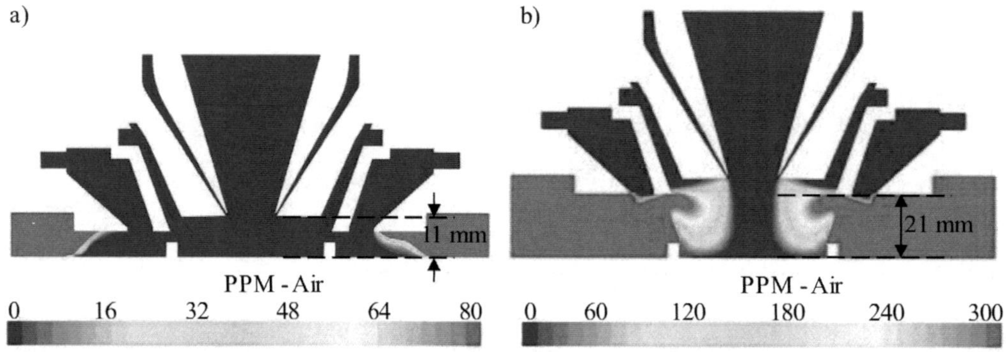

Figure 6. Scalar plot of concentration of air from simulation of the additional shielding gas nozzle at a) a nozzle height of 11 mm and b) a nozzle height of 21 mm.

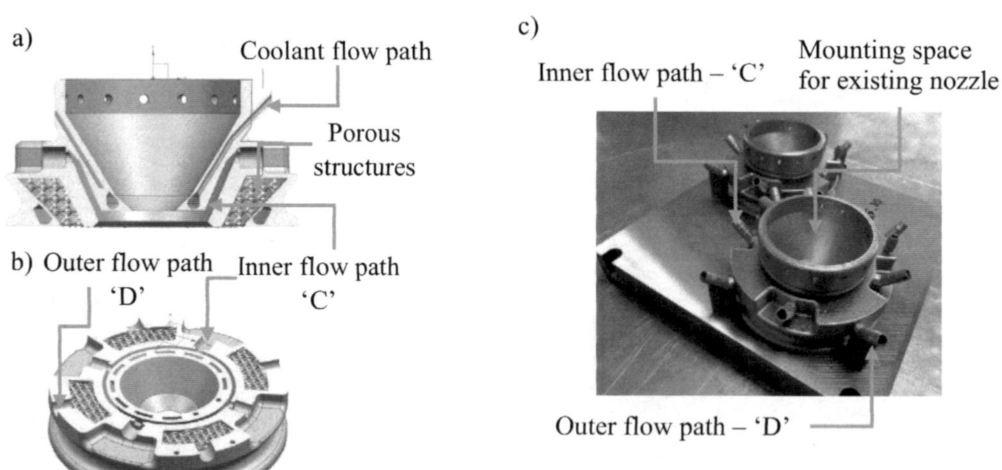

Figure 7. a) Vertical section and b) horizontal section of the 3D model of the additional shielding gas nozzle c) manufactured nozzle by laser powder bed fusion from Inconel 625

The solid part of this nozzle design was built keeping the flow paths similar to that of the simulated design. Figure 7 (a) and Figure 7 (b) shows the two cross-sections of the 3D model of the additional nozzle and Figure 7 (c) shows the manufactured nozzle. Two coolant paths were introduced into the nozzle as it operates at high temperatures around 1000°C. Nozzle and its features were designed to be additively manufactured from Inconel 625 by LPBF.

The manufactured metallic additional nozzle was mounted on the conventional nozzle. Initially, oxygen concentrations were measured in cold conditions without metal deposition process. Recordings were taken for multiple nozzle-to-work distances. It was observed that in cold flow conditions, the spread of shielding gas with the additional nozzle, covered the whole region of substrate below 80 ppm of oxygen as predicted from simulations. Figure 8 shows the experimental recordings of oxygen concentration for the additional nozzle-to-work distance of 11 mm.

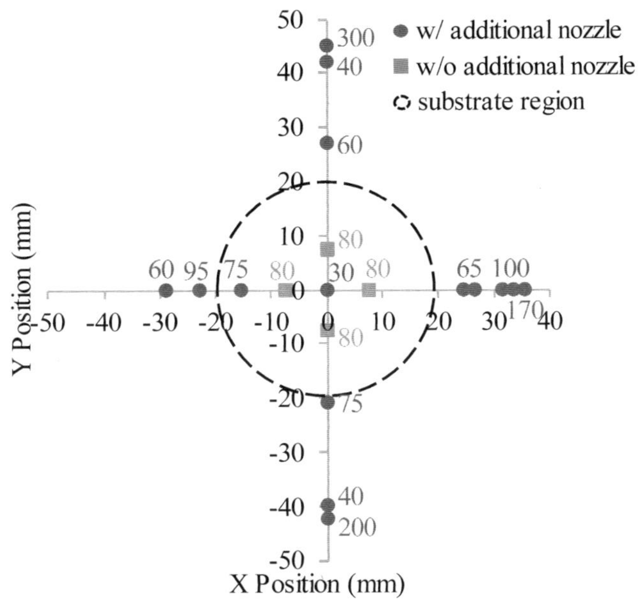

Figure 8. Experimental recordings of residual oxygen concentration (ppm) with and without additional nozzle at a nozzle distance of 15 mm from the material surface.

The labels in the image depict the concentration of oxygen in terms of ppm at respective coordinates relative to the centre of the process zone (0, 0). It was observed that the ppm values increased rapidly, approximately at the radial distance of outer flow path 'D'. During actual processing, the substrate was heated up to 1000°C with an induction ring and material was deposited.

Figure 9 (a) shows the part built of Mo-Si-B by using the additional nozzle. A cube with side length of 10 mm was successfully deposited on the substrate. Figure 9 (b) shows the cross-section of the deposited cube. In the process without additional shielding gas nozzle, the substrate showed a high level of oxidation and decolouration after preheating to 1000°C and it was not possible to proceed processing with these oxidation effects. No cracks were observed inside the cube when additional nozzle was used to provide shielding gas as seen from Figure 9 (b).

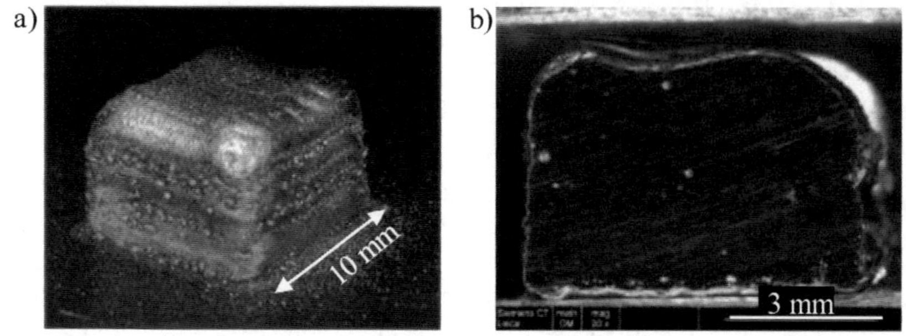

Figure 9. a) Deposited structure when using the additional shielding gas nozzle. b) Cross-section of the printed Mo-Si-B cube.

4. Discussion

4.1. Conventional nozzle

The conventional compound nozzles are used to deliver shielding gas, and material powder onto the substrate. The cover of shielding gas provided by this nozzle system was insufficient to deposit Mo-Si-B and even cover the entire substrate region with less than 80 ppm of oxygen. The reason for insufficient coverage of shielding gas even after increasing flow rate can be attributed to the nature of jet flow and jet mixing. Figure 10 shows some of the important parts of a jet emerging out of a round nozzle.

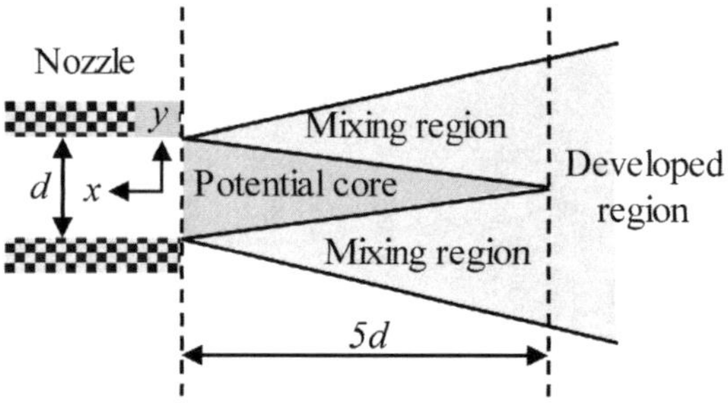

Figure 10. Depiction of potential core and parts of jet as based on [13].

In our case, the outer nozzle which consists of flow path 'B' from Figure 3, has a very small outlet diameter as it is designed to deliver material powder. Therefore, it has a smaller potential core which does not support in shielding the process zone. The potential core of the inner nozzle with flow path 'A' from Figure 3 provides the overall shielding effect.

In addition to diameter '*d*' of the nozzle exit as shown in Figure 10, the jet mixing also depends on the velocity of the flow. It was found in studies that a higher flow rate results in higher Reynolds number (Re) of the flow which enhances the mixing of the jet with ambient atmosphere [15, 16]. The dependence of velocity profiles of a jet flow on Reynolds number can be seen from Figure 11. The velocity profiles are plotted for a H/D ratio of 10 [14]. Velocity profiles depict how quickly the jet disperses into the ambient atmosphere. The centre-line velocity decays very rapidly at higher Re number. It can be inferred therefore that the jet mixing would increase with increase in flow rate. The reason for reduction of '*R*' with increase in flow rate as shown in Figure 5 can be attributed to the dependence of jet mixing on Reynolds number as shown in Figure 11.

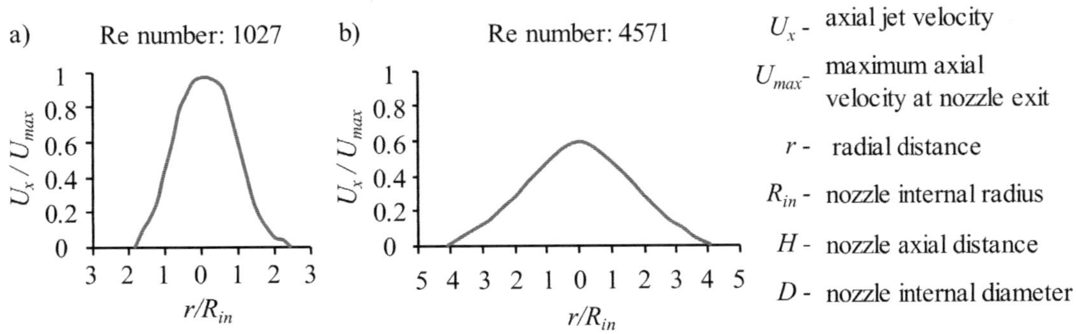

Figure 11. Velocity profiles of jets at two different Reynolds numbers [14].

4.2. Additional Nozzle

4.2.1. Potential core of annular nozzle. The issue of having a limited potential core with the conventional nozzle was addressed by the additional nozzle. Flow from additional nozzle acted as an annular shielding wall around the processing zone thereby separating the inert gases delivered by conventional nozzle from the ambient atmosphere as can be seen from Figure 12. A wider annular flow path was designed for flow path 'D' as can be seen from Figure 3 to have a wider non-mixing zone.

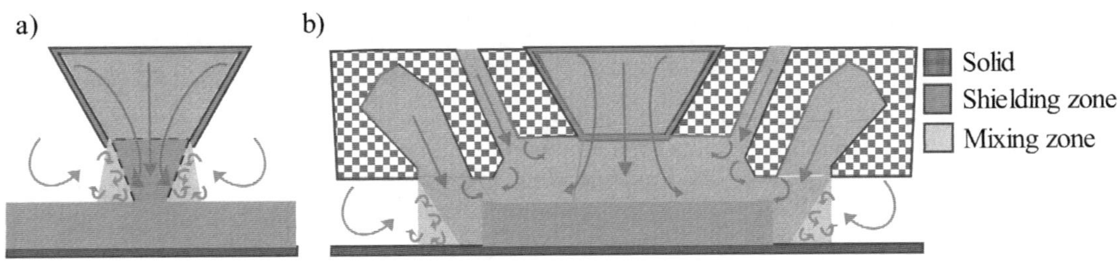

Figure 12. Depiction of mixing zones a) w/o additional nozzle and b) with additional nozzle.

4.2.2. Lattice structures. Lattice structures reduced the effective flow volume inside the nozzle and because of its porous nature, aided in distributing the flow inside the nozzle. A larger annular radius of additional nozzle demanded a very high flow rate of shielding gas. The shielding gas was delivered through pipes which have a very small internal diameter compared to the nozzle. Because of its higher flow rate and sudden change in flow areas, from pipes to nozzle, the shielding gas did not fill the nozzle completely leaving some pockets of ambient air. Therefore, lattice structures were designed to diffuse the shielding gas inside the nozzle and completely displace ambient air from the nozzle. In addition to distributing the flow, porous nature of lattice structures helped in reducing the turbulence of the flow.

5. Conclusions

To reduce the jet mixing with ambient atmosphere during powder DED and enable stable processing without oxidation effects, the gas flow should be maintained in the laminar regime of the flow. It was shown that the shielding gas flow can be stabilized by additionally applied gas flows, while the super positioning of the potential core of the jets was found to have the main impact on the gas mixing. Increasing annular radius of additional nozzle demands an increase in the flow rate of shielding gas to completely fill the nozzle without any pockets of ambient air inside the nozzle. Properly designed lattice structures improved the flow distribution inside the nozzle and provided a better flow uniformity at nozzle exit. CFD simulations were used to understand the flow behaviour of the gases. Lattice structures were modelled as porous region for the simulations. Fibre based prototypes of the additional nozzles were built to validate the simulations and provide an agile feedback in the design loops. An Inconel based prototype was built using LPBF for the final design of the additional nozzle. The newly developed additional nozzle system was able to prevent extensive oxidation during DED of Mo-Si-B deposition. A cube of 10 mm edge length was successfully deposited by using the additional nozzle. No cracks were observed from a cross-section of the cube.

6. Acknowledgements

We extend our sincere gratitude to the German Federal Ministry of Education and Research (Bundesministerium für Bildung und Forschung – BMBF) for providing the financial backing for the research project LextrA (project number 03XP0094).

7. ORCID iDs

Yogesh Nalam https://orcid.org/0000-0002-1993-8459
Tobias Hauser https://orcid.org/0000-0002-3403-5602
Philipp Peter Breese https://orcid.org/0000-0001-9205-0053
Tobias Kamps https://orcid.org/0000-0002-6855-913X
Yves Küsters https://orcid.org/0000-0002-4299-8206
Rebar Hama-Saleh Abdullah https://orcid.org/0000-0002-7970-7329
Joerg Volpp https://orcid.org/0000-0003-0194-9018
Alexander F H Kaplan https://orcid.org/0000-0002-3569-6795

8. References

[1] Hauser T, Breese PP, Kamps T, Heinze C, Volpp J and Kaplan AF 2020 Material transitions within multi-material laser deposited intermetallic iron aluminides *Additive Manufacturing* **34** pp 101242
[2] Ruiz J E, Cortina M, Arrizubieta J I and Lamikiz A 2018 Study of the influence of shielding gases on laser metal deposition of inconel 718 superalloy *Materials* **11** pp 1388
[3] Cortina M, Arrizubieta JI, Ruiz JE, Lamikiz A and Ukar E 2018 Design and manufacturing of a protective nozzle for highly reactive materials processing via Laser Material Deposition *Procedia CIRP.* **68** pp 387-92
[4] Ferreira E, Dal M, Colin C, Marion G, Gorny C, Courapied D, Guy J and Peyre P 2020 Experimental and numerical analysis of gas/powder flow for different LMD nozzles *Metals* **10(5)** pp 667
[5] Li G, Zhang J, Shi T, Shi J, Cheng D, Lu L and Shi S 2020 Experimental investigation on laser metal deposition of Ti-6Al-4V alloy with coaxial local shielding gas nozzle *Journal of Materials Engineering and Performance* **29(12)** pp 7821-9
[6] Yamauchi A, Yoshimi K, Murakami Y, Kurokawa K and Hanada S 2007 Oxidation behavior of mo-si-b in situ composites *Solid State Phenomena* **127** pp 215–20
[7] Anderson J 1995 *Computational Fluid Dynamics.* McGraw-Hill Education.
[8] Hauser T, Reisch RT, Breese PP, Nalam Y, Joshi KS, Bela K, Kamps T, Volpp J and Kaplan AF 2021 Oxidation in wire arc additive manufacturing of aluminium alloys *Additive Manufacturing* **41** pp 101958
[9] Zastawny M and Lardeau S 2020 Application of Simcenter STAR-CCM+ for analysis of CFD sensitivities in NASA Juncture Flow Simulation *AIAA Aviation 2020 Forum* **2020** pp 2736
[10] Davis P L, Rinehimer A T and Uddin M 2012 A comparison of RANS-based turbulence modeling for flow over a wall-mounted square cylinder *20th Annual Conference of the CFD Society of Canada* **8**
[11] Dukhan N, Bağcı Ö and Özdemir M 2014 Experimental flow in various porous media and reconciliation of Forchheimer and Ergun relations *Experimental Thermal and Fluid Science* **57** pp 425-33
[12] Van der Sman RG 2002 Prediction of airflow through a vented box by the Darcy–Forchheimer equation *Journal of Food Engineering* **55(1)** pp 49-57

[13] Kubler R, Rotinat R, Badreddine J and Puydt Q 2020 Experimental analysis of the shot peening particle stream using particle tracking and digital image correlation techniques *Experimental Mechanics* **4** pp 429-33

[14] Labus L T and Symons P E 1972 Experimental investigation of an axisymmetric free jet with an initially uniform velocity profile *NASA Technical Note*

[15] Teng Z, Larsson S, Lundström T, and Marjavaara B 2018 The Effect of Reynolds Number on Jet in Asymmetric Co-Flows: A CFD Study *International Journal of Chemical Engineering* **2018**

[16] Fonte CP, Sultan MA, Santos RJ, Dias MM and Lopes JC 2015 Flow imbalance and Reynolds number impact on mixing in Confined Impinging Jets *Chemical Engineering Journal* **260** pp 316-30

Influence of an external applied AC magnetic field on the melt pool dynamics at high-power laser beam welding

Ömer Üstündağ[1], Nasim Bakir[1], Andrey Gumenyuk[1,2], Michael Rethmeier[3,1,2]

[1]Bundesanstalt für Materialforschung und -prüfung (BAM), Unter den Eichen 87, 12205 Berlin, Germany
[2]Fraunhofer Institute for Production Systems and Design Technology, Pascalstraße 8-9, 10587 Berlin, Germany
[3]Institute for Machine Tools and Factory Management, Technische Universität Berlin, Pascalstraße 8-9, 10587 Berlin, Germany

oemer.uestuendag@bam.de

Abstract. The study deals with the determination of the influence of an externally applied oscillating magnetic field on the melt pool dynamics in high power laser beam and hybrid laser arc welding processes. An AC magnet was positioned under the workpiece which is generating an upward directed electromagnetic force to counteract the formation of the droplets. To visualise the melt flow characteristics, several experiments were carried out using a special technique with mild steel from S355J2 with a plate thickness of up to 20 mm and a quartz glass in butt configuration. The profile of the keyhole and the melt flow were recorded with a high speed camera from the glass side. Additionally, the influence of the magnetic field orientation to the welding direction on the filler material dilution on laser hybrid welding was studied with variating oscillation frequency. The element distribution over the whole seam thickness was measured with X-ray fluorescence (XRF). The oscillation frequency demonstrated a great influence on the melt pool dynamics and the mixing of the elements of the filler wire. The high speed recordings showed, under the influence of the magnetic field, that the melt is affected under strong vortex at the weld root, which also avoids the formation of droplets.

1. Introduction

Despite the availability of high-power laser systems within the range beyond 100 kW on the market, the use of high-power laser beam welding (LBW) for thick metal sections at industrial scale are still under discussion. Its application for plate thicknesses greater than 15 mm has certain technological limitations. One of the limiting factors is the formation of sagging due to the gravitational forces or the hydrostatic pressure often observed during welding in flat position (1G). For that reason, the arc-based welding processes are usually implemented for welding of thick-walled structures with wall thicknesses above 15 mm. However, these are less productive as compared with high-power LBW cause of lower penetration depth. Hence, multi-layer technology is used for welding of thick metal plates. It leads to high heat inputs and distortion or thermal induced residual stress of the plates as well as reworks hindrances such as flame straightening thus expend more time and cost.

Several investigations were conducted to identify the maximum boundaries of the single-pass high power LBW or hybrid laser-arc welding (HLAW) and their challenges and physical backgrounds. Similar challenges or defects are to be expected for high-power LBW or HLAW. Single-pass HLAW was produced up to 28 mm metal thickness using a laser power of just 19 kW with an electromagnetic weld pool support system [1] or HLAW of 25 mm thick plates using cut-wire, which was filled within an air gap between workpieces, with ceramic or flux backing and also up to 50 mm depth by double sided welds approach showed successful outcomes [2]. Multilayer high-power LBW, especially HLAW technique was demonstrated in [3] for one sided steel welds ranging in thickness from 28 mm to 32 mm in two to five layers. Laser beam welding process under vacuum conditions [4,5], laser submerged arc hybrid welding [6], or narrow-gap laser-arc hybrid welding [7] were conducted successfully for deep penetration weld joints. As reported in the studies dealing with single-pass high-power LBW, the formation of sagging or drop-outs is a major challenge during welding. Guidelines for preventing the drop-outs were reported in [8], where the increase of the welding speed or laser power is recommended for the prevention of those defects. The main physical background of this recommendation lies in the fact, that with increasing welding speed the geometrical sizes of the seam is changed to thinner seam width. This leads to the increase of the surface tension which counteracts the hydrostatic pressure on the root part. Hence, the stability criterion is dependent on geometrical dimensions of weld seams and is defined in conformance with [9] as shown in Equation 1:

$$h \times w < 2\, l_{cap}^2, \tag{1}$$

where h is the plate thickness, w is the root width, $l_{cap} = [\gamma/(\rho g_0)]^{1/2}$ is the capillary length, γ is the surface tension coefficient, ρ is the density of the melt, and g_0 is the gravitational acceleration. The geometrical sizes on the root side of the weld pool and the stability criterion for liquid steel is shown in Figure 1 according to [9].

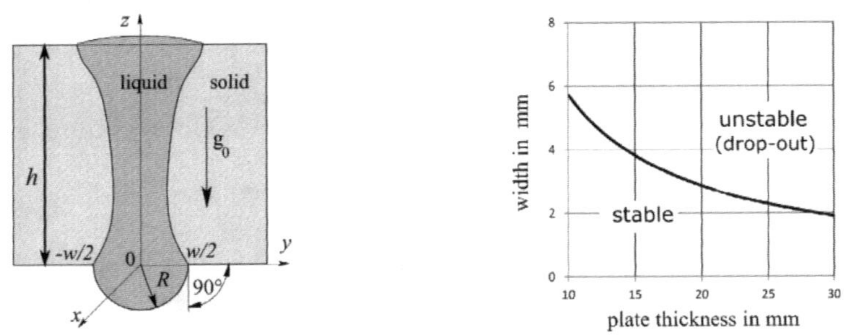

Figure 1. Geometrical sizes on the root side of the weld pool (left) stability criterion for liquid steel (right) [9].

Another typical problem associated with the application of high-power HLAW or wire feed LBW at deep penetration is the inadequate and non-uniform distribution of filler wire elements over the entire seam depth which may deteriorate mechanical properties especially in the root part. The high-power LBW seams are characterized by a high cooling rate, why deteriorated mechanical properties can be observed, especially at LBW of high-strength steels. So, the use of a filler wire and a homogeneous mixing of the filler wire plays an integral role to achieve the required mechanical properties. In [10] it was reported that the Charpy impact toughness of a single-pass hybrid laser-arc welded 25 mm thick structural steel plate is decreased up to 60 % in the root part (laser-dominated zone) compared to the top part (arc-dominated zone). Thereby, the decrease in cooling time over depth, the resulting microstructure and grain size, together with the inhomogeneous mixing have a great impact on the mechanical properties. Studies dealing with the improvement of the dilution are already known. In [11] the influence of the arc mode during HLAW on the dilution was studied, where a pulsed or modified spray modes were recommended. Another effective method to improve the filler wire dilution is to weld thick plates with a small air gap of 0.3 mm to 0.4 mm as reported in [12,13]. The effectiveness of trailing GMAW configuration in the mixing of the melt during the welding of 10 mm thick plates was noted in [14]. The same approach was shown in [15]. Additionally, the shielding gas containing more than 2 % O_2 produces a more homogenous distribution of alloying elements [15]. The impact of the welding parameters such as laser beam power, wire feed speed and resulting arc power, distance between the two heat sources and the configuration was studied in [16]. It was found that with increasing laser power and arc current, the filler wire mixing was improved. With a distance of 2 mm to 6 mm between the wire tip extension and the laser beam, promising results could be achieved.

As the state-of-the-art shows, the challenges during high-power LBW or HLAW of thick-walled steels, especially the formation of drop-outs and inhomogeneous filler wire mixing is well known. Methods with the adaptation of the welding parameters for the prevention of an inhomogeneous mixing were developed for welds with a thickness of up to 15 mm. For thicker plates an additional external force may be needed to influence the melt flow dynamics for a better mixing behaviour. Therefore, external magnetic fields were applied in some studies such as in [17-21]. A magnet with a low frequency in the range of 10 Hz to 20 Hz was applied for the improvement of the element distribution for LBW of 3 mm thick aluminum plate [17-19]. In [20] an external magnetic field was applied for wire-feed LBW of 10 mm thick austenitic stainless steel to improve the mixing behaviour. It was found that a bulging region is formed, which narrows the metal transfer channel from the top to the bottom region. A change of the magnetic field orientation of 10° to 20° to the welding direction was recommended numerically as well as experimentally to eliminate the bulging phenomenon, thus providing a downward transfer channel for the melt flow. An approach to prevent sagging and to improve the filler wire mixing at single-pass HLAW of steel plates with a wall thickness of 20 mm was demonstrated in [21]. Therefore, an AC magnet was positioned under the workpiece and operated with a frequency of approx. 1.2 kHz.

The goal of this study is to describe the melt flow characteristics in detail and to show the influence of the magnetic field orientation and oscillation frequency on the filler wire mixing during high-power LBW and HLAW.

2. Experimental Setup

2.1. Laser beam welding experiments in steel-glass configuration

The high-power fibre laser IPG YLR-20000 with a maximum output power of 20 kW was used as the laser beam source. The emission wavelength and beam parameter product were 1070 nm and 11 mm x mrad, respectively. The laser radiation was transmitted through an optical fibre with a core diameter of 200 µm. A laser-processing head BIMO HP has been selected, which provides a magnification of 2.8 so that the laser beam can be focused into a spot with a diameter of 560 µm. To detect the keyhole during high-power laser beam welding a special setup was necessary. Welding trials in butt joint configuration of 25 mm thick structural steel plate (S355J2) and quartz glass were conducted. A groove with the dimensions of 80 mm x 8 mm x 0.5 mm was milled on the steel plate and filled with an austenitic powder 316L-Si with Ni as tracing element for the later evaluation. Side views of the molten pool were taken with help of a highspeed camera Fastcam 1024PCI and interference band-pass filter at 808 nm and band width of 20 nm. The frame rate and the frame size were 2000 fps and 1200 pixels to 1200 pixels, respectively. An oscillating magnetic field generated by an AC electromagnet was applied to the root side of the weld specimen, where the magnetic field was perpendicular and induced electric current parallel to the welding direction. The schematic representation of the experimental setup is shown in Figure 2. The LBW experiments on steel-glass configuration were performed using a laser beam power of 18.7 kW at a welding speed of 0.9 m min^{-1} and a focal position of -7 mm. The AC magnet was operated at an oscillation frequency of 1.2 kHz and a magnet power of 2.1 kW ± 200 W. To protect the arc on the top side from a deflection due to the oscillating magnetic field, the frequencies have been selected in the kHz range, where the skin layer depth is less than the plate thickness. For the evaluation of the high-speed recordings the displacements in the melt-glass interface were estimated and then the velocities were calculated using the optical flow algorithm according to Lucas-Kanade method.

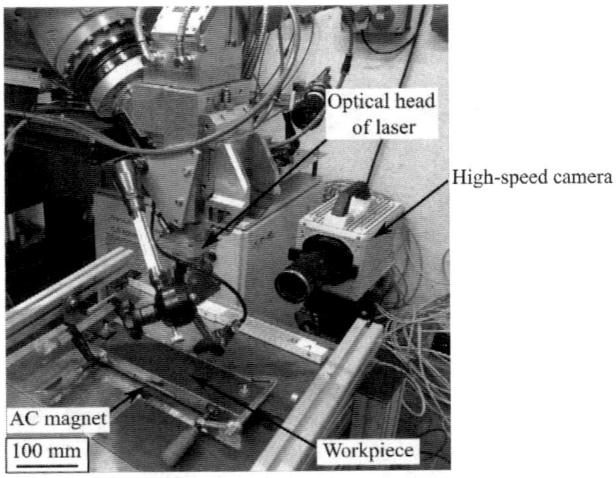

Figure 2. Experimental setup for LBW in steel-glass configuration.

For metallographic inspection, longitudinal sections of the laser beam welded samples were cut. After polishing and etching the samples using 2 % nital solution, the mixing of the powder over the thickness was evident.

2.2. *Hybrid laser-arc welding experiments*

The HLAW experiments were performed using the same laser source as described in Section 2.1. Additionally, a welding machine Cloos Quineo with a maximum current of 600 A was used as an arc power source. The laser optics and GMAW torch were mounted on the robot arm, where the laser axis was positioned 90° to the weld specimen surface and the GMA torch was tilted 25° relative to the laser axis. The processing head and the magnet remained in a fixed position during the welding, where the specimens were moved by an external axis. The experiments were carried out on 20 mm thick structural steel plates (S355J2) in butt joint configuration with an arc leading position and a distance of 4 mm between the two heat sources and under the following welding parameters: laser beam power of 17.7 kW; welding speed of 1.3 m min^{-1}; focal position of the laser beam of -5 mm; wire feed speed of 13 m min^{-1}; stick-out of 18 mm; shielding gas mixing consisted of 18% CO_2 in Ar with a volume flow rate of 20 m min^{-1}. A Ni-based solid wire Thermanit625 (ERNiCrMo-3 according to AWS A-5.14) with a diameter of 1.2 mm was used as a filler wire. Figure 3 shows the setup for the HLAW experiments. Different magnetic field orientation and oscillating frequencies were tested to determine the influence of the magnetic parameters on the filler wire mixing.

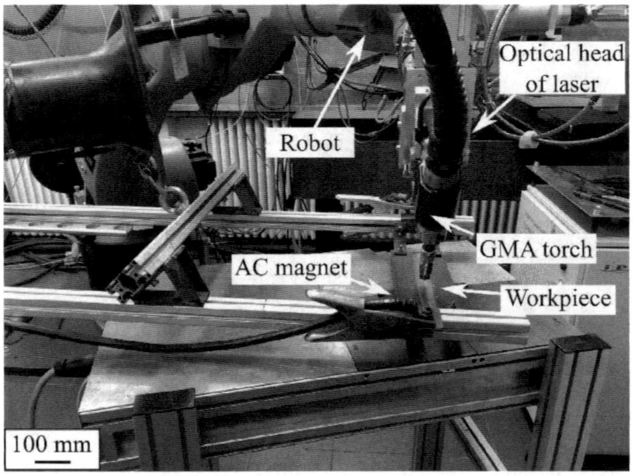

Figure 3. Experimental setup for HLAW of 20 mm thick S355J2 in butt joint configuration with 15°-turn of the magnetic field orientation.

To evaluate the metal mixing, the Ni-content was measured then with the X-ray fluorescence method (XRF) over the whole seam thickness.

2.3. Welding materials
For the experiments structural steel plates of S355J2 with a thickness of 20 mm and 25 mm were used. The filler wire diameter was 1.2 mm. The particle size distribution of the powder was between 53 µm and 106 µm. The chemical composition of the materials used are listed in Table 1.

Table 1. Chemical composition of materials used, shown in wt-%.

Material/Element	C	Mn	Si	P	S	Cr	Ni	Mo	Cu	Nb	Fe
S355J2	0.08	1.3	0.29	0.019	0.004				0.08		bal.
Thermanit 625	0.03	0.2	0.25			22	bal.	9		3.6	< 0.5
316L-Si powder	0.01	0.5	2.1	0.02	0.01	16.9	11.9	3			bal.

3. Results and discussion

3.1. Laser beam welds
The LBW experiments in steel-glass configuration were conducted to observe the prevention of the drop-outs and to evaluate the melt flow characteristic in dependence on the external applied AC magnetic field. First, the magnetic field orientation was unchanged, so that the oscillating magnetic field was perpendicular to the welding direction, while the generated Lorentz force is directed upwards to counteract the hydrostatic pressure. The laser beam was positioned on the front edge of the magnet poles (see Figure 4), so that the drop, which is formed at the rear side of the molten pool is in the middle of the magnet poles, where the magnetic field density is high enough to suppress the drop-out pushing the melt upwards. Figure 4 shows three images of the high-speed records, where the start of the formation of a drop-out (t_0) can be seen clearly. After a time-step of $\Delta t = 5$ ms (10 frames) the volume of the droplet is increased. In the middle of the magnet poles, the electromagnetic pressure compensates the droplet, which is visible on the right-hand side of Figure 4.

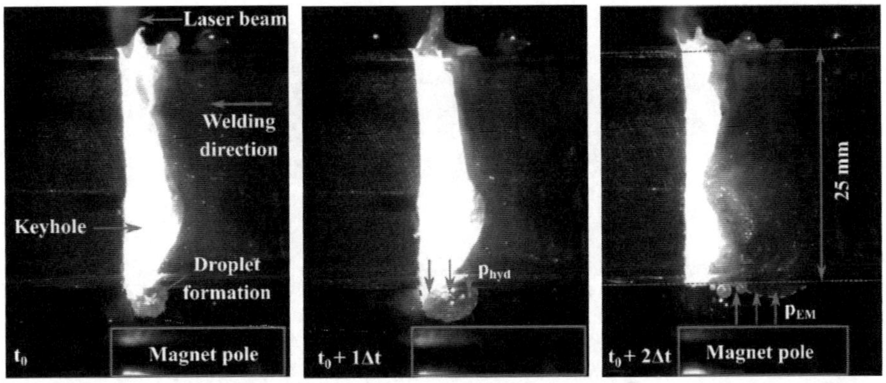

Figure 4. Highspeed images of LBW steel-glass configuration during droplet formation (left and middle) and its prevention through the electromagnetic pressure (right).

With a turn of the magnet poles 45° to the initial state, a vortex on the root part is formed due to the inhomogeneous course of the electric current and its concentration in the weld pool. By turning the magnetic field orientation, a rotational component of the Lorentz force is formed. This vortex is visible on the Figure 5. Additionally, the melt pool geometry can be recognized in terms of a liquid movement. Due to the Marangoni convection on the free surfaces the weld pool length in the top surface as well as in the root is large. The melt pool geometry can also be observed on the welded specimen, especially in the end crater area, see Figure 6. It correlates with the high-speed images and the assumption that this area actually reflects the melt pool geometry. The seam profile at the start due to the ramp time of 450 ms for the laser power from 0 kW to 18.7 kW can be seen clearly.

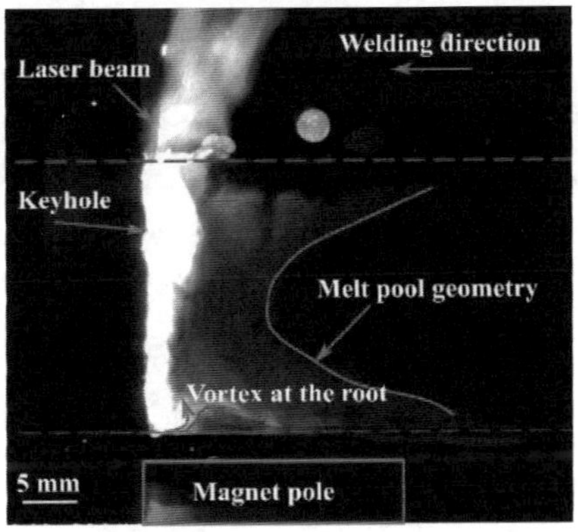

Figure 5. Melt pool geometry of LBW sample with electromagnetic support with turn of the magnetic field orientation of 45°.

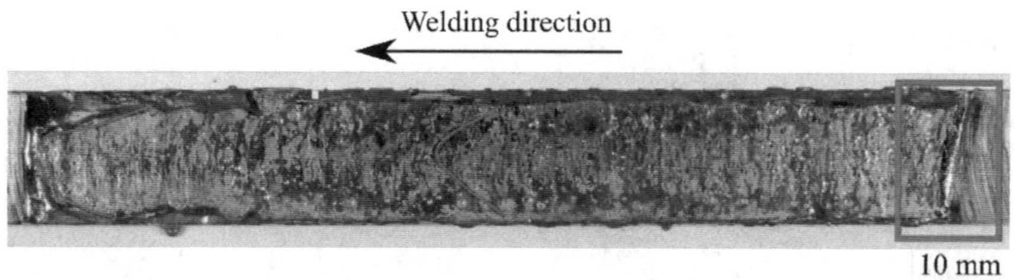

Figure 6. Laser beam welded specimen in steel-glass configuration with the marked melt pool geometry.

Figure 7 shows the velocity vectors for a sequence of LBW sample with electromagnetic support. These analyses can clearly show the flow directions and their velocities. Also, the shape of the molten pool can be estimated as a virtual transition boundary between non-zero and nearly zero displacements.

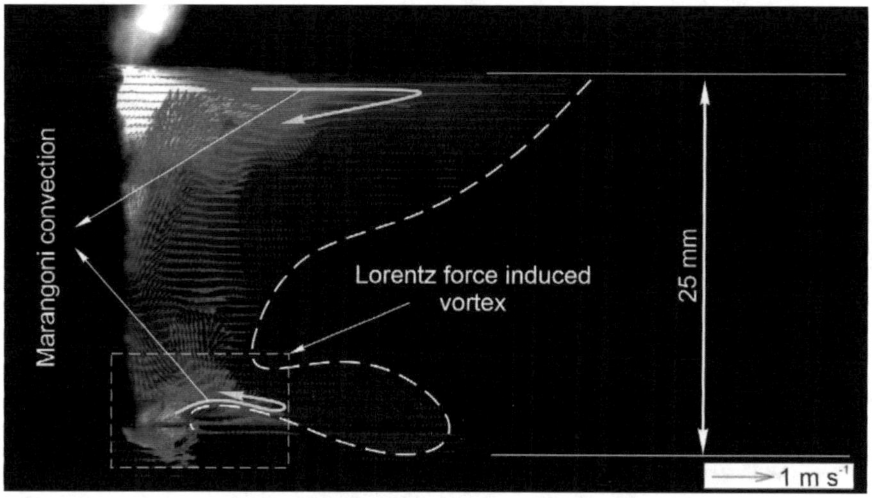

Figure 7. Velocity vectors estimated using optical flow algorithm of LBW sample with electromagnetic support with turn of the magnetic field orientation of 45°.

The velocities vary between 0.01 m s^{-1} and 1 m s^{-1} depending on the distance from the keyhole and the temperature as well as the viscosity. The flows that can be on the laser beam path e.g. on the keyhole wall have two directions, downwards and upwards. The upward and downward flows are mainly driven by the vapor velocity in the keyhole. The Marangoni flow can also be observed on the surface of the molten pool, as shown by the vectors on the surface from left to right and are marked with the yellow lines. The influence of the magnet induced Lorentz force can be observed in the weld root. A drop formation can be seen in Figure 7. The Lorentz force acts against the gravitational force and hydrostatic pressure and pushes the melt again in the weld pool. This leads to the formation of a vortex in the weld root region. It is evident, that two different vortexes are formed on the top side and weld root respectively. On the top surface, the additional elements from the powder are mixed well, where the vortex in the root leads to a material flow within the base material. As it can be seen in Figure 7, a necking region with a low melt flow velocity is formed between the two vortices. This phenomenon appears to the separation in the material transport from the top to the weld root due to the opposite vortices which are formed in the top part and root part and are separated by a necking region. The melt pool geometry correlates with the geometry in Figure 6.

As it can be seen on Figure 8, a weld-through was guaranteed and the drop-outs were prevented by the externally applied magnetic field in every case. Furthermore, the distribution of the austenitic powder

316L-Si which was filled into the prefabricated groove before welding, can be seen. For this analysis the magnetic field orientation was variated between 0° and 45° in steps of 15°. With an orientation of the magnetic field transverse to the welding direction (0°-case), the powder can be measured at a depth up to 13 mm. With a turn of the magnet poles of 15°, the depth of the mixing area decreases to 12.3 mm. For a 30° and 45°-turn of the magnetic field orientation, the depth of mixing decrease to 11.5 mm and 9.7 mm respectively. These new findings contradict with the results published in [21], where a turn of the magnetic field orientation was recommended. Although, there is vortex in the root part which may be helpful for a better mixing or homogenous distribution of the additional wire or powder, but because of the separation of the melt flow movement and the necking in between the two regions, which is described above, the effectiveness of the filler wire or powder transport to the root part is decreased. As it can be seen in Figure 7, a high vortex is formed due to the rotation of the magnetic field orientation especially in the root part. A slight vortex across the entire material thickness would be more helpful for the element transport and mixing behaviour.

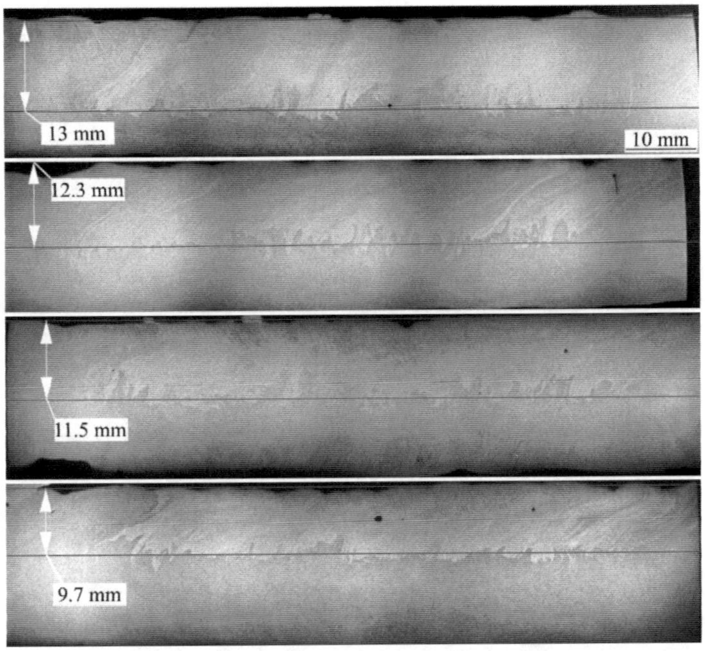

Figure 8. Longitudinal sections of the laser beam welded specimens in steel-glass configuration, filled with an austenitic powder 316L-Si: variation of the magnetic field orientation of 0°, 15°, 30° and 45° perpendicular to the welding direction (from to top to bottom).

3.2. Hybrid laser-arc welds

An example of HLAW plate with electromagnetic weld pool support is shown in Figure 9. The visual test of the welded joint did not reveal imperfections, such as incomplete fusion or lack of penetration. The root quality meets the requirements related to quality level B according to ISO 12932. A weld-through is guaranteed over the entire seam length. The AC magnet was operated with an oscillation frequency of 1.2 kHz at a power of 2.1 kW and a magnetic flux density of 157 mT.

IOP Conf. Series: Materials Science and Engineering 1135 (2021) 012017 doi:10.1088/1757-899X/1135/1/012017

Figure 9. The top surface and the root of the hybrid laser arc welded joint with electromagnetic weld pool support technique using 17.7 kW laser power and 9.8 kW arc power at a welding speed of 1.3 m min^{-1}.

The influence of the oscillation frequency on the element mixing was investigated. Therefore, the magnetic field orientation was turned 30° transverse to the welding direction and the frequency was variated between 0.8 kHz, 1.2 kHz and 1.7 kHz. According to the skin layer theory, the oscillating frequency has a major influence on the depth, where the magnetic field and the induced eddy currents have an impact on the melt flow. With increasing oscillation frequency, the skin layer depth decreases where a strong Lorentz force is formed in a thin layer. With decreasing oscillation frequency, the filler wire mixing was improved as the formed Lorentz forces are distributed over the entire thickness of the plate. A homogenous distribution of the additional filler wire is reached to a depth of approx. 16 mm. It is known that with decreasing oscillating frequency the skin layer depth increases. For a frequency of 800 Hz, the estimated skin layer depth corresponds to 20.5 mm and reaches the entire seam thickness, so that a better mixing takes place. The results of the Ni distribution over the entire seam thickness by a turn of the magnetic field orientation of 30° and variation of the oscillating frequency are shown in Figure 10.

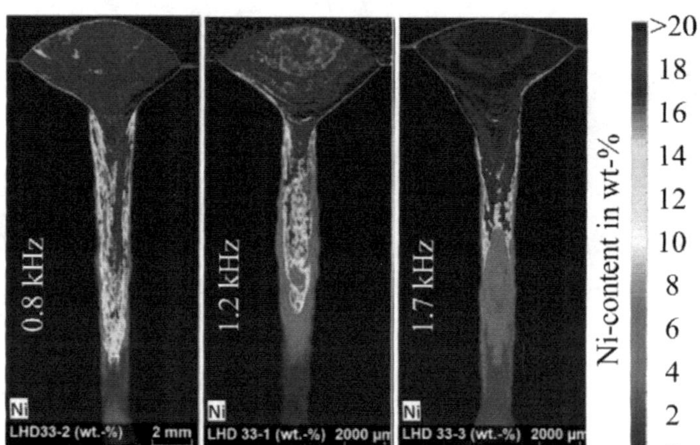

Figure 10. Ni distribution of a single-pass HLAW with electromagnetic weld pool support by a turn of the magnetic field orientation of 30° and variation of the oscillating frequency, measured by XRF

4. Conclusion

Full penetration LBW in steel-glass configuration for 25 mm thick S355J2 were performed for different designs of the orientation of the external oscillating magnetic field. Thereby, high-speed images were taken with a high-speed camera. The formation of the drop-outs on the weld seam root and their prevention due to the electromagnetic pressure provided with an AC magnet could be recorded. The root quality could be classified in the highest evaluation group B according to EN ISO 12932. By turning the magnetic field, a rotational component of the Lorentz force was generated, which forms a strong vortex in the weld seam root. The velocity vectors were estimated using an optical flow algorithm and confirm the vortex in the root. Nevertheless, a strong vortex in the root was not helpful for a better mixing of the additional austenitic powder throughout the entire seam thickness. However, the oscillating frequency shows a high impact on the mixing behaviour due to the change of the skin layer depth. A low frequency of just 0.8 kHz was helpful to transport the filler wire into the seam root part at HLAW of 20 mm thick S355J2 plates.

Funding
The research project was carried out in the framework of the Industrial Collective Research programme (IGF No.20.827N). It was supported by the Federal Ministry for Economic Affairs and Energy (BMWi) through the AiF (German Federation of Industrial Research Associations e.V.) based on a decision taken by the German Bundestag. Financial funding is gratefully acknowledged.

ORCID iDs
Ö. Üstündağ https://orcid.org/0000-0002-7352-429X
N. Bakir https://orcid.org/0000-0001-6847-0365
A. Gumenyuk https://orcid.org/0000-0002-8420-5964

M. Rethmeier https://orcid.org/0000-0001-8123-6696**References**

[1] Üstündag Ö, Avilov V, Gumenyuk A and Rethmeier M 2018 *J. Phys.: Conf. Ser.* **1109** 012015
[2] Wahba M, Mizutani M and Katayama S 2016 *Mater. Des.* **97** 1
[3] Rethmeier M, Gook S, Lammers M and Gumenyuk A 2009 *Q. J. Jpn. Weld. Soc.* **27** 74
[4] Katayama S, Yohei A, Mizutani M and Kawahito Y 2011 *Phys. Procedia* **12** 75
[5] Reisgen U, Olschok S and Longerich S 2010 *Proceedings of 29th International Congress on Applications of Lasers & Electro-Optics ICALEO* Anaheim USA, 638
[6] Reisgen U, Olschok S and Jakobs S 2016 *Phys. Procedia* **56** 653
[7] Zhang C, Li G, Gao M and Zeng X 2017 *Materials* **10** 106
[8] Frostevarg J and Heussermann T 2015 *The IIW International Conference on High Strength Materials–Challenges and Applications* Helsinki Finland
[9] Avilov V, Gumenyuk A, Lammers M and Rethmeier M 2012 *Sci. Technol. Weld. Join.* **17** 128
[10] Üstündağ Ö, Gook S, Gumenyuk A and Rethmeier M 2018 *Procedia Manuf.* **36** 112
[11] Gook S, Gumenyuk A and Rethmeier M 2014 *Sci. Technol. Weld. Join.* **19** 15
[12] Bunaziv I, Wenner S, Ren X, Frostevarg J, Kaplan A F and Akselsen O M 2020 *J. Manuf. Process.* **54** 228
[13] Karhu M, Kujanpaa V, Gumenyuk A and Lammers M 2013 *Proceedings of 32nd International Congress on Applications of Lasers & Electro-Optics ICALEO* Miami USA252
[14] Sohail M, Karhu M, Na S J, Han S W and Kujanpaa V 2017 *J. Laser Appl.* **29** 042009
[15] Zhao L, Sugino T, Arakane G and Tsukamoto S 2009 *Sci. Technol. Weld. Join.* **14** 457
[16] Suder W, Ganguly S, Williams S and Yudodibroto B Y 2021 *J. Mater. Process. Technol.* **291** 117040
[17] Gatzen M, Tang Z and Vollertsen F 2011 *Phys. Procedia* **12** 56
[18] Tang Z and Gatzen M 2010 *Phys. Procedia* **5** 125
[19] Vollertsen F and Thomy C 2006 *J. Laser Appl.* **18** 28
[20] Meng X, Bachmann M, Artinov A and Rethmeier M 2021 *J. Mater. Process. Technol.* **294** 117135
[21] Üstündağ Ö, Avilov V, Gumenyuk A and Rethmeier M 2019 *Metals* **9** 594

IOP Conf. Series: Materials Science and Engineering 1135 (2021) 012018 doi:10.1088/1757-899X/1135/1/012018

The use of non-spherical powder particles in Laser Powder Bed Fusion

Tatiana Fedina[1], Jesper Sundqvist[1], Alexander F. H. Kaplan[1]

[1]Luleå University of Technology, Department of Engineering Sciences and Mathematics, 971 87 Luleå, Sweden

tatiana.fedina@ltu.se

Abstract. Laser powder bed fusion (LPBF) generally involves the use of near-spherical powders due to their smooth morphology and enhanced flowability that allow for easier powder layering and laser processing. Non-spherical powders, on the other hand, are more cost-efficient to manufacture, however, the underlying mechanisms of their movement and interparticle interaction on the powder bed are still unclear. Thus, this study reports on the use of irregular iron-based powder material in LPBF, with a specific focus on particle motion and interaction behavior on the powder bed. The powder morphology, sphericity and particle size were analysed using X-ray computed microtomography and scanning electron microscopy. Based on the acquired data and by using a simplified analytical calculation, the influence of the particle shape/size on the particle movement in LPBF was established. High-speed imaging was employed to investigate the particle flow dynamics in the process zone, as well as the powder entrainment phenomenon. Particle entrainment and entrainment distances along the scanning direction were measured for near-spherical and non-spherical powders. The obtained results were compared between the powders, revealing a dissimilar particle transfer behavior. Non-spherical powder had a shorter entrainment distance partly attributed to the weaker drag force acting on these particles.

1. Introduction

Laser powder bed fusion (LPBF) is currently one of the most widely applied techniques in additive manufacturing (AM) [1]. The process offers various advantages over conventional subtractive methods, such as manufacturing parts of complex geometry, weight and material waste reduction, and processing of new materials [2,3]. Many studies on laser processing of powders in LPBF have been carried out before [4-8]. However, due to the complexity of the process, it is still challenging to understand the movement behavior of powder particles on a powder bed. In most cases, spatter ejection and particle entrainment followed by denudation are the two phenomena which cause powder motion near a solidified track [9, 10]. Spatter ejection describes the formation of spatters, originating in the melt pool [11-15]. Entrainment of particles by an induced gas flow causes the motion of particles in the vicinity of the melt pool which leads to the creation of a denudation zone. The denudation zone refers to the particle depletion area, occurring near the solidified track due to the laser beam-powder interaction [16]. The described phenomena have a direct impact on the process dynamics by activating the movement of powder particles which leads to disturbances in the powder bed [17, 18]. Therefore, it is critical to investigate the particle movement behavior and the mechanisms behind particle motion.

In AM, particle morphology plays an important role as it influences flowability and powder distribution on a build plate [19]. Powders of high sphericity and good flowability such as gas atomized

(GA) powder are normally employed in laser powder bed fusion as these powders have high packing densities and are easy to distribute over a build platform [2]. Heretofore, the movement behavior of particles in a powder system and their interaction with the laser beam have been described mainly for powders of near-spherical shape [4, 9-11, 20, 21]. The influence of laser radiation pressure on the dynamics of powder particles in the powder bed was previously investigated [22]. The authors reported that the radiation pressure can have as significant impact on the particle motion behavior as the aerodynamic forces. Furthermore, the drag force, exerted on powder particles, may dominate the radiation pressure in the vapor plume region. A simulation study by Chen and Yan [23], in which the powder particle-gas phase interaction was discussed, revealed that the metal vapor flow coupled with the vortex flow of inert gas caused spatter formation and denudation in the process zone.

The use of water atomized (WA) powder as a feedstock has hardly been investigated before [24], especially with relation to its movement behavior during laser processing. The morphology of WA powder is dominated by highly irregular particles with a tendency to agglomeration. This is due to the more rapid cooling and solidification rates during water atomization [24, 25]. However, the shape of WA powder can be altered by adjusting the process parameters or by applying thermal or mechanical post-treatment to the powder. Such a shape improvement may promote the application of WA powder in AM and, as a result, significantly decrease the cost of producing final components. Yet additional research is required to understand how non-spherical powder particles behave when the powder bed is exposed to laser irradiation.

This is addressed in the current manuscript with a special focus made on the particle movement behavior of powders of arbitrary shape. The influence of particle morphology and position on the powder bed on the process behavior of gas and water atomized low alloy powders was discussed. High-speed imaging (HSI) was utilized to measure particle entrainment distances in a radial direction for both powders and to identify variations in the powders' behavior under changing process parameters. To evaluate the powders' properties, X-ray computed microtomography (μCT) was employed. Based on the acquired data, the drag force, exerted on the arbitrary-shaped powder particles, was calculated.

2. Materials and methodology

2.1. Feedstock

The experiments were carried out using GA and WA 4130 low alloy steel powders with a particle size of 20-53 μm, provided by Höganäs AB. The powders' bulk composition was analyzed by inductively coupled plasma mass spectrometry (ICP-MS). Carbon and sulphur content was measured using combustion analysis. Inert gas fusion was used to determine nitrogen and oxygen contents. The results are present in table 1.

Table 1. Chemical compositions of gas and water atomized steel powders (wt. %).

Powder	Fe	C	O	N	S	Si	Mn	Cr	Mo
GA 4130	Balance	0.3	0.05	0.0075	0.0067	0.30	0.60	1.0	0.20
WA 4130	Balance	0.3	0.29	0.0030	0.0071	0.45	0.044	1.1	0.28

A Zeiss Xradia 510 Versa μCT system was employed to determine particle size distributions of the powders. To measure flowability and apparent density of the powders, the Hall flowmeter funnel method was used. Table 2 shows the powders' properties.

Table 2. Technological properties of the 4130 low alloy steel powders.

Powder	Flowability (s/50g)	Apparent density (g/cm^3)	Particle mean diameter D_{50} (μm)
GA 4130	22	4.09	37.70
WA 4130	30	3.27	27.44

It can be observed that the mean particle size diameter D_{50} of water atomized powder (D_{50}=27.5 μm) was slightly lower compared with gas atomized powder (D_{50}=37.7 μm). Moreover, prior to the experiment, to improve flowability and apparent density of the WA powder, the material was mechanically treated by the powder supplier. This treatment resulted in a slight increase of fine particle fraction in the material. The particle shape, observed using an FEI Magellan 400 extreme high-resolution scanning electron microscope, is demonstrated in figure 1, featuring various morphologies present in GA (figure 1 a) and WA (figure 1 b) powders. In addition, the particles' cross-sections can be seen (figure 1 c, d), displaying a minor porosity that exists in both materials.

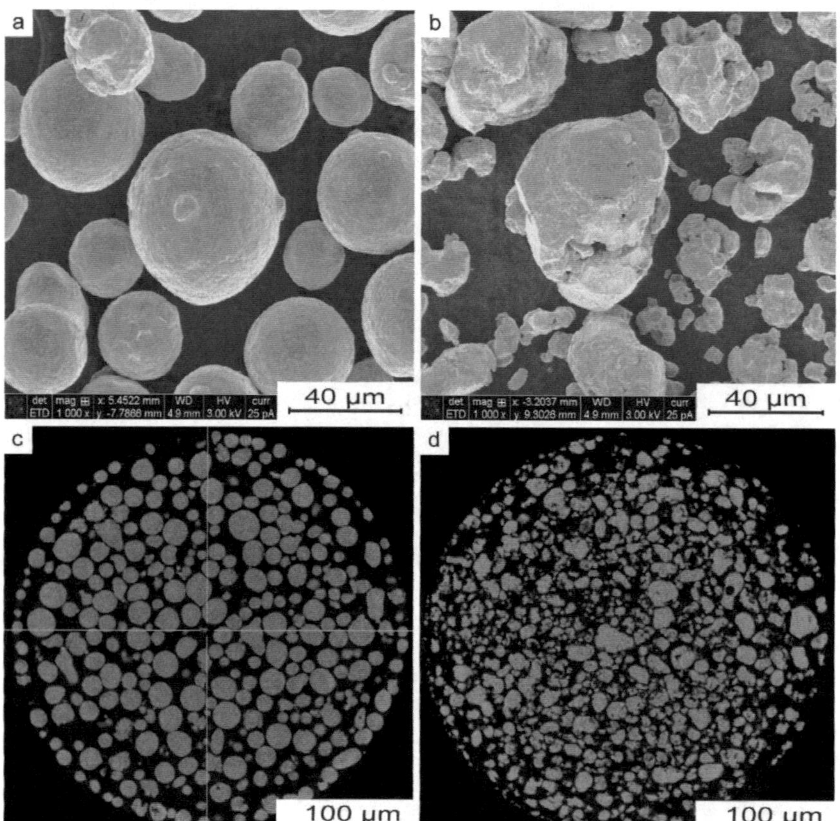

Figure 1. Morphology of GA (a) and WA (b) 4130 steel powders. GA (c) and WA (d) powder cross-sections can also be observed.

2.2. X-ray computed microtomography

The powder properties were evaluated using a Zeiss Xradia 510 Versa μCT machine with 7 W and 80 kV accelerating voltage. To hold the powders, a tube made of X-ray transparent polycarbonate with a 0.8 mm inner diameter was used. The powder specimens were located 6.5 mm from the CCD detector and 11.5 mm from the X-ray source. An objective lens with a 20 × magnification was utilized to allow the sample analysis within an 842 × 842 μm field of view and with a 0.85 μm spatial resolution. The exposure time was between 25 to 29 s. For the powder analysis, around 950 projections of each powder sample were captured. Dragonfly 2020.1 software [26] by Object Research Systems was used for the image analysis.

2.3. Processing of preplaced gas and water amortized powders

The powder processing was carried out in a laboratory LPBF system, equipped with a continuous wave IPG Yb-doped fiber laser (with a 1070 nm laser wavelength). The laser optics had a 250 mm focal length collimator and a 150 mm focal length focusing lens which allowed a 75 μm laser beam spot diameter at the focus. To prevent back reflections that can occur during processing, the optics head was inclined at 7°. The experiment was carried out using following parameters: 33 mm/s scanning speed and 60 μm powder layer thickness. The laser power was varied between 150 and 300 W with a step of 50 W. To prevent oxidation in the process zone, a gas tube with a diameter of 20 mm was placed above the powder bed. Throughout the experiment, argon was used as a shielding gas with a flow rate of 22 L/min. The flow rate was tested prior to the laser processing to ensure a laminar flow regime in the process area. The powders were preplaced on a powder bed, dividing it into two halves. The setup details are demonstrated in figure 2.

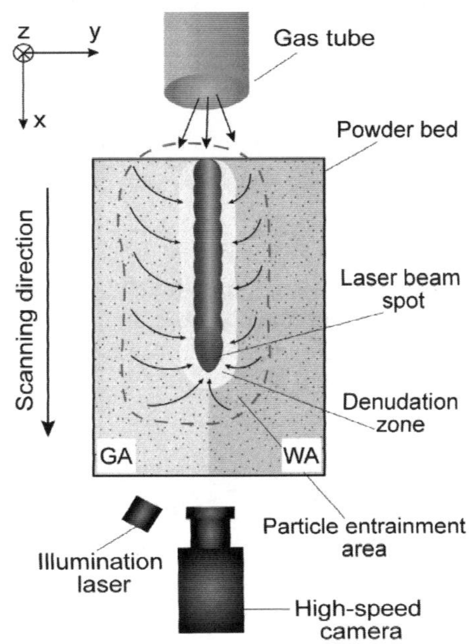

Figure 2. A schematics of the laboratory LPBF setup, featuring the processing of individual tracks when using GA and WA powders. The red dashed line highlights the powder entrainment area. The camera is positioned with an inclination of 39° to the surface horizontal.

Depending on the system supplier, process parameters and conditions can vary. Normally, commercial LPBF machines are equipped with a scanning optics, operate at higher speeds than the value used in this study, and the processing is carried out in a sealed chamber filled with an inert gas. In the current work, despite a lower speed limit, an open linear axes laser system was utilized in order to provide more experimental flexibility and record the process with high-speed imaging. Therefore, although the powders were processed at a lower scanning speed compared to the values used in industrial LPBF systems, the authors believe the results are still consistent and representative.

2.4. High-speed imaging analysis

A Photron FASTCAM Mini UX100 high-speed imaging camera, coupled with an illumination laser CAVILUX CW (fiber-coupled outputs with a CW power of 50 W, 810 nm wavelength) provided by CAVITAR were utilized to observe the motion of variously-shaped powder particles and measure particle entrainment distances in a radial direction as shown in figure 2. To enable a clearer observation of particle dynamics in the powder bed, a 20 mm macro ring was attached to the camera lens. To remove the process light, a narrow band pass filter was used over the lens. The high-speed imaging videos were captured at 8000 frames per second (fps) and an exposure time of 80 μs. The camera was positioned with an inclination of 39° to the surface horizontal (figure 2). The HSI videos were analyzed using a Photron FASTCAM Viewer software. The particle velocity and entrainment distances for both powders were also measured based on the data acquired from the HSI camera. The velocity measurement was carried out on the GA and WA powder particles before their detachment, moving radially, on the sides of the solidified tracks.

2.5. Drag force calculation

Due to the complexity of a powder system which consists of arbitrary-shaped particles, no research has been reported on the movement behavior of such powder materials in LPBF. Nor was it needed as most of LPBF systems operate with near-spherical powders. This section interprets the physical mechanisms behind particle movement of various shape based on the drag force calculation made in this study. The drag force acting on near-spherical GA and non-spherical WA powder particles will be calculated in two regions: in the vapor jet area and in the vicinity of the melt pool.
The following assumptions were made to simplify the calculation:
(1) The powder bed was assumed to be flat. No surface roughness of the outer layer was accounted for.
(2) The powder motion was described only for individual particles which were not in contact with any other surfaces/particles.
(3) The particle velocity was determined by HSI in the area near the edge of the melt pool in a radial direction. No particle rotation was accounted for.
(4) The gas velocities were1 m/s in the area of 250 μm from the melt pool and 150 m/s in the vapor flow region. The value was taken from a simulation study carried out by Chen and Yan [23].

To determine the drag force F_d, a Reynolds number Re and a drag coefficient C_d need to be calculated. The Reynolds number Re is defined as

$$Re = \frac{\rho_g U D_p}{\mu_g} \tag{1}$$

where ρ_g is the gas density; U is the relative velocity between the gas flow and the particle given as $|U_g - U_p|$, where U_g and U_p are the gas and particle velocities, respectively. D_p is the mean particle diameter; μ_g is the gas viscosity [22].

The drag coefficient C_d was calculated using a formula [27] that can be applied for arbitrary-shaped particles. It is also is valid in a wide range of Re numbers. The coefficient C_d is then defined as

$$C_d = \frac{8}{Re}\frac{1}{\sqrt{\varphi_{\parallel}}} + \frac{16}{Re}\frac{1}{\sqrt{\varphi}} + \frac{3}{\sqrt{Re}}\frac{1}{\varphi^{\frac{3}{4}}} + 0.4210^{0.4(-\log\varphi)^{0.2}}\frac{1}{\varphi_{\perp}} \tag{2}$$

where φ is the overall sphericity, $\varphi_{\parallel}$ is the lengthwise sphericity and $\varphi_{\perp}$ is the crosswise sphericity. φ refers to the ratio between the surface area of a volume equivalent sphere and that of the considered particle. $\varphi_{\parallel}$ is the ratio between the cross-sectional area of a volume equivalent sphere and the difference between half the surface area and the mean projected longitudinal cross-sectional area of the considered particle. $\varphi_{\perp}$ is expressed as the ratio between the cross-sectional area of a volume equivalent sphere and the projected cross-sectional area of the considered particle.

Equation 3 describes the drag force F_d, exerted on powder particles of various morphology:

$$F_d = \frac{1}{2} C_d \rho_g A_{p\perp} U^2 \tag{3}$$

where C_d is the drag coefficient; ρ_g is the gas density; U is the relative velocity between the gas flow and the particle. $A_{p\perp}$ is the projected cross-sectional area of the particle, subjected to the flow.

3. Results and discussion

The motion of powder particles on the powder bed is largely affected by their shape. To understand the influence of powder morphology on the particle movement in the powder bed, particle entrainment distances were measured for GA and WA powders. Figure 3 shows a correlation between the radial entrainment distances generated by the induced gas flow and the varying energy densities E defined as

$$E = \frac{P}{U \cdot l \cdot w} \tag{4}$$

where P is the laser power; U is the scanning speed; l is the powder layer thickness and w is the beam size diameter [28].

As demonstrated in figure 3, the entrainment of gas and water atomized powder particles increased with the increase of volumetric energy density. The observed behavior can be explained by the rapid increase in the pressure gradient in the process area as more energy is deposited across the laser scanned area. However, the processing of GA powder resulted in a 28 % larger entrainment area compared to WA powder (1.5 mm and 1.17 mm at E=150 J/mm^2, respectively). This is likely due to the difference in particle morphology between the powders. Being irregularly shaped, the WA powder has a tendency to developing contacts with neighboring particles which leads to mechanical interlocking and reduced mobility of the particles on the powder bed.

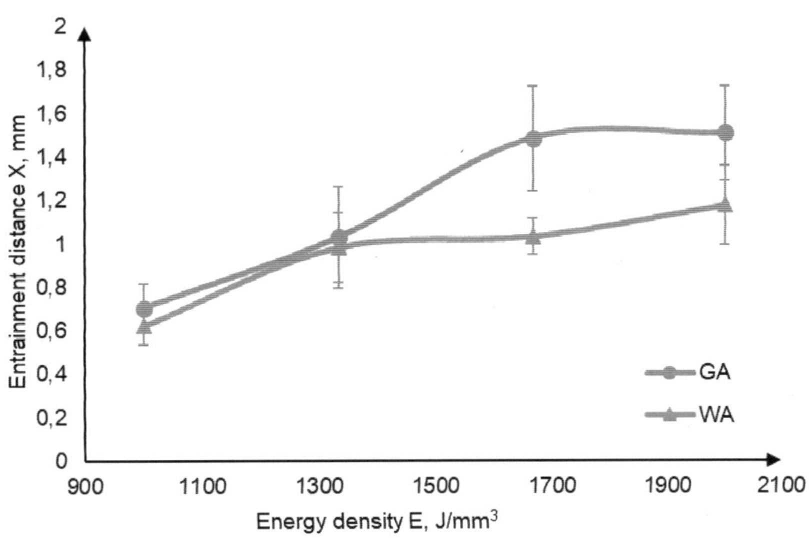

Figure 3. Entrainment distances vs energy densities given for GA and WA powders.

Furthermore, such behavior could also be caused by the difference in the magnitude of the drag force exerted on near-spherical GA and irregularly shaped WA powder particles. To confirm the assumption, the drag force, experienced by both powders, was calculated in two regions on the powder bed: in the vapor region (this implies in the region next to the melt, not above it as only the particle movement before its detachment was studied) and in the proximity of the melt pool, approximately 250 μm from the melt, presumably in the denudation zone. To simplify the description, let us call the mentioned regions as region 1 (vapor) and region 2 (250 μm from the melt pool). The values acquired from μCT and the study by Yan et al. [23] were used to aid the calculation. In region 1, the drag force experienced by both powders was approximately 3 orders of magnitude stronger than in the region near the melt pool. This can be explained by the strong radial drag in the vapor plume region that pulls the particles towards the melt pool.

In the region 2, as expressed in equation (1), the Re number was calculated, resulting in 4.42 and 3.37 for GA and WA powders, respectively. The drag coefficient C_d was determined using equation (2) which was derived for arbitrary shaped particles by accounting for particles' shape based on their sphericity and orientation to the gas flow. The C_d calculation yielded 7.32 for GA and 10.65 for WA powders. Finally, the drag force for both materials was determined using equation (3), demonstrating a significant difference in the drag force acting on the powders. The drag force exerted on GA powder particles (2.8×10^{-8} N) in region 2 was 70 % greater in comparison with WA powder (1.66×10^{-8} N), which supports the above mentioned assumption. This finding explains the difference in the GA and WA powder particle transfer behavior. The WA powder's propensity to interact with nearby particles could promote agglomeration and high interparticle friction which both would lead to a development of a high counter force (s) exerted in the opposite direction to the drag force.

The movement of particles in the powder bed differs depending on the distance from the laser process area. Figure 4 highlights a sequence of images with different sizes of the particle entrainment area captured by HSI. The laser beam position was precisely between the two powder beds which allowed to study the particle motion behavior simultaneously. The red and black arrows present in figure 4 (a), (d) show the direction of particle motion. The reason for adding up the arrows only to the two snapshots (figure 4 a, d) was to improve the process visualization and, at the same time, to not distract the reader's attention from the particle entrainment areas. The powder bed half containing GA powder exhibits a larger area of particle entrainment in comparison with the WA powder half, regardless of the surface energy density. The observed difference stems from the dissimilar morphology of the powders and their ability to flow, suggesting that a number of particles in the WA powder bed half form mechanical bonds with their neighbors and remain in their initial position.

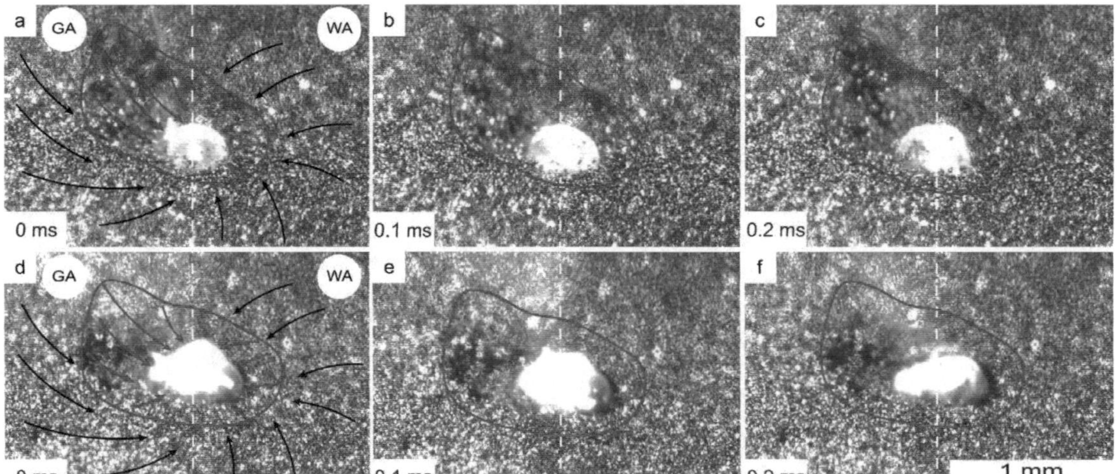

Figure 4. A series of high-speed imaging snapshots, highlighting the areas of particle entrainment near the melt pool, created at surface energy densities E = 125 J/mm^2 (a, b, c) and E = 150 J/mm^2 (d, e, f). The active entrainment zone is encircled in red, displaying the particles next to the melt pool moving rapidly towards the powder-laser interaction area. The white dashed line divides the powder bed into 2 halves, with GA powder on the left and WA powder on the right. The red and black arrows demonstrate the direction of particle motion inside the active (encircled in red) and passive (marked with black arrows) entrainment zones.

It can also be seen in figure 4 (a-f) that regardless of the particle shape, the entrainment zone can be divided into 2 domains: the area of active entrainment (encircled in red) where the particles in the vicinity of the melt pool are rapidly entrained due to the Bernoulli effect [5] and a so called "passive" entrainment zone (black arrows outside the laser exposed area) where individual particles outside the process area are pulled in the direction of the melt pool.

It is worth mentioning that the area of particle entrainment fluctuates depending on the position of the vapor which, in turn, depends on the geometry of the melt pool.

4. Conclusions

This study has investigated the influence of powder shape on the particle movement behavior when processing near-spherical and irregular powder particles in LPBF. The following conclusions can be reported:

- Particle movement behavior changes depending on the particle morphology and position on the powder bed.
- Non-spherical WA powder particles demonstrated a restricted mobility on the powder bed which was attributed to their developed surface area and reduced flowability.
- The influence of particle shape and position on the dynamics of gas and water atomized powders was studied. The drag force exerted on GA powder particles near the melt pool was 70 % greater in comparison with WA powder. In addition, in the vapor plume region, the drag force experienced by both powders was approximately 3 orders of magnitude stronger than in the region near the melt pool.
- Regardless of the powder morphology, the entrainment of particles in a radial direction increased with the increase of surface energy density. Furthermore, when processing GA powder, the particle entrainment area was 28 % larger compared to WA powder. Such behavior is likely a combined result of the stronger drag force, exerted on near-spherical gas atomized powder particles, and the tendency of non-spherical water atomized powder to mechanical interlocking.

The results from this study demonstrate that the use of non-spherical powder particles leads to a narrower particle entrainment area compared with near-spherical powders. The observed behavior indicates that less denudation should occur around the solidified track. These findings contribute to the current limited knowledge about the application of non-spherical powder particles in additive manufacturing and their behavior.

Acknowledgments

The authors express their gratitude to the SPAcEMAN project and wish to acknowledge support from EIT Raw Materials (grant agreement No.17070), and the CINEMA project funded by EU-ERDF Interreg Nord program (grant agreement No.304-7463-2018). A special thanks goes to Anna Larsson and Sven Bengtsson at Höganäs AB and Henrik Lycksam and Fredrik Forsberg at LTU for their cooperation and help with conducting this research work.

ORCID IDs

T Fedina https://orcid.org/0000-0003-4443-3097
J Sundqvist https://orcid.org/0000-0002-9010-1555
A F H Kaplan https://orcid.org/0000-0002-3569-6795

References

[1] Kruth J P, Froyen L, Van Vaerenbergh J, Mercelis P, Rombouts M and Lauwers B. Selective laser melting of iron-based powder 2004. *Journal of Materials Processing Technology* ***149*** (1–3):616-622.

[2] Gorsse S, Hutchinson C, Gouné M and Banerjee R. Additive manufacturing of metals: a brief review of the characteristic microstructures and properties of steels, Ti-6Al-4V and high-entropy alloys 2017. *Science and Technology of Advanced Materials* ***18*** (1):584-610.

[3] Herzog D, Seyda V, Wycisk E and Emmelmann C. Additive manufacturing of metals 2016. *Acta Materialia* ***117***:371–392.

[4] Scipioni Bertoli U, Guss G, Wu S, Matthews M J and Schoenung J M. In-situ characterization of laser-powder interaction and cooling rates through high-speed imaging of powder bed fusion additive manufacturing 2017. *Mater. Des* ***135***:385–396.

[5] Leung C L A, Marussi S, Atwood R C, Towrie M, Withers P J and Lee P D. In situ X-ray imaging of defect and molten pool dynamics in laser additive manufacturing 2018. *Nat. Commun.* ***9*** 1–9.

[6] Taheri Andani M, Dehghani R, Karamooz-Ravari M R, Mirzaeifar R and Ni J. A study on the effect of energy input on spatter particles creation during selective laser melting process 2018. *Addit. Manuf.* ***20*** 33–43.

[7] Ye D, Zhu K, Fuh J Y H, Zhang Y and Soon H G. The investigation of plume and spatter signatures on melted states in selective laser melting 2019. *Opt. Laser Technol.* ***111*** 395–406.

[8] Le Dantec M, Güniat L, Leistner M, Figi R, Bleiner D, Leparoux M and Hoffmann P. Impact of oxygen content in powders on the morphology of the laser molten tracks in preparation for additive manufacturing of silicon 2020. *Powder Technol.* ***361*** 704–710.

[9] Guo Q, Zhao C, Escano L I, Young Z, Xiong L, Fezzaa K, Everhart W, Brown B, Sun T and Chen L. Transient dynamics of powder spattering in laser powder bed fusion additive manufacturing process revealed by in-situ high-speed high-energy x-ray imaging 2018. *Acta Mater.* ***151*** 169–180.

[10] Bidare P, Bitharas I, Ward R M, Attallah M M and Moore A J. Fluid and particle dynamics in laser powder bed fusion 2018. *Acta Mater.* ***142*** 107–120.

[11] Ly S, Rubenchik A M, Khairallah S A, Guss G and Matthews M J. Metal vapor micro-jet controls material redistribution in laser powder bed fusion additive manufacturing 2017. *Sci. Rep.* ***7*** 1–12.

[12] Wang D, Wu S, Fu F, Mai S, Yang Y, Liu Y and Song C. Mechanisms and characteristics of spatter generation in SLM processing and its effect on the properties 2017. *Mater. Des.* ***117*** 121–130.

[13] Liu Y, Yang Y, Mai S, Wang D and Song C. Investigation into spatter behavior during selective laser melting of AISI 316L stainless steel powder 2015. *Materials and Design* ***87***:797–806.

[14] Esmaeilizadeh R, Ali U, Keshavarzkermani A, Mahmoodkhani Y, Marzbanrad E and Toyserkani E. On the effect of spatter particles distribution on the quality of Hastelloy X parts made by laser powder-bed fusion additive manufacturing 2019. *Journal of Manufacturing Processes* ***37***:11–20.

[15] Ashcroft I A, Clare A T, Johns D, Gasper A N D, Wang X, Szost B and Sharma S. Spatter and oxide formation in laser powder bed fusion of Inconel 718 2018. *Additive Manufacturing* ***24***: 446–456.

[16] Matthews M J, Guss G, Khairallah S A, Rubenchik A M, Depond P J and King W E. Denudation of metal powder layers in laser powder-bed fusion processes 2017. *Addit. Manuf.* ***114*** 677–693.

[17] Dal M, Gunenthiram V, Fabbro R, Koutiri I, Coste F, Peyre P and Schneider M. Experimental analysis of spatter generation and melt-pool behavior during the powder bed laser beam melting process 2017. *J. Mater. Process. Technol.* ***251*** 376–386.

[18] Lutter-Günther M, Bröker M, Mayer T, Lizak S, Seidel C and Reinhart G. Spatter formation during laser beam melting of AlSi10Mg and effects on powder quality 2018. *Procedia CIRP.* ***74*** 33–38.

[19] Tan J H, Wong W L E and Dalgarno K W. An overview of powder granulometry on feedstock and part performance in the selective laser melting process 2017. *Addit. Manuf.* ***18*** 228–255.

[20] Martin A A, Calta N P, Hammons J A, Khairallah S A, Nielsen M H, Shuttlesworth R M, Sinclair N, Matthews M J, Jeffries J R, Willey T M and Lee J R I. Ultrafast dynamics of laser-

metal interactions in additive manufacturing alloys captured by in situ X-ray imaging 2019. *Mater. Today Adv.* ***1*** 100002.

[21] Tran H C and Lo Y L. Heat transfer simulations of selective laser melting process based on volumetric heat source with powder size consideration 2018. J. *Mater. Process. Technol.* ***255*** 411–425.

[22] Haeri S, Haeri S, Hanson J and Lotfian S. Analysis of radiation pressure and aerodynamic forces acting on powder grains in powder-based additive manufacturing 2020. *Powder Technol.* ***368*** 125–129.

[23] Chen H and Yan W. Spattering and denudation in laser powder bed fusion process: multiphase flow modelling 2020. *Acta Mater.* ***196*** 154–167.

[24] Hoeges S, Zwiren A and Schade C. Additive manufacturing using water atomized steel powders 2017. *Met. Powder Rep.* ***72*** 111–117.

[25] Thummler F and Oberacker R 1993. *Introduction to Powder Metallurgy.* The Institute of Materials, London.

[26] Dragonfly 2020.1 computer software. Object Research Systems (ORS) Inc, Montreal, Canada, 2018.

[27] Hölzer A and Sommerfeld M. New simple correlation formula for the drag coefficient of non-spherical particles 2008. *Powder Technol.* ***184*** 361–365.

[28] Willy H J, Li X, Chen Z, Herng T S, Chang S, Ong C Y A, Li C and Ding J. Model of laser energy absorption adjusted to optical measurements with effective use infinite element simulation of selective laser melting 2018. *Mater. Des* ***157***:24–34.

IOP Conf. Series: Materials Science and Engineering 1135 (2021) 012019 doi:10.1088/1757-899X/1135/1/012019

Structural performance of laser welded edge joints made of high-strength steel

Juha Peippo[1], Antti Ahola[2], Timo Björk[2], Antti Salminen[1]

[1]University of Turku, Department of Mechanical and Material Engineering, Finland
[2]Lappeenranta-Lahti University of Technology LUT, LUT School of Energy Systems, Finland

juha.peippo@utu.fi

Abstract. Laser welding is a widely-used fusion welding process in industry. However, laser welding is not a common welding process in the manufacture of industrial crane structures. There has been a remarkable increase in the available strength classes of steel grades over the last 10 years, such that strengths of up to 1200 MPa are now commercially available. This enables the use of thinner materials in welded products and at the same time, has opened up new possibilities for using laser welding more widely in the manufacture of steel structures. This study focuses on the static and fatigue strength of the laser-welded joints. Details investigated are edge joint with flange preparation between two rectangular tubes and edge joint between flat bar and rectangular tube. A novel fatigue strength assessment concept, the FAT_{mod} method, is applied to assess the theoretical fatigue performance of the joint in comparison with the effective notch stress method with a FAT630 design curve. The FAT_{mod} method is based on the local stress ratio at a fatigue-critical point of the joint and the analysis considers the strength of the material, surface quality and applied stress ratio in the assessment of fatigue. The study shows that the samples failed from the base material side in static tests and the FAT_{mod} method developed was found to agree well with the test results.

1. Introduction

Laser welding is a widely-used fusion welding processes in various industrial applications. However, laser welding is not a common welding process in the manufacturing of industrial crane structures. The most frequently-used processes are submerged arc welding (SAW) and gas metal arc welding (GMAW). There has been a remarkable increase in the available strength classes of steel grades over the last decade and strengths of up to 1200 MPa are now commercially available. This has enabled the use of thinner materials in welded structures and consequently, has opened up wider possibilities of utilizing the laser-welding process in the fabrication of steel structures. This has provided not only a change in plate thickness but typically, has stimulated the whole construction process with fresh ideas on how to produce the required performance of the crane. This investigation focuses on the static and fatigue strength of laser welds. The details investigated are edge joint with flange preparation I-type joint between two rectangular tubes and edge joint between flat bar and rectangular tube.

2. Experimental tests

2.1. Test specimens

Two different specimen configurations were used in this study: the first test specimen type consisted of a rectangular hollow section with flat bar and the second consisted of two rectangular hollow sections welded together. The thickness of the flat bars was 25 mm and the steel grade used was normal construction steel S355J2. The rectangular hollow sections were manufactured using a laser-welding process. The material thickness of the rectangular profile was 3 mm and the steel was direct quenched steel S900. Fig. 1 shows the main dimensions of the test samples and the nominal material properties from the literature are shown in Table 1. The conventional carbon equivalent values (CEV) and the other carbon equivalents (PL) for laser weld were calculated [1]. The maximum carbon equivalent values available from steel suppliers are shown in Table 2.

Table 1. Chemical compositions [%] of used steel [1] [2] [3]

	C	Mn	Cr	Mo	V	Cu	Ni	S	Si	P	B
S355J2	0.2	1.6	0	0	0	0	0	0.025	0.55	0.025	0
Optim900QC	0.20	1.6	0.8	0.7	0	0.3	2	0.01	0.05	0.03	0

Table 2. Mechanical properties of used steel. [3] [4]

	f_y [MPa]	f_u [MPa]	A_5 [MPa]	CEV/CEVmax	LP
S355J2	355	460	23	0.47/0.45	0.27
Optim 900 QC	900	950	7	0.92/0.58	0.27

The test pieces were 100 mm long and were cut from longer sections, about 1000 mm long see Fig. 1. All welds were produced without filler material.

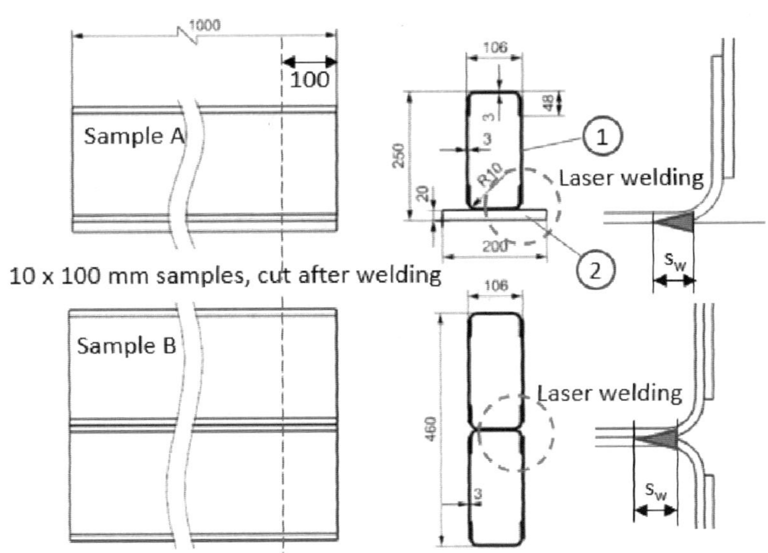

Fig. 1. The shape and main dimensions of the test samples. a) laser weld joint between S355J2 rectangular profile and HS-steel profile and b) laser weld joint between two HS-steel hollow-profile. (red arrows show laser-weld seam positions; thickness of part 1 is 3 mm and part 2 is 20 mm; width of part 2 is 200 mm)

The welding processes was carried out using a CO2-laser welding machine. The welding power was 4.4 kW and welding speed was ether 1.50 m/min or 2.50 m/min. The average width of the weld was about 0.9 mm and penetration s_w was about 4.7 mm in edge joint seams; see Fig. 1 . The laser machine type used was a CO2-Laser Rofin Sinar 4.5 kW and the shielding gas used was helium. Gas flow was 10 l/min. All weldings were performed at normal room temperature of about 20 °C without preheating the welding. The welding parameters are seen in Table 3. Focus = 0 means the laser beam was adjusted so that its focal point was on the surface of test specimen.

Table 3. Used welding parameters and approximation of heat input values Q.

Joint	Focus [mm]	Power [kW]	Speed [mm/min]	Gas	Q [kJ/mm]
Plate and O-profile		4.4	2500	Helium/10 l/min	0.07
Two O-profiles joint	0	4.4	2500	Helium/10 l/min	0.07
Lap joint		4.4	1500	Helium/10l/min	0.12

Heat input values Q were calculated based on Eq.(1) that is given in standard SFS-EN 1011-1

$$Q = k \cdot \frac{P}{v \cdot 1000} \tag{1}$$

where k is thermal coefficient, P is power [W] and v is welding speed [mm/s]. The Eq. (1) gives results in unit kJ/mm. The thermal coefficient, which SSAB proposed to use for laser weld [5], was selected 0.7.

2.2. Static tensile tests

In order to validate the ultimate performance of the structure, static tests were carried out. The tensile test setup is presented in Fig. 2. The samples were clamped to the table of the test ring and external force was applied to the test pieces through the rigid yoke.

Sample A Sample B Potential failure locations

Fig. 2. Test setups for samples A and B. The red arrows show cylinder force directions in the test ring. Potential failure locations.

There were six tensile tests in total. The tensile strength results and typical failure modes are shown in Fig. 3. The average depth s_w of the laser-weld seams was about 4 mm. The ultimate load carrying capacity was 139 kN for sample A and 114 kN for sample B, respectively. The rupture location was at the weld toe outer side of the beam and the fracture passed perpendicularly through the wall thickness of the base material. The nominal tensile capacity of the webs of the hollow profile is 540 kN (f_y·area). This means that the capacity of the formed and welded area related to the capacity of the webs is 0.26 for sample A and 0.21 for sample B.

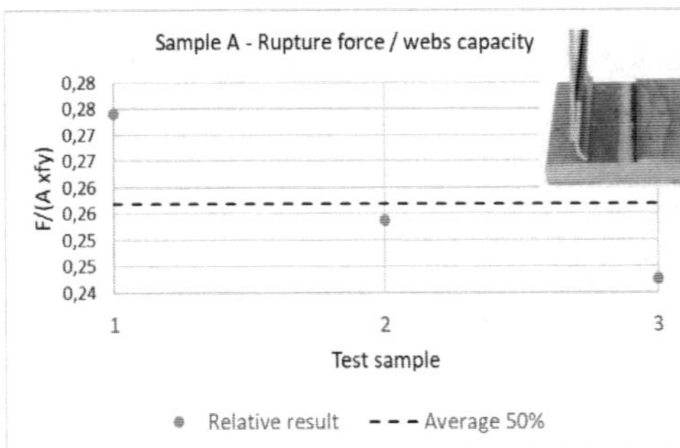

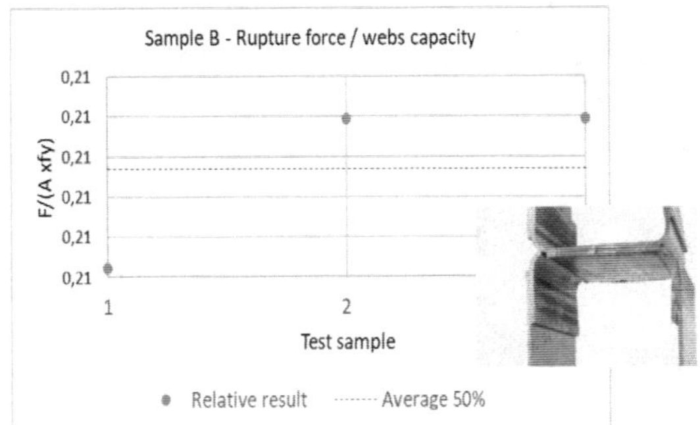

Fig. 3. The tensile strength test results for samples A and B.

2.3. Fatigue tests

Fatigue tests were carried out in the steel structures laboratory at LUT University. The tests were performed using constant amplitude loading, and the stress ratio R was set at close to 0.1, as shown in Table 4. The test matrix and test results are presented in Table 4. The nominal membrane stress range in the webs, equal to F / A_w, is presented in Table 4. The reference area used is $A_w = 600$ mm^2.

The locations of the fatigue failures are shown in Fig. 2. Point 1 illustrates the crack through the base material, Point 2 the crack at the weld toe propagating in the base material and Point 3 the crack at weld root propagating in the base material, respectively.

Table 4. Test matrix and the results of fatigue tests.

Sample	F_{max} / F_{min} [kN]	R [-]	$\Delta\sigma$ at web [MPa]	N_f [cycles]	Failure
A-KO_4	10.40 / 0.20	0.02	17.1	102000	Point 1, at end of profile
A-KO_5	35.66 / 0.20	0.01	59.1	1438	Point 1, at end of profile
A-KO_6	9.93 / 0.20	0.02	16.2	25440	Point 1, at end of profile
A-KO_7	15.30/5.00	0.33	17.2	63452	Point 1, at end of profile
A-KO_8	12.06 / 0.00	0.00	20.1	13920	Point 1, at end of profile
A-KO_9	12.35 / 0.31	0.03	20.1	13560	Point 1, at end of profile
A-KO_12	93.13 / 0.85	0.01	153.8	180	Point 1, at end of profile
B-KOA1_1	10.34 / 0.17	0.02	16.9	139962	Point 1, at end of profile
B-KOA1_2	10.28 / 0.24	0.02	16.7	129686	Point 1, at end of profile
B-KOA1_3	10.34 / 0.28	0.03	16.8	122854	Point 1, at end of profile
B-KOA1_4	12.36 / 0.25	0.02	20.2	36478	Point 1, at end of profile
B-KOA1_5	12.35 / 0.24	0.02	20.2	34730	Point 1, at end of profile
B-KOA1_6	12.50 / 0.39	0.03	20.2	91500	Point 1, at end of profile
B-KOA1_7	12.17 / 0.20	0.02	19.9	71492	Point 1, at end of profile

The depth of the weld seams was measured after the failure for four test samples. The selected samples were assumed to present a typical cross section of weld seam shapes, and two joints were investigated in detail, as illustrated in Figure 7. The measured mean penetration s_w was 4 mm for sample A and 4.55 mm for sample B. These values were subsequently used in this investigation and are slightly different than the values obtained from the test samples used in the static tests.

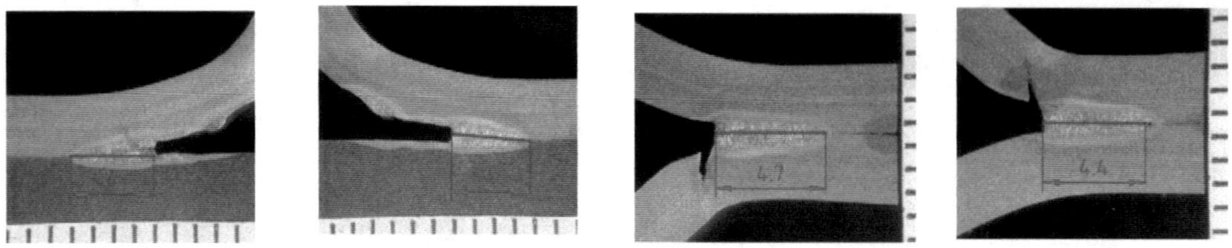

Sample A Sample B

Fig. 4. Typical depth of weld seams.

3. FE-models

The linear-elastic stress concentration factors K_t are needed in fatigue strength assessment of the joints using the notch stress concepts. These were determined with the FE method at both the weld toe and weld root sides of the laser weld in the flange groove. Parabolic tetra elements were used. The effective notch stress approach was applied and a fictitious notch radius of $r = 0.05$ mm was applied in the model in accordance with the recommendations for a thin-walled joint ($t < 5$mm) [6]. Six elements over the 45-degree arc were used in the models, fulfilling the mesh density requirement proposed by Fricke [7] .

Double symmetry was exploited and a linear material model was used. Young's modulus $E = 210000$ MPa and Poisson ratio $\nu = 0.3$ were used. The gap between connected palates was assumed to be zero, and no contact configuration was applied to this gap.

The external tensile force was applied to the models through a yoke and the value of the force was 3750 N in one quarter FE model. The yoke was connected to the rectangular hollow section using a contact boundary condition. The friction value was set to be 0.2 in this contact area. The measured average values for weld dimensions were used in FEA. Fig. 5 presents the mesh and boundary conditions used in FEA.

The stress concentration factors were calculated in three different locations. The factors $K_{t,w,t}$ and $K_{t,w,r}$ were calculated based on stress distribution through the weld seam, while the stress concentration factor $K_{t,b,t}$ was determined based on stress distribution through the plate thickness at the boundary line between the weld and the base plate, corresponding to the potential fatigue failure paths in the studied laser welds. The calculated stress distributions $\sigma(x)$ were divided in membrane-, bending- and non-linear stress components, as shown in Fig. 5. Separation was carried out by applying the widely-used Eqs presented in reference [8].

Linear-elastic notch stress concentration factors were calculated using the following formula

$$K_t = \frac{\sigma_{memb} + \sigma_{bend} + \sigma_{nlp}}{\sigma_{memb} + \sigma_{bend}} \quad (2)$$

The stress components and absolute stress concentration factors are given in Table 5. Stresses were calculated with external force 3750 N in test setup.

Table 5. Selected stresses and the stress concentration values. K_t . Membran stress component is compress, because the yoke does not be rigid and external force effects on the middle of the yoke.

	σ_{memb}	σ_{bend}	σ_{nlp}	$\lvert K_t \rvert$	Stress path (see Fig. 8)
$K_{t,w,t}$	-69	143	1112	16.0	sample A, Line 1
$K_{t,w,r}$	-69	-143	1002	3.7	sample A, Line 1
$K_{t,b,t}$	-14	515	1775	4.5	sample A, Line 2
$K_{t,w,t}$	-120	42	1161	13.9	sample B, Line 1
$K_{t,w,r}$	-120	-42	-552	4.4	sample B, Line 1
$K_{t,b,t}$	-28	491	1159	3.5	sample B, Line 2

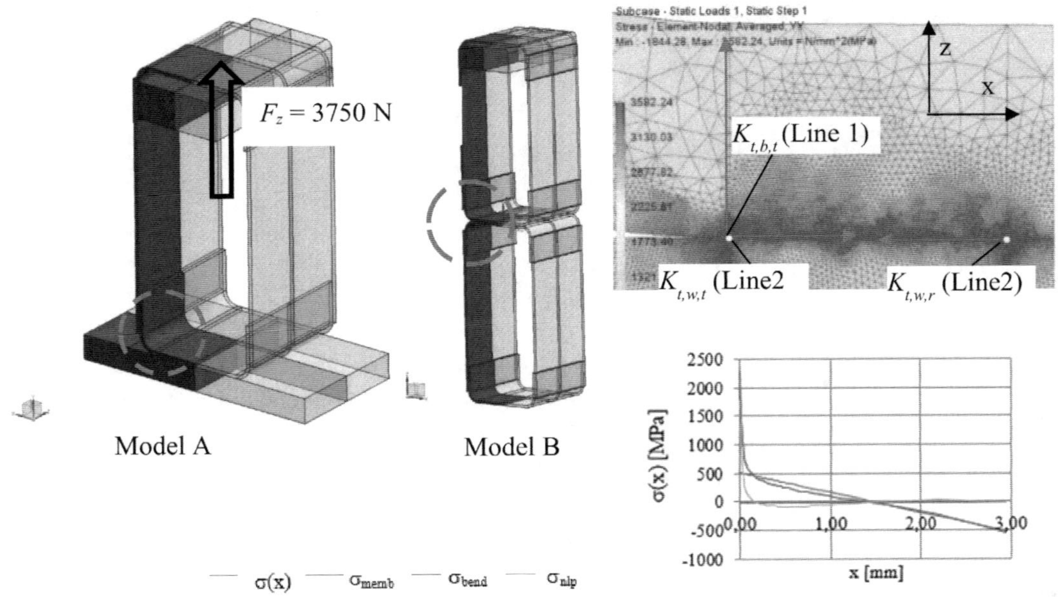

Fig. 5. The type of test samples and detail of mesh in FE models that were used in the determination of stress concentration factors. Lines 1 and 2 show the stress plotting lines that were used in subsequent calculations. The stresses were read in nodes on the outer surface part. Example stress distribution from line 1 and stress components (x is distance from critical point).

4. Fatigue strength assessment of the joints

4.1. Modified notch stress approach

The fatigue strength of laser-welded joints is evaluated by means of a modified local stress approach, with reference to Nykänen et al. [9] [10] [11], Björk et al. [12] and Peippo et al. [13].

The linear-elastic stresses and strains were transformed to elastic-plastic (σ_1, ε_1) components by using Eq. (5)

$$\left.\begin{aligned} \varepsilon_1 \cdot \sigma_1 &= \frac{\left(K_{fmod} \cdot \sigma_{nom,max} + \sigma_r\right)^2}{E} \\ \varepsilon_1 &= \frac{\sigma_1}{E} + \left(\frac{\sigma_1}{K'}\right)^{1/n'} \end{aligned}\right\} \rightarrow output\ \sigma_1\ and\ \varepsilon_1 \tag{3}$$

where $\sigma_{nom,max}$ is the maximum nominal stress, σ_r is residual stress, E is Young's modulus, n' is the cyclic strain hardening exponent, and K' is the cyclic strength coefficient. In this investigation n' is 0.164 and $K' = 1.65 \cdot f_u$ [14]. K_{fmod} is a notch-sensitive factor, that is calculated according to Eq. (6). The effect on surface quality is taken into account in Eq. (6) through the term $C_{\sigma R}$, that is, the roughness correction factor for axial stress. Factor $C_{\sigma R}$ is calculated using Eq. (7).

$$K_{fmod} = (K_f + \frac{1}{C_{\sigma R}} - 1) \tag{4}$$

$$C_{\sigma,R} = 1 - a_R \cdot log(R_z) \cdot log(\frac{2 \cdot f_u}{S_{u,min}}) \tag{5}$$

where a_R is a roughness constant (for steel a_R is 0.22), $S_{u,min}$ is minimum ultimate tensile strength (for all structural steel $S_{u,min}$ is 400 MPa [15]), R_z is the roughness value of the surface in [μm] and f_u is the ultimate strength of the material.

Elastic-plastic stress and strain ranges are calculated ($\Delta\sigma$, $\Delta\varepsilon$) by using Eq. (8). This equation is based on Neuber's rule and kinematic hardening rules

$$\left.\begin{aligned} \Delta\varepsilon \cdot \Delta\sigma &= \frac{\left(K_{fmod} \cdot \Delta\sigma_{nom}\right)^2}{E} \\ \Delta\varepsilon &= \frac{\Delta\sigma}{E} + 2\left(\frac{\Delta\sigma}{2 \cdot K'}\right)^{1/n'} \end{aligned}\right\} \rightarrow output\ \Delta\sigma\ and\ \Delta\varepsilon \qquad (6)$$

where $\Delta\sigma_{nom}$ is nominal stress range in the location under investigation. True stress ration R_{true} is calculated by the equation (9). The stress ratio R_{true} values were limited between -2 to 0.5.

$$\left.\begin{aligned} \sigma_2 &= \sigma_1 - \Delta\sigma \\ R_{true} &= \frac{min(\sigma_1, \sigma_2)}{max(\sigma_1, \sigma_2)} \end{aligned}\right\} \rightarrow output\ \sigma_2\ and\ R_{true} \qquad (7)$$

All stresses are commeasured into a reference stress system, corresponding to the true stress ratio of $R_{true} = 0$. A Smith-Watson-Topper (SWT) mean stress correction is applied to define a new stress range $\Delta\sigma_{ref,R=0}$ referring the equivalent stress range, when $R = 0$, according to Eq. (10)

$$\Delta\sigma_{ref,R=0} = \frac{K_{fmod} \cdot \Delta\sigma_{nom}}{\sqrt{1 - R_{true}}} \qquad (8)$$

4.2. Stress ranges

The fatigue test results are tabulated in Table 6 and the stress ranges calculated using the linear stress concentration factor are given in Table 5. Thus, the linear stress calculation equations can be carried out according to Eqs. (9) – (11).

$$\Delta\sigma_1 = K_{t,b,t} \cdot \Delta\sigma_{nom} \qquad (9)$$

$$\Delta\sigma_2 = K_{t,w,t} \cdot \Delta\sigma_{nom} \qquad (10)$$

$$\Delta\sigma_3 = K_{t,w,r} \cdot \Delta\sigma_{nom} \qquad (11)$$

The stress range is calculated using Eq. (12). The calculated stress range can be positive or negative depending on the stress component values.

$$\Delta\sigma_{nom} = \sigma_1 - \sigma_2 \qquad (12)$$

Table 6. Fatigue test results and stress ranges based on linear stress calculation and linear stress concentration factors

Test sample	Point 1 $\Delta\sigma_1$ [MPa]	Point 2 $\Delta\sigma_2$ [MPa]	Point 3 $\Delta\sigma_3$ [MPa]	N_f [cycles]	Failure location	Test sample	Point 1 $\Delta\sigma_1$ [MPa]	Point 2 $\Delta\sigma_2$ [MPa]	Point 3 $\Delta\sigma_3$ [MPa]	N_f [cycles]	Failure location
A-KO_4	1553	811	-829	102000	Point 1	B-KOA1_1	1099	-733	-484	139962	Point 1
A-KO_5	5374	2804	-2869	1438	Point 1	B-KOA1_2	1085	-724	-478	129686	Point 1
A-KO_6	1475	770	-788	25440	Point 1	B-KOA1_3	1087	-725	-479	122854	Point 1
A-KO_7	1562	815	-834	63452	Point 1	B-KOA1_4	1309	-874	-577	36478	Point 1
A-KO_8	1826	953	-975	13920	Point 1	B-KOA1_5	1308	-874	-577	34730	Point 1
A-KO_9	1826	953	-975	13560	Point 1	B-KOA1_6	1308	-874	-577	91500	Point 1
A-KO_12	13992	7302	-7470	180	Point 1	B-KOA1_7	1293	-863	-570	71492	Point 1

4.3. Modified fatigue strength FAT_{mod} using the reference stress ratio

The mean FAT_m classes were calculated for Point 1, where the fatigue failures occurred. The reference stress component $\Delta\sigma_{ref,R=0}$ is used in regression analyses. The slope of the S-N curve is set to be $m = 3$. For comparison, the regression analyses were also conducted by using free slope S-N curves. The calculated mean reference $FAT_{m,Rtrue=0}$ and characteristic $FAT_{c,Rtrue=0}$ values are given in Table 7 and the S-N curves are plotted in Fig. 7. In the calculations it was assumed that no residual stresses exist and that surface quality $R_z = 100$ (the surface quality was only an estimate).

The effect on the residual stress assumption to deviation value in regression analyses is estimated in Fig. 11. The value of the residual stress was varied in relation to material yield strength, initial residual stress values were calculated with Eq. (13), and the results were applied in Eq. (5).

$$\sigma_r = f \cdot f_y, \tag{13}$$

where f is scaling factor that vary from -1 to 1 with interval 0.2.

The minimum deviation was given with the residual stress assumption that f is -0.4, as shown in Fig. 6. The regression analyses were repeated using that initial residual stress assumption. The results are given in Table 7 and in Fig. 7.

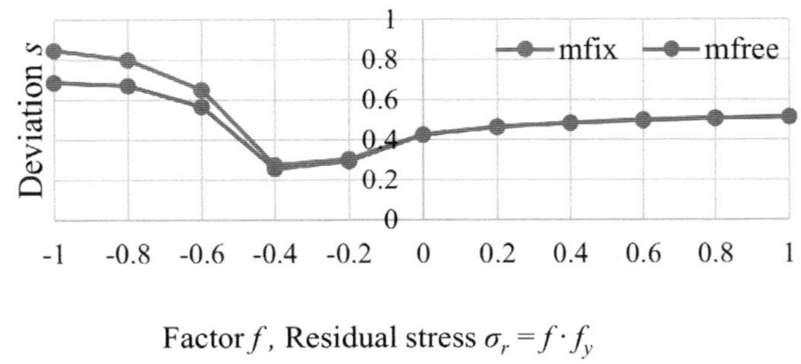

Fig. 6. Deviation vs. residual stress assumption in regression analyses.

Table 7. Reference FAT-classes and deviations.

Test sample	Residual stress	Location	m_{fixed}	$FAT_{m,Rtrue=0}$ and deviation	$FAT_{c,Rtrue=0}$	m_{free}	$FAT_{m,Rtrue=0}$ and deviation	$FAT_{c,Rtrue=0}$
A+B	0	Line 1	3	137 0.427	53	2.9	131 0.426	49.0
A+B	$-0.4 \cdot f_y$	Line 1	3	107.7 0.276	58	2.5	83.2 0.236	44.8

The comparison between fatigue test results and calculated fatigue life estimations were given in Fig. 8. The residual stress assumption was varied between 0 (A) and $-0.4f_y$ (B). Residual stresses were calculated by using the smallest yield strength of the detail. Dotted lines were plotted by using 95% confidential level.

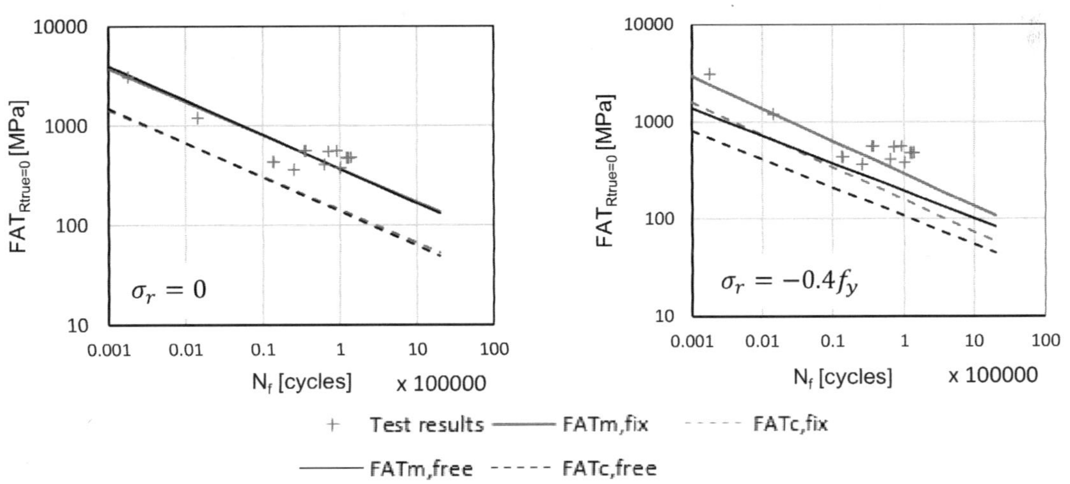

Fig. 7. S-N-curve based assumption of non-existing residual stresses and residual stress assumption $-0.4 \cdot f_y$, and surface roughness of R_z =100.

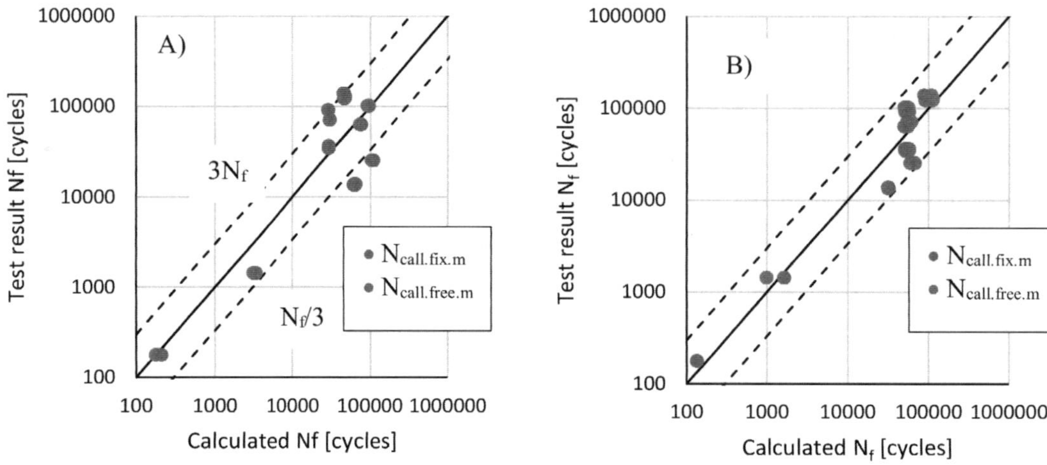

Fig. 8. Tested fatigue time vs. calculated fatigue time with a 95% confidence level (dotted lines) and with residual stress assumption $\sigma_r = 0$ MPa in figure A and $\sigma_r = -0.4f_y$ MPa in figure B

4.4. Comparison of the modified fatigue strength FAT_{mod} and effective notch stress methods

The effective notch stress method gives a constant FAT class of 630 MPa, when the effective notch radius is 0.05 mm [7]. Linear notch stress is used and the slope of the SN-curve is m = 3. Fatigue life predictions generated using ENS and FAT_{mod} methods are presented in comparison with the test results in Fig. 9. Calculated lives were obtained with the characteristic values. Results show that the ENS results are unconservative (below the black line of the constant) while FAT_{mod} results are conservative (located above the black line). Furthermore, it is worth mentioning that the FAT_{mod} method is able to predict, for example, the effect of different parameters, such as applied stress ratio, see Eq. (5), on fatigue strength, and thus enabling more accurate estimations of fatigue life.

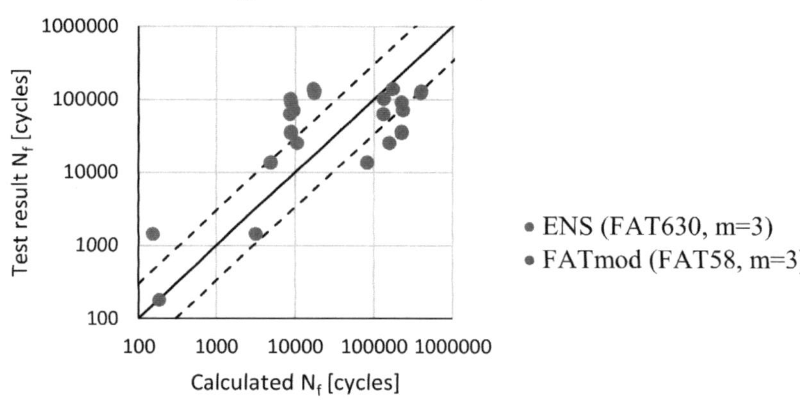

Fig. 9. Calculated fatigue time estimations.

5. Conclusion

In this research work static and fatigue experimental tests for laser-welded joints made of normal and ultra high strength structural steels were carried out. The results were compared with numerical values obtained through a modified local strain approach. Based on these results the following conclusions can be drawn:

- Nominal static performance of joint compared to capacity of wall thickness in pure tension is rather low, equal to 0.21–0.26 of the web capacity.
- The results indicate that the FAT_{mod} method has the capability to describe stress states and it gives a reasonably accurate fatigue time estimation for the details that were investigated. However, the number of test samples was limited and to obtain greater reliability, further tests are needed.
- The fatigue life comparisons indicated that the ENS method with FAT630 ($r = 0.05$ mm) gives an unconservative fatigue life estimation in comparison with the investigated FAT_{mod} method for the laser welds studied.
- Deviation in the statistical analyses was lowest when the residual stress assumption in the equations was equal to $\sigma_r = -0.4f_y$.

6. Acknowledgments

This work was funded by the Research Fund. The Research Council is further acknowledged for support.

ORCID IDs

A Ahokas https://orcid.org/0000-0001-8706-5680
A Salminen https://orcid.org/0000-0002-9071-3682
J Peippo https://orcid.org/0000-0001-5681-3877
T Björk https://orcid.org/0000-0001-8870-5678

7. List of references

[1] "B2B Metel. Online metalll marketplace," [Online]. Available: http://www.b2bmetal.eu/en/pages/index/index/id/149/. [Accessed 23 10 2016].

[2] "Steel Express," [Online]. Available: http://www.steelexpress.co.uk/structuralsteel/S355-S275-structural-steel.html. [Accessed 23 10 2016].

[3] "Strength 900 datat sheet," SSAB, [Online]. Available: http://www.ssab.fi/tuotteet/brandit/strenx/tuotteet/strenx-900. [Accessed 23 10 2016].

[4] "Grand steel," [Online]. Available: http://gangsteel.net/product/alloy/en100252//s355j2.html. [Accessed 23 10 2016].

[5] SSAB, "WeldCalc 2.2 calculation program," SSAB, 2012.

[6] C. Sonsino, W. Fricke, F. Bruyne, A. Ahmadi and G. Zhang, Notch stress concepts for the fatigue assessment of welded joints - Background and applications, International Journal of Fatigue, 34 (2012) 2-16.

[7] W. Fricke, "IIW-Doc. XIII-2240r2-08/XV-1289r2-08 IIW Recommendations for the Fatigue Assessment by Notch Stress Analysis for Welded Structures," 2010.

[8] A. Hobbacher, "IIW document XIII-1965-03 / XV-1127-03 Recommendations for fatique design of welded joints and components".

[9] T. Nykänen and T. Björk, "A new proposal for the assessment of fatigue strength of steel butt-welded joints improved by peening. IIW International Conference. High-Strength Materials – Challenges and Applications, 2-3 July 2015, Helsinki, Finland".

[10] T. Nykänen and T. Björk, "A new proposal for assessment of the fatigue strength of steel buttwelded joints improved by peening (HFMI) under constant amplitude tensile loading," in *FFEMS Fatigue & Fracture of Engineering Materilas & Structures* , 2016, pp. 566-582.

[11] T. Nykänen and T. Björk, "Assessment of fatigue strength of steel buttwelded joints in as-welded condition - Alternative approaches for curve fitting and mean stress effect analysis," in *Marine Structures 44*, 2015, pp. 288-310.

[12] T. Björk, H. Mettänen, A. Ahola and M. Lindgren, "Fatigue strerength assessment of duplex and super-duplex stainless steel by 4R method," in *IIW : XIII-2683-17*, 2017.

[13] J. Peippo, T. Björk and T. Nykänen, "A novel method for fatigue assessment of steel plates with thermally, Welding in the World (2018) 62:105-115, https://link.springer.com/article/10.1007/s40194-017-0529-7," 2017.

[14] D. Radaj, C. Sonsino and W. Fricke, "Fatigue assessment of welded joint by local approaches. 2nd edition. Cambridge: Woodhead Publishing," 2006.

[15] B. Hänel, E. Haibach, T. Seeger, G. Wirthgen and H. Zenner, "FKM-Guidline. Analytical strength assessment. 5th Edition. ISBN 3-8163-0425-7," 2003.

[16] "Ruukin kuumavalssatut erikoisteräkset," Ruukki Oy, 16 5 2012. [Online]. Available: https://wiki.oulu.fi/.../Ruukin+kuumavalssatut+erikoisteräkset+-+Olli+Vähäkainu.pdf. [Accessed 9 10 2016].

[17] J. Peippo, A modified nominal stress method for fatigue assessment of steel plates with thermally cut edges, Lappeenranta: Acta Universitatis, 2015.

[18] M. Ysuyama, H. Gujimoto, M. Uchihara and K. Fukui, IIW: III-1380-06. The Laser welded joint stregth of advanced automotive high strength steel sheets.

Surface roughness variance on different levels of surface inclination of powder bed fused tool steel 1.2709

Jussi Karlsson[1], Aki Piiroinen[1], Markus Korpela[2], Antti Salminen[3]

[1]Machine Technology Centre Turku Ltd., Finland,
[2]LUT University, Laboratory of Laser Materials Processing and Additive Manufacturing
[3]University of Turku, Department of Mechanical and Materials Engineering, Finland,

jussi.karlsson@koneteknologiakeskus.fi

Abstract. Evolution of additive manufacturing has allowed increased flexibility and complexity of designs over conventional manufacturing e.g. formative and subtractive manufacturing. Restricting factor of laser powder bed fusion of metals (PBF-LB/M) additive manufacturing is the as-built surface quality. To promote an understanding of the surface roughness and suitable surface measuring technologies octagon shaped tool steel 1.2709 samples was developed and manufactured. Different surface measuring technologies was also literary reviewed. Studied samples were manufactured with commercially available laser-based powder bed fusion system using standard parameter set provided by the system manufacturer. Surface roughness was measured from top and down skins from multiple different building angles in a way that process specific effects, such as direction of movement of the powder re-coater, was considered. Based on these measuring results of the samples the effect surface inclination are discussed. The results show that building angle strongly affects to surface roughness of laser-based powder bed fused parts. Surface roughness was measured to be more than five times worse in unsupported angle manufactured down facing surfaces when compared with vertical walls.

Keywords: Additive Manufacturing, Surface roughness, tool steel (1.2709), PBF-LB/M, L-PFB, SLM

1. Introduction

The driving force of ever grooving success of additive manufacturing is enabled allowed increased flexibility and complex geometry over conventional manufacturing e.g. formative and subtractive manufacturing. As-built surface quality is restricting factor of most popular additive manufacturing technique of metals, the laser-based powder bed fusion (PBF-LB/M) of metals using a laser based system. The quality or need of post processing is heavily involved by the use of support structure. The use of support structure automatically leads to poor surface quality and need of excessive post processing to finalize surface quality to application specific surface finish. The cost vice optimum would be to always manufacture so-called self-supporting shapes that do not need support structure. Unsupported down facing surfaces have partially fused particles which are lowering the surface quality comparing surface quality on the top up facing surfaces. Often when additive manufacturing surface roughness values are mentioned they are top surface and side surface quality values not the overall average quality values including also down facing surfaces. The factors affecting overall surface roughness in PBF-LB/M manufacturing process can be divided on process parameters, material properties and design aspects of the work piece. Effecting process parameters are layer thickness, spot overlap, energy density, laser beam velocity and spot angle. Material effects are powder grain size and chemical composition. Design effects are part building orientation, placement of the work piece, surfaces inclination angles. Overall process parameters for smoother surface have direct impact on additive manufactured part fatigue performance and need for further post processing steps. [1,2,3,4,5]

There are several physical phenomena in the surface formation of metal bodies produced by AM techniques that affect the surface shape as well as interfere with surface measurement. The laser, the powder layer, and the interaction between the layers beneath it significantly affect the surface details, i.e., the topography. Typical SLM printed part surface defects can be seen in Figure 1 [6]. The weld tracks are the most evident features, appearing as ridge-like formations indicating the path followed by the processing laser while traversing the powder bed. Smaller-scale ripples on the weld tracks are formed as a result of the cyclical process of liquefaction and solidification of the melt pool as the laser moves across the surface [7]. Partially melted powder particles as well as splashes cause unevenness on the surfaces of the body. The recesses on the surfaces may be due to discontinuity of the welding track, incompleteness of adjacent welds, and porosity due to gas adhesion [8].

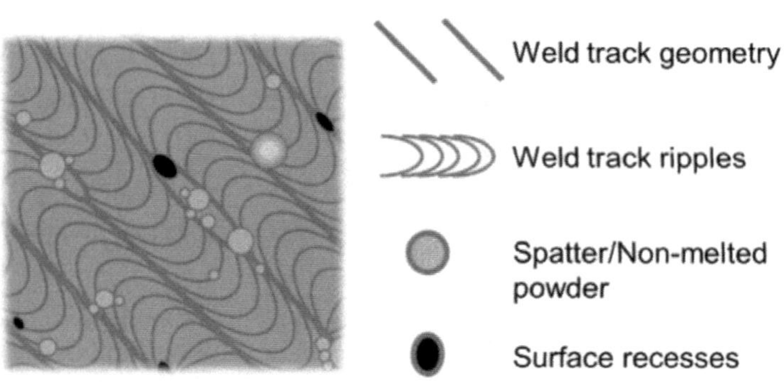

Figure 1 Topographic features relevant to investigation of surfaces, as they appear on a layer of an SLM metallic part. [6]

The purpose of this study is to promote an understanding of PBF-LB/M manufacturing process incline angel effect to surface roughness. Octagon shaped tool steel 1.2709 samples was designed and manufactured. Octagon shape was selected, because of prediction that shape would support itself and less support would be needed. Shape would also reveal if powder spreading (re-coater angle) has on effect on surface quality. Surface roughness was measured on samples rising and falling unsupported surfaces using optical profilometer and stylus profilometer. Different surface measuring technologies was also literary reviewed.

2. Surface measuring metrology

In general, surface metrology is defined as the measurement and characterization of a surface topology that broadly describes the shape and roughness of a surface. In this study, we are only interested in surface texture.

Lou et al. (2019) have specifically investigated methods for measuring the surface area of bodies made by additive manufacturing (AM) and the PBF (Powder Bed Fusion) method in comparison with conventionally made methods for measuring bodies. According to them, traditional measurement arrangements are not suitable for AM products due to its particularly different manufacturing method. In traditional manufacturing, regular patterns were formed on the machining surfaces, but with the AM technique, the surface formation is different. Partially melted particles and recesses in the surface distort the result and the results obtained do not sufficiently describe the functionality of the product. The surface of the print and even the shape of the surface change in the post-treatment, in which loose or partially melted particles can be removed. Prior to post-treatment, measuring the surface with a stylus-based device can be challenging if there are large partially melted particles on the surface. The stylus may not be able to measure surface depressions, particles will distort the result, the stylus may get stuck in the particles, or in the worst case the stylus will be damaged during measurement. [10]

ISO 4286 defined the surface structure parameters e.g. Ra and Rq values, while the surface texture parameters of the area are defined in ISO 25178-2 with e.g. Sa and Sq values. In industry, these values are mainly used to describe the texture of the surface, but these values do not provide enough information to control the production of additional manufacturing. [3]

Townsend et al. (2016), based on a review of surface roughness studies of AM products, the most commonly used surface roughness measurement method is a stylus-based tangential measurement. Optical measuring devices have also been used and their use is becoming more common. The optical measuring device gives a wider view of the surface by measuring the area at once. In this study, we have used a stylus device and an optical measuring device. [10]

3. Experimental setup

In order to get more surface quality data on powder bed fusion process, octagon shaped samples covering a range of surface angles were built with EOS M290 a laser powder bed fusion system. EOS Maraging Steel MS1 material and MS1 performance 2.0 parameter settings with 40 μm layer thickness was used on build processes. MS1 have a chemical composition corresponding to US classification 18% Ni Maraging 300, European 1.2709 and German X3NiCoMoTi 18-9-5. Figure 2 shows the detailed geometry of the total set of 6 octagon pieces has been made on our experimental setup. Octagons plane inclination angels vary from rising top surfaces 90°, 60°, 50°, 45°, 40°, 30° angles to falling unsupported down facing surfaces 60°, 50°, 45°, 40°, 30°, 20° angles. Octagon shape was selected, because of prediction that shape would support itself and less support would be needed. Shape would also reveal if powder spreading (re-coater angle) has on effect on surface quality. Placement of pieces on a powder bed is shown in Figure 3.

Surface quality was measured first with Keyence VX-3000 optical profilometer with surface area of 2 x 2 mm^2 rising surfaces and 4 x 4 mm^2 falling surface. Arithmetic mean surface roughness for surface value Sa and sum value Sz of largest peak height value and the largest pit depth value within measured surface area was measured. Second measuring method was stylus profilometer (contact measurement) MarSurf PS10 with 5 µm measuring stylus. With stylus profilometer, every octagon rising inclination plane surface quality was measured five times from different locations randomly over plane surface area. Final Ra and Rz results which is used in this study was average of these five measurements. Arithmetic mean surface roughness parameter Ra / Sa and sum value of highest peak to the lowest measurements Rz / Sz is used here as a representation of surface roughness.

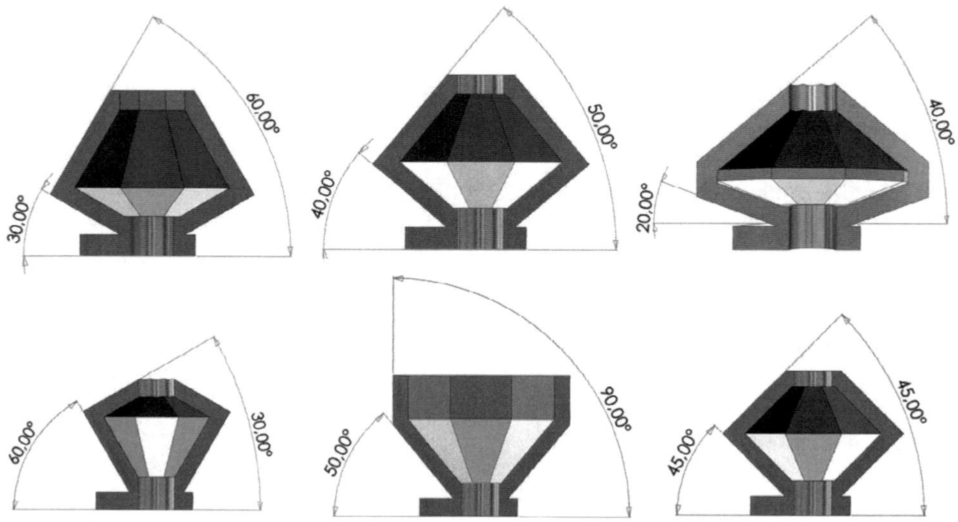

Figure 2. Detailed geometry of the total of 6 octagon pieces has been made on our experimental setup.

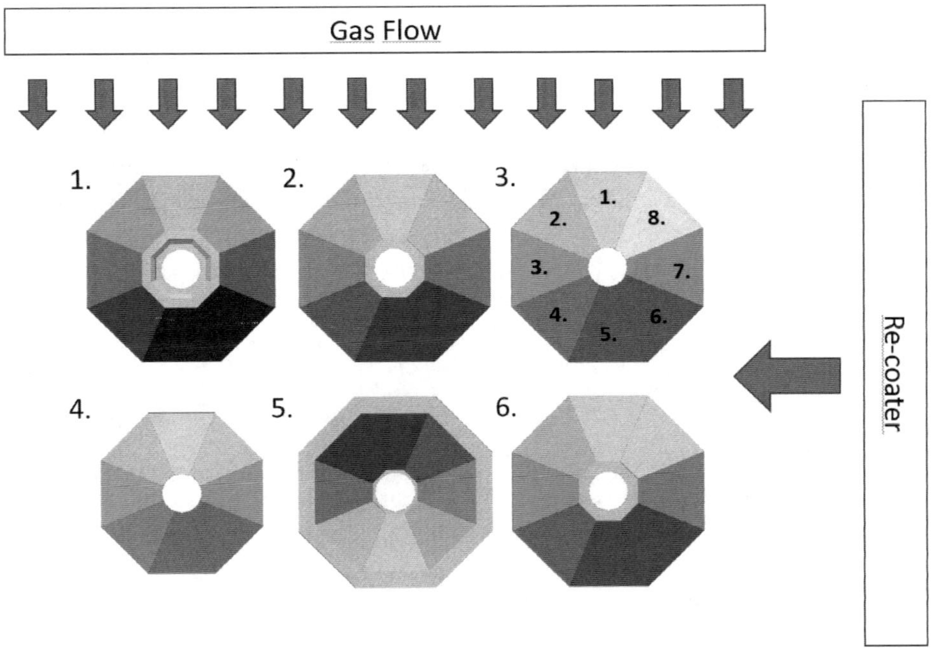

Figure 3. Detailed placement of 6 octagon pieces.

4. Results and discussion

Six octagon shape sample pieces was produced using PBF-LB/M additive manufacturing process. Octagon with 20 degrees falling surface fail on manufacturing without support structures. All other test pieces was possible to print without support structures. Due to poor surface quality on down facing surfaces stylus profilometer measuring tip was stuck on partially fused particles which result stylus profilometer to fail measure surface quality. On top facing surfaces measurements variance between optical and contact measuring methods was insignificant. One effect on small difference on values between measuring methods could be explained with measuring order where optical measurements was done first and contact stylus measurement second. Some of the partially fused particles could be fallen of between measuring.

The dependence of the surface roughness on the manufacturing inclination angle is illustrated on fig. 4 drawing and in line graph format on fig. 5 Sa and Ra - values and fig 6. Sz and Rz values. Values are average value of all eight top or down facing surfaces of one work piece. Top and down facing surface shown same trend on surface quality where surface quality deteriorated as the incline angle decreased toward the horizontal level. Top surface lowest surface roughness values on was measured on 90° angel (vertical side wall) Ra 6,7 µm and highest value Ra 11,0 µm on 30° angel. Down facing surface lowest surface roughness values on was measured on 30° angel Ra 17,9 µm and highest value Ra 58,0 µm on 60° angel. On average values shows that roughness values was three times higher on down facing surfaces comparing to top surface values. There is also significant difference of quality between down skin surfaces on 30° (Ra 58,0 µm) and 40° (Ra 36,0 µm) inclination angels. This information gives one more reason to consider not to design AM part overhangs smaller incline angels than 45° comparing to horizontal base. Values also show that ssurface roughness was measured to be more than five times worse in unsupported angle-manufactured down facing surfaces when compared with vertical walls.

The dependence of the surface roughness on the manufacturing re-coater angles is illustrated in in line graph format on fig. 7 and fig. 8. Octagon shape gives eight re-coater angles to investigate and given result values are average values from all work pieces with same re-coater angle. Re-coater angle 0° (surface 1, fig. 3) are towards gas flow and angle 270° (surface 7, fig. 3) are towards re-coater. Surface roughness measurements values variance on different re-coater angle was insignificant on incline surfaces. Top facing surface has slight increase in Rz surface roughness value between 180° to 315°, but same effect is not noticeable on other measurement values. Ra and Za top facing surface roughness values was between 7 – 11 µm on all re-coater angles and Sa surface roughness values for unsupported down facing surfaces was between 30 – 39 µm. Top facing Sz surface roughness values had biggest variance but there was no clear pattern for the variation. Calculating average value for all the surface roughness measurement values the calculation gives us Sa value 22,4 µm and Sz value 210,5 µm for overall surface quality to our test pieces outer surface.

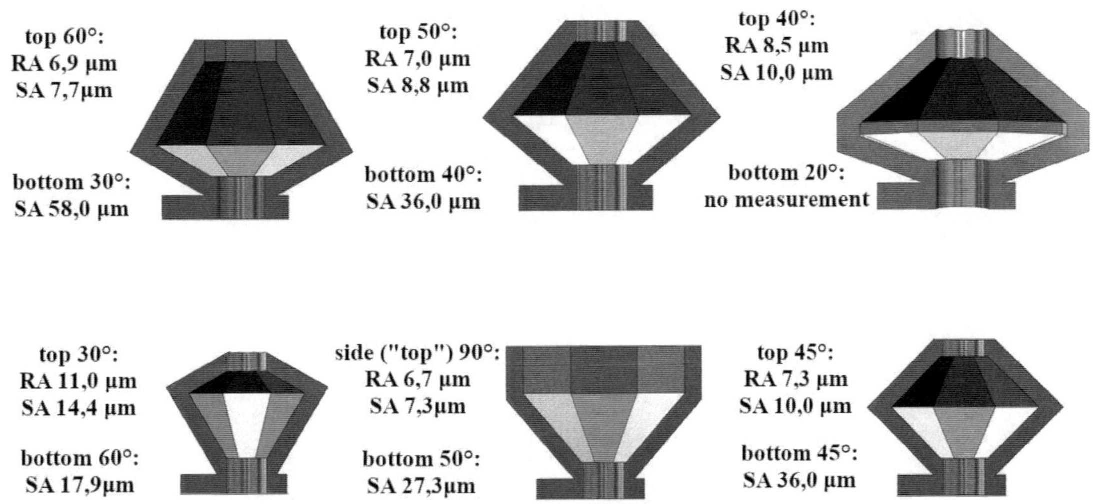

Figure 4. Surface roughness measurement results with different plane inclination angels, SA values are measured with optical profilometer and Ra values are measured with stylus profilometer. Values are average of all surfaces of one work piece.

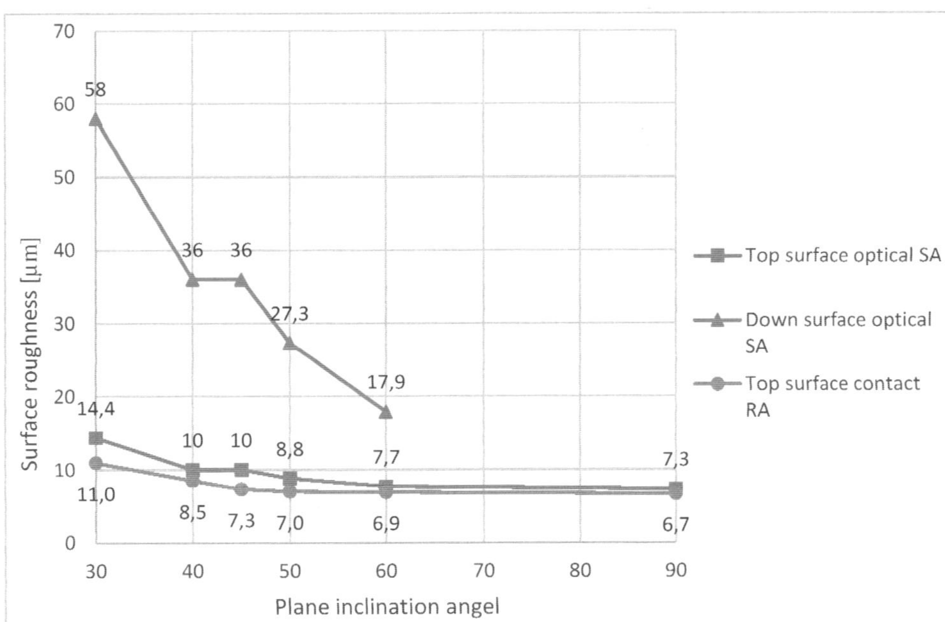

Figure 5. Line graph of surface roughness measurement results with different plane inclination angels, SA values are measured with optical profilometer and Ra values are measured with stylus profilometer. Values are average of all surfaces of one work piece.

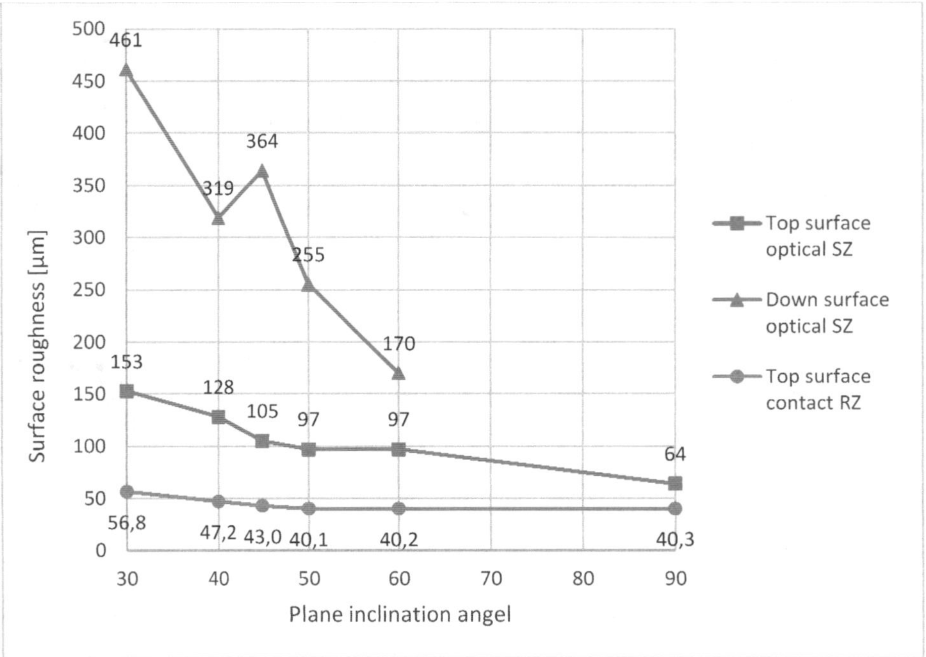

Figure 6. Line graph of surface roughness measurement results with different plane inclination angels, SZ values are measured with optical profilometer and Rz values are measured with stylus profilometer. Values are average of all surfaces of one work piece.

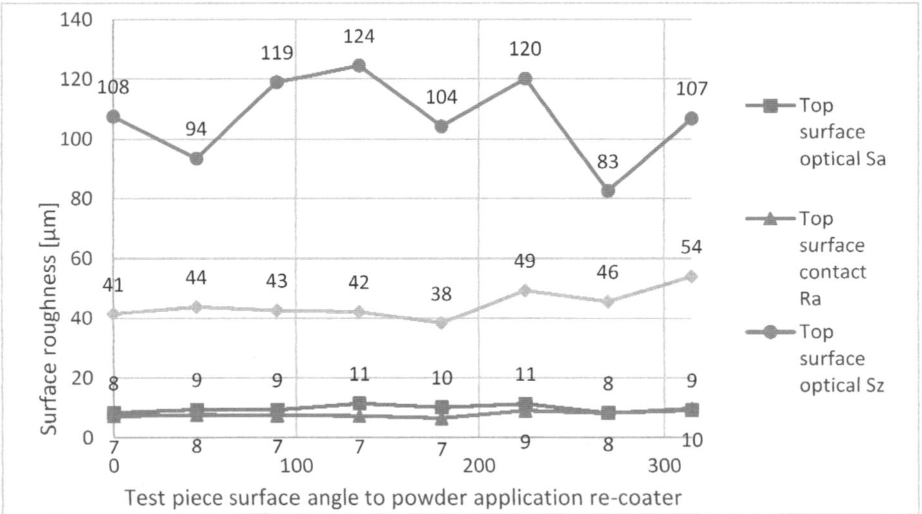

Figure 7. Line graph of top surface roughness measurements mean value from results of all test pieces with different re-coater angles, Sz / Sa values are measured with optical profilometer and Rz / Ra values are measured with stylus profilometer. Values are average of all work pieces.

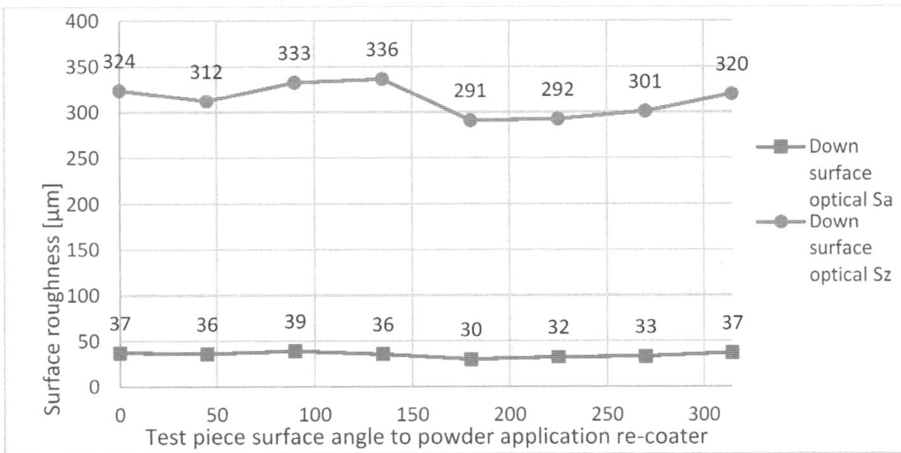

Figure 8. Line graph of down facing surface roughness measurements mean value from results of all test pieces with different re-coater angles, Sz / Sa values are measured with optical profilometer. Values are average of all work pieces.

5. Conclusions

This study designed and investigated PBF-LB/M manufacturing process surface roughness quality for several incline angels and powder spreading angles. Octagon shaped tool steel 1.2709 samples were designed and additively manufactured using EOS M290 laser powder bed fusion system and commercial EOS parameter sets MS1 performance 2.0 parameter. The effect of the surface roughness was analysed multiple inclinations. Octagon shape of the samples gives eight re-coater angles to investigate. Main findings was concluded:

1. Surface roughness on unsupported down facing surfaces are so low quality that stylus profilometer with contact measuring tip was unapplied to measure surface quality.

2. Top and down facing surface shown same trend on surface quality where surface quality deteriorated as the incline angle decreased toward the horizontal level.

3. The surface roughness values show that roughness values was three times higher on down facing surfaces comparing to top surface values.

4. Surface roughness was measured to be more than five times worse in unsupported angles manufactured down facing surfaces when compared with vertical walls.

5. Surface roughness measurements values variance on different re-coater angle was insignificant on incline surfaces. Sample set was small and horizontal surfaces was not measured.

Further research and development is needed to find uniform standard ways measure and compare additive manufacture parts over all surface quality. Furthermore, considering issues could be the effect of post-manufacture polishing on uniformity of overall surface quality, part location and shielding gas effect on build plate and inner surfaces quality measuring technics.

IOP Conf. Series: Materials Science and Engineering 1135 (2021) 012020 doi:10.1088/1757-899X/1135/1/012020

ORCID iDs

J Karlsson https://orcid.org/0000-0003-1665-8010

A Piiroinen https://orcid.org/0000-0002-8473-126X

M Korpela https://orcid.org/0000-0003-4936-1440

A Salminen https://orcid.org/0000-0002-9071-3682

References

[1] Jason C. Fox, Shawn P. Moylan, Brandon M. Lane, "Effect of Process Parameters on the Surface Roughness of Overhanging Structures in Laser Powder Bed Fusion Additive Manufacturing", Procedia CIRP, Volume 45, 2016, Pages 131-134, ISSN 2212-8271, https://doi.org/10.1016/j.procir.2016.02.347

[2] K.A. Mumtaz, N. Hopkinson, "Selective Laser Melting of thin wall parts using pulse shaping", Journal of Materials Processing Technology, Volume 210, Issue 2, 2010, Pages 279-287, ISSN 0924-0136, https://doi.org/10.1016/j.jmatprotec.2009.09.011.

[3] S. Lou, X. Jiang, W. Sun, W. Zeng, L. Pagani, P.J. Scott, " Characterisation methods for powder bed fusion processed surface topography" Precision Engineering, Volume 57, 2019, Pages 1-15, ISSN 0141-6359, https://doi.org/10.1016/j.precisioneng.2018.09.007.

[4] Fatemi A., Molaei R., Simsiriwong J., Sanaei N., Pegues J., Torries B., Phan N., Shamsaei N.," Fatigue behaviour of additive manufactured materials: An overview of some recent experimental studies on Ti-6Al-4V considering various processing and loading direction effects", Journal of Fatigue & Fracture of Engineering Materials & Structures, Volume 42, Issue 5, 2019, https://doi.org/10.1111/ffe.13000

[5] Calignano, F. Investigation of the accuracy and roughness in the laser powder bed fusion process. Virtual and Physical Prototyping, 2018, Journal, 13:2, 97-104, https://doi.org/10.1080/17452759.2018.1426368

[6] Richard Leach, Adam Thompson, Nicola Senin, "A metrology horror story: The additive surface" , Manufacturing Metrology Team, Faculty of Engineering, University of Nottingham, NG7 2RD, Nottingham, UK

[7] Mazumder J. Overview Of Melt Dynamics In Laser Processing. Opt. Eng. 1991; 30:1208–19.

[8] Simonelli M., Tuck C., Aboulkhair N. T., Maskery I., Ashcroft I., Wildman R. D., Hague R. A Study On The Laser Spatter And The Oxidation Reactions During Selective Laser Melting Of 316L Stainless Steel, Al-Si10-Mg, And Ti-6Al-4V. Metall. Mater. Trans. A Phys. Metall. Mater. Sci. 2015; 46:3842–51

[9] Whitehouse, David J.. Surfaces and Their Measurement. Oxford: Elsevier Science & Technology, 2004.

[10] Townsend A., Senin N., Blunt L., Leach R.K., Taylor J.S., "Surface texture metrology for metal additive manufacturing: A review", Precision Engineering, Volume 46, 2016, pp. 34-47, ISSN 0141-6359, https://doi.org/10.1016/j.precisioneng.2016.06.001.

A study on use of wobble features in laser welding of low alloy steel joint with butt joint configuration

Timo Kankala[1] Antti Salminen[1, 2]

[1] Machine Technology Center Turku Ltd, Turku, Finland
[2] University of Turku, Department of Mechanical and Materials Engineering, Turku, Finland

timo.kankala@koneteknologiakeskus.fi

Abstract. Laser welding is modern digital welding process, which thanks to several advantages over traditional welding processes, is gaining ever growing role in manufacturing. The process has still some weaknesses. The better the beam quality the smaller the focal point, the actual welding tool, diameter is. Typically, because of this the welding of joints with lesser quality e.g. larger air gap is difficult or even impossible. So-called beam manipulation opens opportunities to deal with the problem. The dynamic beam manipulation gives opportunities to control the weld dimensions during the welding process by the requirements of individual locations of weld joint. This study used the two dimensional scanner to manipulate beam during welding with so called wobble function. Four different wobble configurations were tested in welding of low-alloyed steel with different joint qualities. The wobble typically made the welds wider, provided typically higher heat input and thus lowered the hardness of the joint. Wobble increased typically the root quality, but there are differences between different wobble parameters. It was possible to weld joints with wider air gaps in the selected material thickness, but the wider air gap and wobble caused finally, when wide enough the sagging of the joint.

1. Introduction

1.1. Laser welding production case

Machine Technology Center Turku Ltd (MTC) has been laser welding of special dimension rectangular hollow section steel beams for maritime industry for several years. These steel beams are used in cruise ship staircase construction as sidewall beams, and are later GMAW-welded together with steps, painted and installed to the ship blocks by the Turku shipyard's subcontractors (Fig.1 and 2). About 2000 beams have been laser welded to the nine different cruise ships during the years.

IOP Conf. Series: Materials Science and Engineering 1135 (2021) 012021 doi:10.1088/1757-899X/1135/1/012021

Fig. 1 and 2. Staircases under installation in Meyer Turku shipyard.

These steel beams are normally laser welded longitudinally from two C-shape press brake bended plates in horizontal position (PC) one side at the time, in butt joint configuration, as shown in Fig.3. Steel beams are typically from 2 m to 4 m long. At the ends there is typically also added separate bended parts in different angles. These parts are also laser welded together longitudinally and with the long parts crosswise. Special jigs are designed and build to use in laser welding to press these C-shape plates together during laser welding. Pressing force is carried out with pneumatic system.

Fig. 3. Horizontal laser welding position.

Regardless that this laser welding production case has been going on several years, there is still some principal problems in it. MTC's large robotized laser welding system has no seam tracking system, which is normally crucial to any large laser welding work stations. Part preparation or setting into fixtures can be incomplete in different ways, e.g. poor laser cut quality, large air gap, linear misalignment. These shortages in production causes unnecessary trouble, time consuming corrections and unnecessary excessive heat input to the work piece. Therefore new laser welding techniques are always under observation in MTC to make production easier and quality better.

1.2. Laser welding with beam manipulation technique

A major improvement to original welding was achieved by manipulation of the focal point size. Using longer 500 mm focal length in existing ordinary laser welding head (Precitec YW52) has given more tolerance for the welding set up and eased the suitability of the large laser welding robot system in accuracy vice compared to the formerly used 300 mm focal length. This is realized by the larger focal point diameter. Diameter with used optical setup is changed to 1.0 mm from previous 0.6 mm. This has helped to achieve satisfying weld quality more easily.

Anyway, this relatively new laser welding technique to use dynamic beam manipulation, or beam oscillation, seems to be very interesting and promising technique to check out and to solve or at least to relieve these problems mentioned in chapter 1.1. The technique has been used e.g. to change the phase balance in welding of duplex stainless steel [1].

In Turku University of Applied Science's research investment program invested to the new laser welding head with beam oscillation, so called wobble head. Public tendering and purchase were done in 2020 and winter 2021 new "wobble head" was installed to the MTC's large robot welding cell. Together with University of Turku and Turku University of Applied Science, MTC is now studying the use of different wobble functions and parameters to achieve better welds, more easily, compared to the stable laser welding head, generally and specially in this steel beam production case. This study is one of the first test trials done with this new welding head in MTC.

New IPG FLWO D50 wobble head is also additionally equipped with IPG LDD-700 Inline Coherent Imaging (ICI) sensor system, which can be used e.g. as a seam tracking system, keyhole depth measuring system and weld bead profile measuring system. These features were not examined in this study.

Some test welds were also videotaped with Cavitar C300 welding camera, which was purchased to MTC in Interreg Baltic Sea Region program INforM project. These videos were not precisely examined in this study. However, welding camera and ICI -sensor system will open up lot of possibilities for further research studies.

Fig. 4. IPG FLWO D50 wobble head with LDD-700 sensor system.

2. Test material

Test materials are taken from actual production steel plates, which have been delivered by the customer. Flat plate test material pieces are cut out from several C-shaped press brake bended production parts. Test piece edges were laser cut like in the normal production parts also

Material is hot rolled fine grain steel S355MC according to the standard EN 10149-2 [2] of thickness 4 mm. The nominal chemical and mechanical properties of the steel are given in Tables 1 and 2, based on steel manufacturer test report (EN 10204 – 3.1) [3] of the melt analysis and mechanical tests of the cast of the test material. Mechanical properties represent the thermomechanically treated (TM) and pickled condition.

Tab. 1. Chemical properties of S355MC. [3]

	C	Si	Mn	P	S	Al	Nb	V	Ti
%	0.064	0.01	0.72	0.008	0.005	0.034	0.027	0.008	0.001

The carbon equivalent (Ceq) of the steel is 0.20 (Ceq = C+Mn/6+(Cr+Mo+V)/5+(Ni+Cu)/15).

Tab. 2. Mechanical properties of S355MC. [3]

Tensile test		
RP0,2 (MPa)	RM (MPa)	A (%)
418	479	33

Because the test plate materials were cut out of the actual left-over work pieces, the dimensions were not according to standard size welding procedure test plate (e.g. 350x150mm), but only about 250mm long.

3. Laser welding equipment

10kW IPG YLS 10000 fiber laser with four way beam switch was used in laser welding. Linear test welding lines were performed with industrial 6-axis Yaskawa Motoman UR50 –robot.

IPG FLWO D50 wobble head was used with 400 mm focal length and 150 mm collimation length. Connected process fiber has a diameter of 300 µm. Theoretical focal spot diameter is then 800 µm, which was also tested with Primes FocusMonitor - laser beam focus measurement system at 5kW laser power, see figure 5 for details. Focal lens is protected with cover slide window and with air knife, also additional air knife pipe was added to the welding head to cut the metal vapor movement towards welding head and to keep the optics clean from the welding spatter and fume. Additional air knife pipe was installed about 5cm above welding spot.

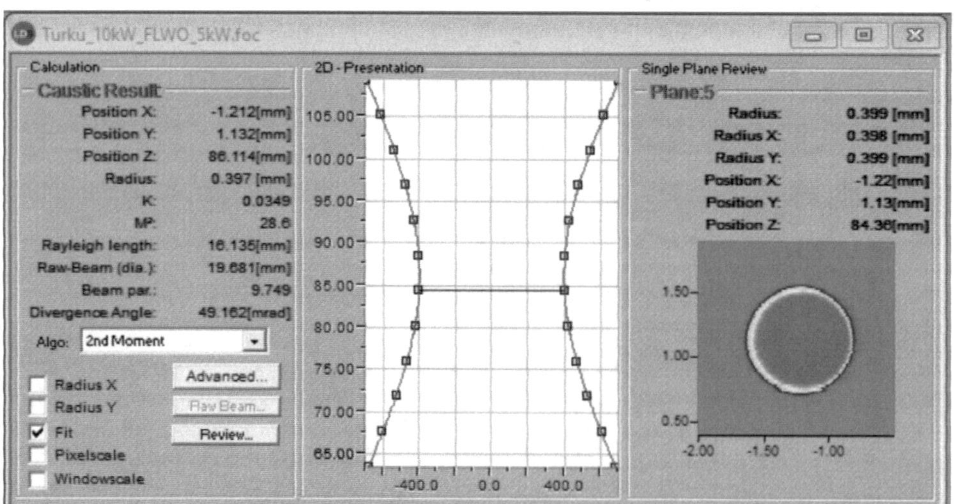

Fig. 5. Result of the focal spot measurement.

4. Laser welding parameters

Main laser welding parameters were kept constant and only some wobble function parameters were varied during test welds. Tested wobble oscillation modes were selected from the basic electric waveform controlled functions of the wobble controller based on previous general laser welding experience. Butt joint I-groove test pieces were laser tack welded with nominal zero air gap.

4.1. Constant test parameters

Laser power and welding speed were kept constant in every test run, 5.7kW and 2.1 m/min were used. No shielding gas or no root gas was used. Focal spot vertical position was at the top surface of the plate, ±0mm. Focal spot horizontal position, and wobble oscillation movement center, was in the middle of the joint, as illustrated in figure 6. Plates were welded at flat position (PA) according to the standard EN ISO 6947. The parts were tack welded at both ends prior welding.

4.2. Wobble parameters

Also, some of the main wobble parameters were kept constant, like amplitude 1 mm and frequency 200 Hz, but oscillation mode was varied. Three different modes of oscillation was used in these tests; line oscillation, infinity oscillation and circle oscillation. 1 mm amplitude designate in wobble controller the full width movement in linear mode and diameter in circle mode. In infinity mode the length of the infinity ∞ -symbol was 0.60 mm and width were 0.24 mm. Also ordinary basic laser weld without any oscillation was made with wobble head as a reference weld.

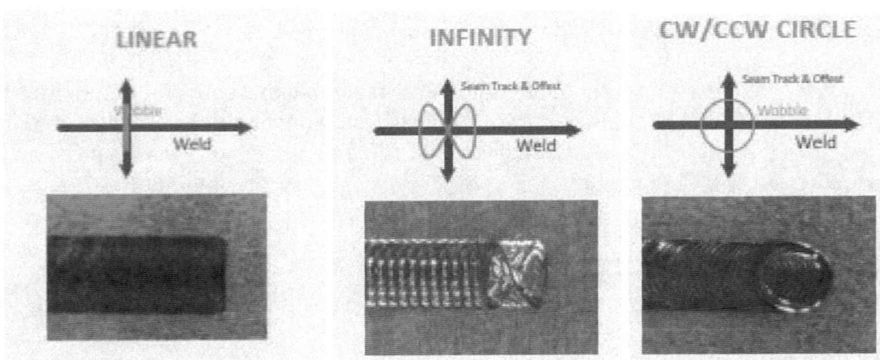

Fig. 6. Wobble head oscillation modes.

These tested wobbling parameters (dimensions and frequency) lead on roughly to the following actual focus spot speeds during the oscillation movements and welding movement; linear mode 435 mm/s, infinity mode 360 mm/s and circle mode 664 mm/s.

5. Test results

Macrographic examination and hardness tests were carried out for test pieces (Fig. 7). Specimens for macrographs and hardness testing were first cut out from welded test plates, then the edges were machined, grinded, polished and etched with 3% nitric acid, so called Nital etchant in room temperature.

Fig. 7. Laser welded test plates

5.1. Macrographs and hardness tests

Hardness tests were made with Innovatest Falcon 600 hardness tester with 1kg force (HV1). EN ISO 22826 [5] standard was used only as a guideline for the hardness tests.

Hardness indentations are from the middle of the plate thickness (t/2), about 2 mm from surface and root, and about 1 mm from surface and root. Except all three base material (BM) hardness indentations are measured from the middle of the plate thickness several millimetres from weld (not seen in macrographs).Weld metal (WM) three hardness indentations are measured in the vertical middle line of the weld and heat affected zone (HAZ) indentations as close as possible from the fusion line. WM and HAZ indentations are shown in the macrograph figures 9, 10, 11 and 12. Measured hardness values are shown in the tables 3, 4, 5 ad 6.

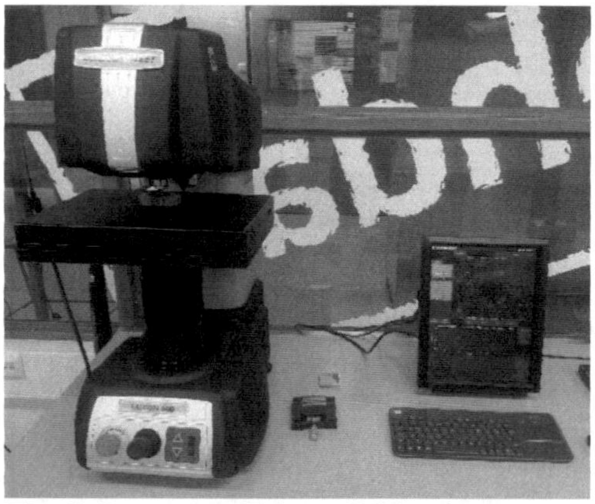

Fig. 8. Innovatest Falcon 600 hardness tester

EN ISO 13919-1 [6] standard was used to evaluate weld imperfections from welds and from their cross-section macrographs. Also weld width from fusion lines was measured from macrographs from three different locations; near surface, middle and near root. Weld width values are shown beside macrograph respectively.

5.1.1. Test results from non-wobbled weld.

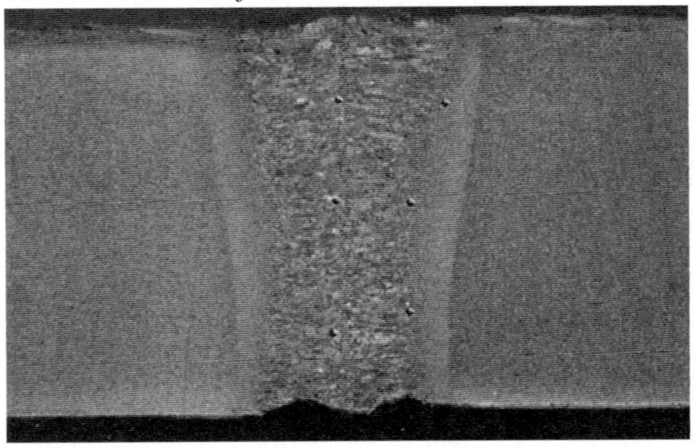

Weld width
2.15mm

1.40mm

1.51mm

Fig 9. No oscillation (ordinary laser weld)

Imperfections; root concavity 0.19mm (level B. ≤ 0.1 t = 0.4mm)

Tab. 3. HV1 hardness from non-wobbled weld.

	BM	HAZ	WM
middle	169.46 166.81 166.87	215.4	249.66
surface		205.36	252.33
root		219.33	258.56
average	168	213	254

5.1.2. Test results from linear wobbled weld.

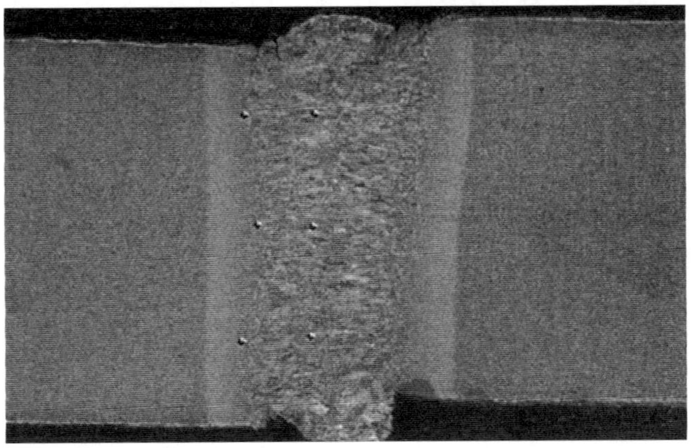

Weld width
1.80mm

1.50mm

1.47mm

Fig. 10. Linear oscillation

Imperfections; cracks on weld surface (not permitted), linear misalignment 0.3mm (level B, ≤ 0.1 t = 0.4mm), excessive penetration 0.43mm (level B, ≤ 0.2mm + 0.15 t = 0.8mm), incorrect weld toe (<90° EN ISO 5817) [7]

Tab. 4. HV1 hardness from linear wobbled weld.

	BM	HAZ	WM
middle	186.03 185.58 185.55	222.79	255.46
surface		206.26	224.33
root		207.40	223.24
average	186	212	234

5.1.3. Test results from infinity wobbled weld.

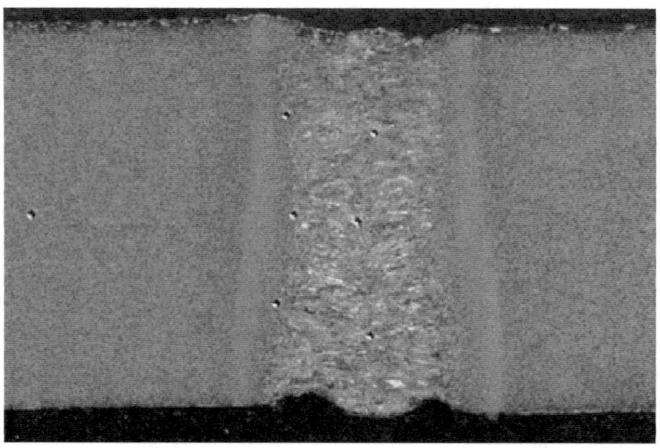

Weld width
1.74mm

1.47mm

1.99mm

Fig 11. Infinity oscillation

Imperfections; root concavity 0.18mm (level B, ≤ 0.1 t = 0.4mm)

Tab. 5. HV1 hardness from infinity wobbled weld.

	BM	HAZ	WM
middle	184.14 179.67 173.6	230.19	253.93
surface		233.63	266.67
root		223.33	252.93
average	179	229	258

IOP Conf. Series: Materials Science and Engineering 1135 (2021) 012021 doi:10.1088/1757-899X/1135/1/012021

5.1.4. Test results from circle wobbled weld.

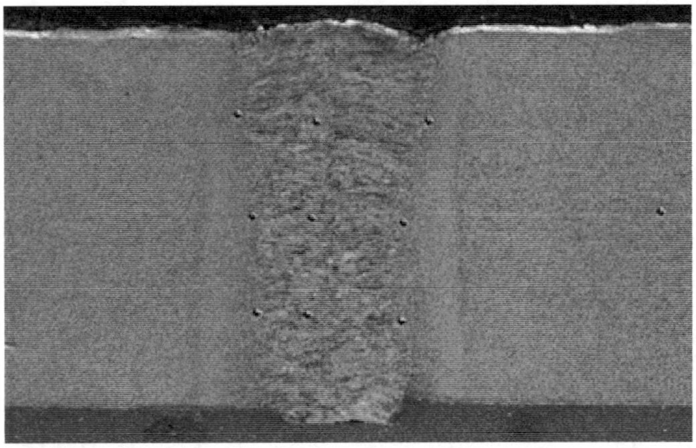

Fig 12. Circle oscillation

Imperfections; crack on weld root (not permitted), excessive penetration 0.25mm (level B, ≤ 0.2mm + 0.15 t = 0.8mm), incompletely filled groove 0.17mm (level B, ≤ 0.1 t = 0.4mm).

Tab. 6. HV1 hardness from circle wobbled weld.

	BM	HAZ1	HAZ2	WM
middle	175.59 179.22 166.27	209.76	210.08	246.08
surface		197.73	202.93	256.02
root		204.3	211.44	248.76
average	174	204	208	250

6. Conclusions

In this study tested wobble features did not affect to the laser weld hardness or weld width. Average weld metal (WM) hardness value of all tests was 249HV1 (-15HV1 to +9HV1) and heat affected zone (HAZ) average value was 213HV1 (-9HV1 to +16HV1). So significant softening or any change of welded material hardness was not discovered. Probably the reason for this is, that base material (S355MC) has a low carbon content and a low carbon equivalent, which makes it less hardenable even with low heat input laser welding. Used laser welding energy input was 0.163kJ/mm.

Weld width values does not show remarkable difference between any wobble mode or even compared to the non-wobbled weld. Reason for this could be that very high focal spot motion speeds, mentioned in chapter 4.2, during oscillation does not give neither enough heat energy nor interaction time to the weld sidewalls further. This makes possible to keep keyhole and melt pool open in the middle of the weld, but not at the sidewalls, because total energy input is always the same in these tests.

Unfortunately some of the imperfections, like linear misalignment especially in linear wobbled weld (Fig. 10), were caused by the imprecise test weld set up. This unintended misalignment was bigger than in other test welds, but size of even this misalignment could actually be a typical case also in production welding. It can been seen from figure 10, that at least these tested linear wobble parameters did not give smooth root side as a result.

IOP Conf. Series: Materials Science and Engineering 1135 (2021) 012021 doi:10.1088/1757-899X/1135/1/012021

Also crack formation in some welds was surprise. It was thought that wobble motion shuffles weld melt pool, and slow down rapid cooling of the laser weld.

Tested wobble parameters did not cause sagging of the weld joint. Mainly because the air gap between the test plates was zero.

At the end it has to be said, that lot of further studies and experiments are needed to achieve all the advantages to the production laser welding what the beam manipulation can offer.

ORCID IDs

T. Kankala https://orcid.org/0000-0001-8432-1765
A. Salminen https://orcid.org/0000-0002-9071-3682

References

[1] Salminen, A., Westin, E., Effect of Heat Input Modification on Duplex stainless Weld Quality. Proc. 29th Int. Conf. on Applications of Lasers and Electro Optics ICALEO 10, Anaheim, CA, USA, September 26-30, Laser Institute of America. Paper 1903. p. 668-675.
[2] EN 10149-2 Hot rolled flat products made of high yield strength steels for cold forming Technical delivery conditions for thermomechanically rolled steels.
[3] Test Report. SSAB Europe Oy. Testing and Inspection, Raahe, Finland (2020)
[4] EN ISO 6947 Welding and allied processes — Welding positions.
[5] EN ISO 22826 Destructive tests on welds in metallic materials — Hardness testing of narrow joints welded by laser and electron beam (Vickers and Knoop hardness tests)
[6] EN ISO 13919-1 Electron and laser-beam welded joints — Requirements and recommendations on quality levels for imperfections — Part 1: Steel, nickel, titanium and their alloys
[7] EN ISO 5817 Welding — Fusion-welded joints in steel, nickel, titanium and their alloys (beam welding excluded) — Quality levels for imperfections

doi:10.1088/1757-899X/1135/1/012022

Melt pool forming a buttonhole in tailored blank welding with multiple laser spots

Alexander F. H. Kaplan[1], Stephanie M. Robertson[1], Joerg Volpp[1], Jan Frostevarg[1]

[1]Luleå University of Technology, Department of Engineering Sciences and Mathematics, Division of Product and Production Development, 971 87 Luleå, Sweden

alexander.kaplan@ltu.se

Abstract. Laser beam welding of tailored blank butt joints of different sheet thickness generates asymmetric melt pool conditions. By employing two, three or four tailored laser beams, additional options for shaping the melt pool conditions can be offered. As observed by high speed imaging, in most multi-spot cases a large stable buttonhole was generated, by the trailing laser beams asymmetrically towards the thinner sheet. Correspondingly, the ablation pressure from the multiple boiling fronts has generated a fast melt jet, particularly along the thicker sheet. In many cases the boiling front kept open to the keyhole rear. The buttonhole differs from the Catenoid-like shape reported earlier. The walls are steeper and the horizontal shape can be asymmetric. The melt pool can switch between different stable modes. Inclined arrangement of three beams enabled even two separate, parallel boiling fronts and melt jets, combining behind the opening. Despite the large buttonhole, sound welds were achieved. Solely for four equal laser beams, arranged as a square, a melt pool without buttonhole was generated. Provided the driving forces from the ablation pressure along with the melt flow are sufficiently explored and understood, new opportunities to optimize the welding process are available.

1. Introduction

In welding processes, the melt pool is the essential central domain determining the energy transfer as well as many aspects of the final quality of the welded joint. For given extension from heat conduction, the shape of the weld pool is governed by the melt flow resulting from the driving forces and the counteracting surface tension forces. In keyhole laser welding of thin sheets, a few authors [1-4] have reported the occurrence of a so-called buttonhole, a hole in the pool much larger than the keyhole that keeps open during the quasi-steady-state welding process because the hole boundary forms a Catenoid-like shape. The Laplace pressure contributions of the vertical and horizontal directions cancel each other out. The adequate Young-Laplace equation writes:

$$p_L = \frac{\gamma}{R_V} + \frac{\gamma}{R_H} \tag{1}$$

where p_L is the resulting surface pressure, γ the surface tension, R_V and R_H the surface curvatures in the vertical and horizontal coordinate, respectively. Non-circular elongated holes with a length-to-width aspect ratio of up to four were observed by high speed imaging of laser welding of thin (0.5 to 2.1 mm)

sheet aluminium [5,6] or mild steel [6], which has led to a calmer process for the mode of elongated holes, as was supported by calculations on the Laplace pressure.
In the most common keyhole laser welding processes, the laser beam forms a quasi-steady keyhole, either of full or partial penetration. The melt passes around it either horizontally or underneath. The convective heat transfer generates an elongated melt pool. The melt pool shape just at the solidification interface determines the resulting weld quality, which can include reinforcement, undercuts or underfill as well as imperfections like porosity, hot-cracking or lack of fusion. Underfill can be promoted by root drop-out or spatter. For deeper understanding of the melt pool flow, experimental high speed imaging observations, HSI, and computational fluid dynamics simulation, CFD [7,8], are suitable methods. Already early CFD simulations as well as semi-analytical models explained how the melt can be horizontally redirected around the keyhole, similar to a cylindrical obstacle, through the laser-induced ablation pressure from boiling at the leading keyhole front. This can even cause strong melt pool jets in the wake of the keyhole. It can be kept in mind that strongly accelerated melt is accompanied by a thinner cross section according to the conservation of mass, as well as by a drop in static pressure, from conservation of momentum – and vice versa when slowing down in the tail or against surface tension forces. In contrast to the self-regulating front wall, the rear wall of the keyhole results from less controlled mechanisms, which are less understood yet, like the forces from the vapour jet against surface tension [9], but also complex additional jets, as more recent sophisticated CFD-simulations show [7,8].

From high speed imaging of the keyhole front at 180 000 fps for fibre laser welding of stainless steel, waves were observed that travel down the front [10]. The melt speed, in the range of 4-18 m/s, increased linearly with the laser power. This confirms the likely occurrence of root spatter for full penetration keyhole welding as well as the favoured melt flow route underneath rather than around the keyhole [11], which can under certain circumstances like high welding speed lead to humping. Mathematical modelling of the absorption of a laser beam at a wavy keyhole front [12] indicated for near infrared lasers (NIR, here 800-1100 nm) the favoured amplification of surface waves, explained by the angle-dependent Fresnel-absorption, while the longer CO_2-laser wavelength (10.6 μm) theoretically does not support the amplification. For fibre laser processing (remote cutting) of aluminium, the authors recently observed that no regular waves were formed. However, for NIR-lasers and steel the presence of the waves down the keyhole front was meanwhile confirmed for many different conditions. Typically the melt pool at the top surface during keyhole laser welding looks quite similar at the root, while CFD-simulations [7] have shown that the tail can be much shorter inside the material, by lack of the convective surface flow to the rear side. Marangoni convection is often only a minor melt flow contribution during laser welding, although present. Instead, the ablation pressure used to be the main driving force.

Butt joints of two sheets with different thickness are usually joined during tailored blank welding, as semi-finished parts particularly for the automotive industry. Welding sheets of equal thickness but different steel grades is also denominated tailored blanks. This study focuses on different thickness, causing the interesting aspect of laterally asymmetric melt pool and keyhole conditions. In particular, the shape of the weld cap from the thicker to the thinner sheet can be optimized, normally placing the step on top while the bottom surfaces are levelled out. Although numerous studies on tailored blank laser welding (TBLW) have been published [13], the melt pool for these asymmetric conditions was hardly studied in depth, neither by HSI nor by CFD. For process monitoring [14], a few photographs are provided that indicate the asymmetric melt pool. The temperature field was recently calculated [15] by finite element analysis, FEA, for TBLW with a dual beam, rotated 45° and 90° to the welding direction.

Laser welding with two laser spots has been occasionally studied. CFD-simulation of the melt pool during tandem welding [16-18] showed for different spot distances how the second beam contributes through a step in the melt profile to a combined, elongated keyhole development. For large distances the keyholes are gradually separated through a melt wall. Certain keyhole locations can experience overheating and violent evaporation events, supporting a transient behaviour [18]. More than two laser beams were seldom studied. The seam cross section from laser keyhole welding was shaped for up to five beam spots lined up perpendicular to the welding direction [19]. Fabbro [11] has explained, via the weld cap appearance when stopped, how the arrangement of two or three (triangle) beams broaden and govern the melt flow and the pool shape. The influence of multiple beams (e.g. as a matrix) or of shaped beams (e.g. C-shape, for battery welding) generated by DOE (diffractive optical elements) on the heat flow conditions during conduction mode laser welding was studied and optimized by heat flow FEA [20], supported by thermal imaging.

Another option for beam shaping is provided by scanner optics, where sufficiently fast beam movement (e.g. circles, ellipses, etc.) can enable quasi-steady state conditions. One strategy studied was to oscillate the laser beam across a filler wire, fed in leading position. This has led to the formation of stable buttonholes in the melt pool, observed by HSI. While the buttonhole mode led to even smoother weld cap surfaces than by the keyhole mode, in the transition regime between keyhole and buttonhole welding the buttonhole periodically detached into larger pinholes in the solidified joint [2]. These buttonholes do not appear as Catenoids but have steeper vertical walls with sharper transitions to the top along with a rounded horizontal shape. For the same case, the buttonhole occasionally formed even during CFD-simulation of the weld pool, including pinhole detachment under certain conditions [3,4].

When the authors carried out a wider study on the influence of multi-spot laser beam patterns on the melt pool during welding of duplex steel, from high speed imaging the formation of buttonholes was noticed, for most of the tailored blank cases. Since these aspects and their combinations were yet hardly studied by high speed imaging, selected results are presented and discussed in the following.

2. Methodology

Tailored blank laser welding was carried out for a butt joint configuration of a 1.8 mm thick steel sheet with a 1.1 mm thick sheet, as illustrated in figure 1(b), for zero gap. Six different multi-spot patterns were arranged according to figure 1(c) and the shown nomenclature. Apart from the different beam spots the experiments had identical parameters. It is worthwhile to mention that these six experiments were part of a wider study of 28 experiments, while all other 22 multi-spot welding experiments were carried out for butt joints of equal thickness, like in figure 1(a) (though also for 1.8 mm thickness), which can serve as reference. These experiments were part of a larger study of steel welding for electric vehicles, particularly based on squared steel tubes as frames. In the project, laser beam welding was compared to electric arc welding. The sheets had a length of 100 mm and a width of 70 mm. The edges were mechanically cut. Duplex steel was applied, grade DP800 for the 1.8 mm thick sheet and the stronger grade DP1000 for the thinner 1.1 mm sheet.

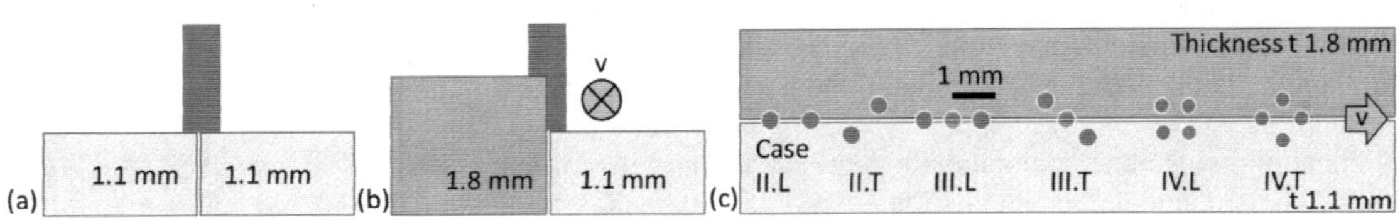

Figure 1. Butt joint configuration: (a) cross section for equal sheet thickness, (b) step joint (tailored blank, here presented results), (c) top view of arrangement of the different multi-spot cases.

As already illustrated in figure 1(c), six different multi-spot arrangement of a fibre laser beam were applied and compared. The multiple spots were formed through a special optics, named *quattroXX*, manufactured by the company *AdlOptica*, Germany. The optics is placed between the collimator and the focusing optics. The *quattroXX*-optics is based on beam splitters. By adjusting two rings that rotate the optical elements, it can be chosen to form one, two, three or four beam spots, which can be numerically simulated beforehand. In addition, within certain limits the distance between the spots, the power and the power density level per spot can be chosen. However, at least in one direction symmetry will be maintained. Another parameter is the arrangement of the beams relative to the travelling direction, i.e. rotation. Lateral off-set would be a further option, not studied here. Corresponding to figure 1(c), the measured spots are shown in figure 2(b), and 3D-visualized in figure 2(a) for Case IV. Case II has a power of 1500 W per spot and an average power density of 1.7 MW/cm^2 (spot diameter 333 μm). Case III: outer spots 1120 W, centre 760 W; Case IV: 840 W and 660 W (= 0.76 MW/cm^2).

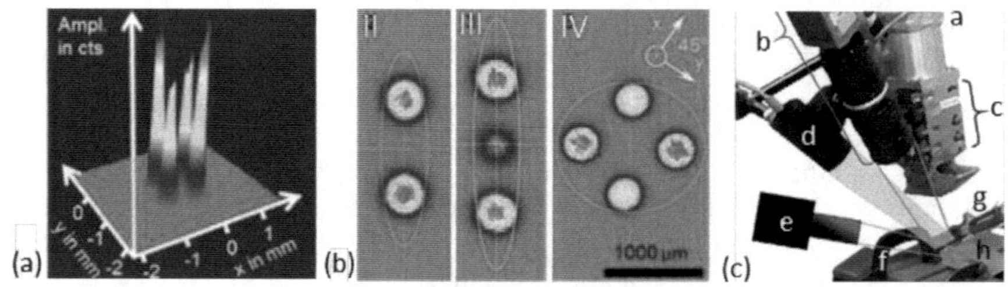

Figure 2. (a) measured lateral power density profile of a shaped four spot laser beam (Case IV); (b) beam profiles for the three beam pattern cases II,III,IV (same scale); (c) experimental laser welding set-up (a: *quattroXX* optics, b: high speed imaging, *Cam1*, c: focusing optics, d: illumination laser, e: *Cam 2*, f: fixtures, g: shielding gas nozzle, h: workpiece).

The choice of the beams and arrangements is based on assumptions for the generation of thermal cycles with a second peak for tandem spot arrangements, an asymmetric melt pool, or distribution of the heat between the thicker and thinner sheet, among other hypotheses for cooperation between the beams.

From the wider experiments of butt joint welding of equal thickness these six spot arrangements have mainly led to the formation of separate keyholes per beam, only occasionally combining. The welds were stable. A buttonhole has never formed. So it was interesting to continue the study for welding two sheets of different thickness, because of the added complexity of asymmetric conditions.

Any other process parameters were kept unchanged for the six beam spot cases. The experimental set-up is shown in figure 2(c). A 5 kW high power Yb:fibre laser was used (*IPG, YLS-5000-S2T-Y16*, wavelength 1070 nm), operated at 3 kW, continuous wave. The process fibre had a core diameter of 200 µm (focal length 250 mm, collimator focal length 150 mm), generating 333 µm-spots (focal plane at the upper surface level of the thicker sheet, Rayleigh length ±6 mm). After the processing fibre, the beam parameter product was 6 mm·mrad. The initially random polarization of the laser beam is expected to be partially altered towards linear polarization, by each of the four beam splitters of the *quattroXX* optics. This requires further investigation. The polarization can affect the Fresnel-absorption behaviour at the boiling front. The optics was mounted to an industrial 6-axis robot (*ABB, IRB 4400 M98A*). The laser beams were moving at a speed of 3 m/min while the clamped workpiece sheets were at rest. Argon was applied as shielding gas from the top (flow rate: 18 L/min), while the root was not shielded.

Two high speed cameras were employed, according to the set-up in figure 2(c), to simultaneously observe the melt pool top surface. Camera *Cam1* (*RedLake NR4-S2*) observed the process from a more vertical position (lateral inclination to the sheets 65°), at 4000 fps, *Cam2* (*Photron Fastcam mini UX100*) more from the side (lateral inclination 45°), at 25 000 fps. *Cam1* was mounted at the robot arm to follow the whole process (~2 seconds), which added some oscillations to the recording. In contrast, *Cam2* was employed in a stationary manner, which avoided shaking but enabled only recording of a half the process (~1 s). An illumination laser beam (diode laser, *Cavitar*, beam power 100 W, wavelength 808 nm) was applied for sufficient brightness while simultaneously blocking the processing laser light and most of the process emissions by a narrow band filter around the illumination laser wavelength. The resulting high speed images were studied and analysed with the respective camera software as well as with *Fiji/ImageJ* software.

3. Results and discussion

A survey of the main observed behaviour of the weld pool for the six beam cases is shown in figure 3. Different operating modes as a function of time during welding can be distinguished. The first period of recording (white bar in figure 3, about 200 ms), when the laser beam is switched on, followed by ramping up the welding speed, starts always with a clear beam-print through the simultaneous initiation of keyholes. Once the constant speed was reached, all processes still required a certain run-in time (black bar in figure 3, lasting 50-500 ms, depending on the beam case) until a stable mode of operation was achieved. Although showing interesting transient melt flow phenomena, these two initial periods will not be discussed below, similarly the final beam switch-off period (second white bar). The yellow bars correspond to the main stable modes of operations achieved, which will be presented and discussed in the following. While Case IV.T developed a typical stable melt pool for laser welding, although with four keyholes, for the other five cases a large stable buttonhole (BH) has formed. For three cases, II.T, III.T and IV.L, a certain change in the conditions has led to switching to a different stable mode (blue bar), which is also interesting to compare, although not deliberately planned, and will be discussed below. Otherwise all modes remained stable until either a change in conditions took place or the laser beam was turned off. Since metal melts usually react in the order of 10 ms to fluidmechanic changes, all operational modes that lasted longer than 100 ms in a similar behaviour and scale were regarded as stable. For some of the run-in periods it was a question of definition whether to already include part of their period into the stable mode, particularly for Case IV.T.

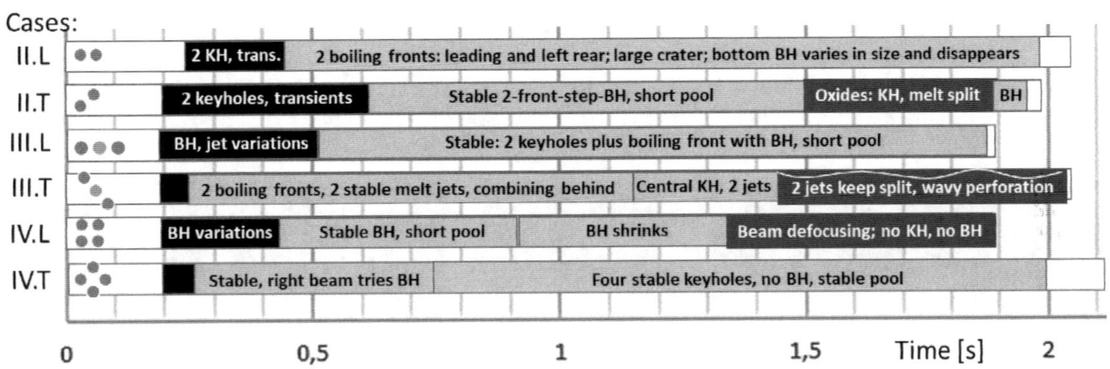

Figure 3. Operating modes of the melt pool as a function of time for the six laser beam cases; white: switch-on/-off periods; black: transient run-in phases; yellow: stable operating modes; blue: different modes through parameter disturbances. (KH: keyhole, BH: buttonhole).

Figure 4 shows typical weld cross sections. For Case IV.T, the common weld pool without buttonhole, a suitable transmission from the thicker to the thinner sheet was achieved, despite some vertical edge mismatch. The usual narrowing of a weld towards the root can be seen. For Case II.T, as one example of welding with a stable buttonhole, the cap became much narrower (top: 45%) and the root is 40% wider than the cap, which will be discussed below. Undercuts have formed for this case. Narrowing of the weld cap can be desirable and hence one motivation for welding with a buttonhole.

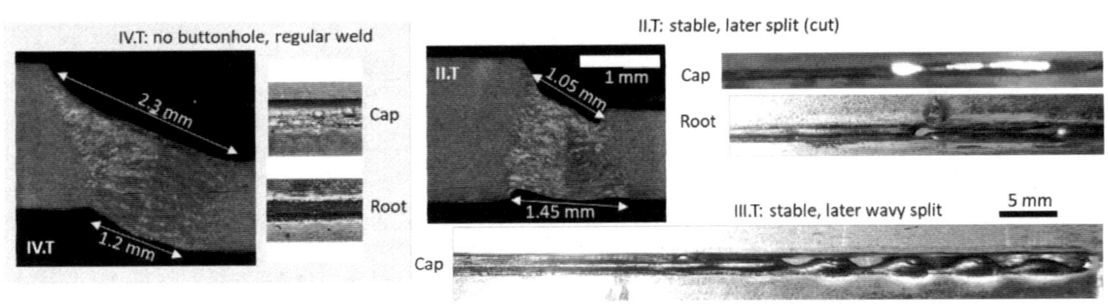

Figure 4. Typical cross section of a welded joint for a case with buttonhole, here II.T, where a more narrow weld seam was generated, while wider for the regular weld pool for Case IV.T. Weld surface appearance (cap and root): Case IV.T: no BH, regular weld, wider. Case II.T: stable weld via buttonhole, fails later (cut). Case III.T: stable weld via buttonhole, fails later (wavy perforation).

Figure 4 also displays examples of resulting weld surface appearances. All welds had a stable appearance throughout, except the final phases of the three cases mentioned where the conditions have changed. For Case II.T it can be seen that the stable weld suddenly changes to a perforated, interrupted track. The mode change was triggered by a short pollution disturbance at the surface. In a short final period the process changed back to a stable buttonhole, hence a stable weld appearance again. For Case III.T the stable weld changed to wavy periodic interruptions and holes. From HSI it became obvious that the boiling fronts became weaker and finally disappeared, which indicates that the beam focus drifted away, or the laser power decayed. Note that the left hand regular weld surface appearances for Cases II.T and III.T are representative for all welds. Case IV.T shows a stable weld when no buttonhole was formed. This weld cap became wider. In most cases the root showed a slight reinforcement, while exceptionally a slight undercut or underfill has formed.

In the following, based on representative high speed images, the six beam cases will be presented, discussed and compared. A scale can be provided horizontally but measurement is limited due to the 3D complex melt surfaces, including the inclined camera views. The 0.7 mm edge step also serves as scale.

Case II.L - Two beam spots, longitudinal (tandem)

First, high speed images for the tandem case II.L are presented in figure 5.

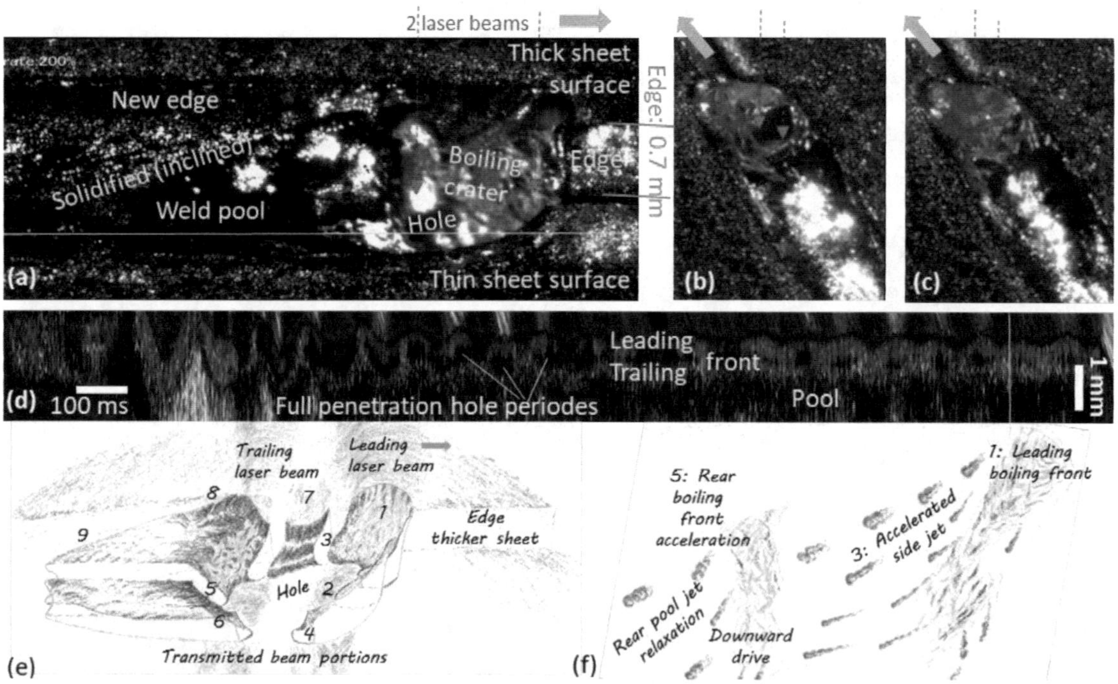

Figure 5. Case II.L, TBLW with two laser beams, longitudinal (tandem) beam arrangement; (a) high speed image of the melt pool, top camera view (Cam1), (b) side camera view (Cam2), showing a hole, (c) Cam 2, no hole; (d) streak image extracted from (a) (line); (e) sketch of the melt pool (1: leading boiling front at the thick sheet, 2: at the thin sheet, 3,4: corresponding fast, thin melt flow channels to left and right, 5,6: slowing down and recombining of the melt, 7: first new edge, 8: second new edge formed, 9: pool solidification); (f) left (thick sheet) melt flow field from first to second boiling front.

As illustrated, the leading beam generates a boiling front, both, on the thick and thin sheet, accelerating melt to the rear, via the sides and underneath. The trailing beam would not be needed. It experiences already an environment where the melt was moved to the side channels. However, the trailing beam generates a second boiling front, now against the travelling direction, which is unusual for laser keyhole welding. It processes melt downwards and determines the rear shape of the large hole, followed by the bigger melt pool tail. This better control of the keyhole rear side can be an advantage, to calm down the melt pool, avoiding defects.

The appearance of holes through the crater melt is fluctuating (for the limited angle of the two cameras), see figures 5(b),(c). The kind of hole can be a large buttonhole, or either of the two beams drills a smaller hole at the crater bottom. In particular, the trailing boiling front accelerates material downwards, for redistribution. However, at the top, the rim of the big crater keeps in a very stable shape, which can be well seen in the streak image, figure 5(d) (the low frequency oscillations induced by the robot movement need to be subtracted/ignored). The trailing beam ensures to keep the boiling crater big and its rear wall stable. The intermittent formation of full penetration holes in this big crater can be seen in the streak image, too.

Case II.T - Two beam spots, tilted

Figure 6(a) shows a high speed image of a buttonhole forming, for Case II.T. As illustrated, also in a sketch, figure 6(c), the leading laser beam to the left generates a deep boiling front at the thicker sheet which also transmits to the thinner sheet. In absence of significant underfill and spatter, the wide hole is evidence of redirection of the melt to the sides. Since the melt film is very thin at the top, for balance of mass the melt has either to achieve very high speed or the film becomes wider at the bottom, or both.

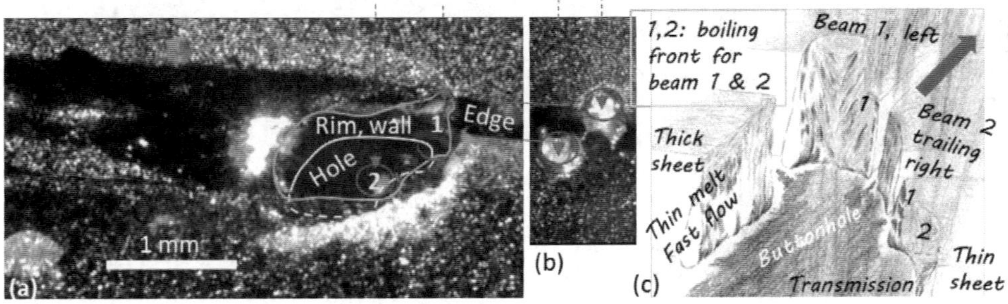

Figure 6. Case II.T, two laser beams tilted; (a) HSI, Cam1; (b) HSI during start, emphasizing the two laser spots; (c) sketch of the left leading (1) and right trailing (2) boiling front.

The root width of the welded joints is slightly wider than the cap (see figure 4), indicating some channel widening, while simultaneously faster melt movement was observed. The significant length of the hole indicates that the melt has no urge to slow down. Probably due to high momentum, the two streams recombine far behind the boiling front. While the side wall appears steep and rather flat, far from a surface tension equalized Catenoid, refer to Eq. (1), the rear shape of the buttonhole tends more to the latter. Compared to the left beam just succeeding through the thicker sheet, the right, trailing laser beam has an excess of beam power for the thinner sheet. This easily enables a second boiling front that widens to a big hole, driving the melt to a channel to the right, while a significant portion of the laser beam can be expected to pass through the hole. The tilted arrangement permits a stable, wide buttonhole.

As indicated in figures 3 and 4, towards the end the laser beams were hitting some oxide flakes at the surface. Within milliseconds the mode of operation has changed, to a stable left keyhole, a right boiling front and splitting of the melt, causing a cut. After 20 mm, the stable buttonhole mode restarted.

Case III.L - Three beam spots, longitudinal

When three spots are arranged in longitudinal direction, along the edge, see figure 7, the first stronger spot and the second weaker spot generate keyholes, which occasionally combine. While those two beams and keyholes start creating a pool similar to single beam welding (though here asymmetric, TBLW), the third (again stronger) laser beam (Case III.L) generates a boiling front and a stable long buttonhole.

As HSI shows and as illustrated, a clear flat wall forms in the thicker sheet, similar to Case II.T, figure 6, but even more stable and pronounced. This corresponds again to a thin, fast melt film as a channel, while the buttonhole closes with a more curved section into the wider pool. The streak image (a camera array recording plotted over time; here selected the horizontal yellow line in Fig. 7(b); for further explanation see [10]), selected across the two keyholes and the buttonhole, confirms that the buttonhole and its boundaries keep very stable (again disregarding the robot oscillations). The two keyholes seem to carry out a preparatory melt formation while the third beam, the open boiling front, easily sustains a big open hole while fuelling the two side channels with accelerated melt. Although the III.L-beams do not widen the boiling front, the third beam was sufficient for a very stable buttonhole.

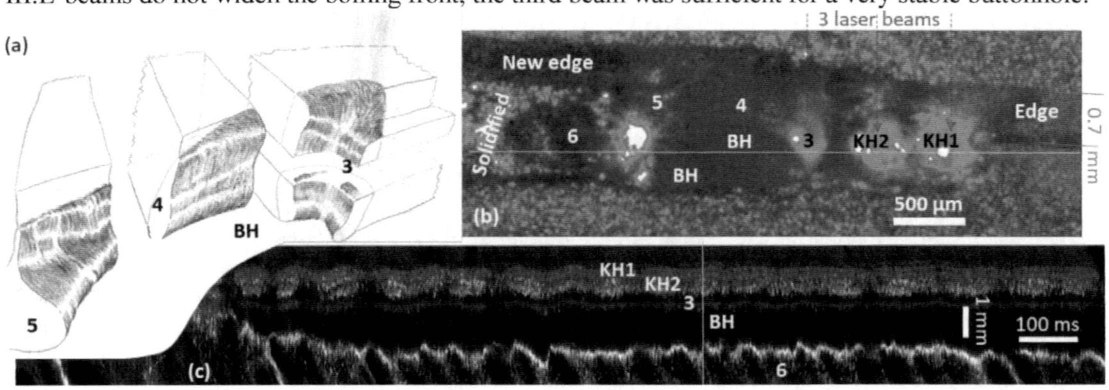

Figure 7. Case III.L; (a) sketch of the boiling front (marked 3), accelerating the melt to a thin side wall (4), slowing down and widening behind (5) to a Catenoid-like surface; (b) HSI, Cam1, showing the two leading keyholes (KH1, KH2) the boiling front (3), the buttonhole (BH) and the pool (6); (c) streak image (array recording over time, see [10]) extracted from the yellow line location in (b).

Case III.T - Three beam spots, tilted

The melt behaved very differently when the three beams were tilted, Case III.T, see figure 8. The leading beam (1) on the thin sheet and the trailing beam (3) on the thick sheet generated two separate boiling fronts and melt jets, which only recombined at some distance behind to a common melt pool. The rear melt part resembles the (torus-like) Catenoid buttonhole curvature, while the two front melt parts are separated (disregarding the solid connection). Initially, figure 8(b), the weaker central beam (2) hardly generated boiling but at a later stage it formed a continuous keyhole, connecting to the leading boiling front (1), figures 8(a),(c), but not to the trailing one (3). At the final stage of the run, the two parts became only periodically welded together, followed by regular wavy gaps, see figure 4. Basically such concept of separately controlled melt jets, which combine behind, can have interesting advantages and applications. Here it was apparently vulnerable to some changing conditions (e.g. lateral distortion of the sheets, resulting from heat accumulation during the welding) that kept the jets separated, see figure 8(d). Note that the left jet was rather thin and did not fully penetrate the thick sheet, which however was succeeded behind, when combining with the other jet.

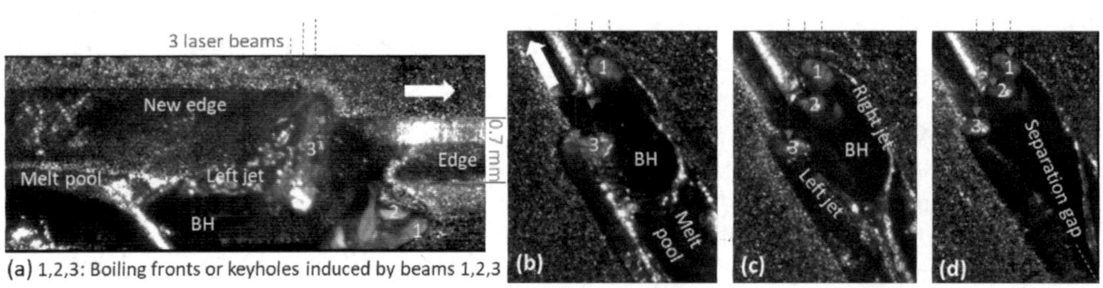

Figure 8. Case III.T, three beams, tilted, HSI; (a) Cam1 and (b) Cam2, two boiling fronts (1,3) and melt jets, forming a buttonhole-like geometry; (c) now also the weaker central beam spot (2) generates a keyhole; (d) later the jets remain separated, causing an intermitting gap instead of a weld seam.

Case IV.L - Four beam spots, longitudinal

According to figures 1,2, four spots were arranged as square, with almost equal power, here in-line with the travelling direction, Case IV.L. High speed images of the weld pool in figure 9 show a stable buttonhole that is quite similar to the Cases II.T and III.L. The beams share the boiling fronts that they develop, as illustrated in figure 9(c). At the left hand side, the leading beam carries out the main melting of the thick sheet into depth, while the trailing beam contributes to a smaller extent, probably accompanied by significant beam transmission through the buttonhole. The melt flows also down to the right hand side where the leading right beam again performs most of the melting, though now at the thinner sheet. There again some excess beam power can transmit through and the buttonhole is easily kept open, a mechanism similar to Case II.T. The boiling front generated by the trailing beam at the thin sheet can hardly be distinguished from the leading beam and again most likely contributes to a smaller extent to the melting and melt acceleration, while part of the beam is transmitted. The four beams ensured a stable large buttonhole, probably with much excess of power. As indicated in figure 3, after a while the buttonhole was shrinking in length but remained stable. Towards the final stage, it seemed as if the beams became weaker, either by some fading of the focal plane position or by a power decay, in an undeliberate but interesting manner. One reason could be heating or pollution of an optical element. Finally, probably because the spot power densities dropped below the boiling threshold, all four keyholes and the buttonhole disappeared and conduction mode welding took place.

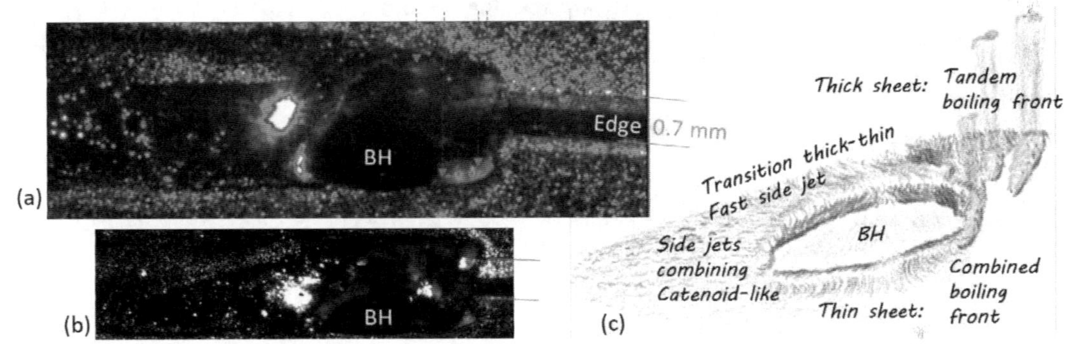

Figure 9. Square of four beams, in-line, Case IV.L; (a),(b) HSI of the weld pool for Cam1 and Cam2, respectively; (c) sketch of the melt pool.

Case IV.T - Four beam spots, tilted

Finally, images of the melt pool for four beams, a square 45° rotated, are shown in figure 10. Notably, this is the only of the six beam cases studied here that did not form a buttonhole. The four beam spots clearly generated four keyholes, although occasionally two of them combined, accompanied by a stable melt pool, appearing to what is common for single spot laser keyhole welding. The weld became much wider than for all the buttonhole welds, and narrowed to the bottom, see figure 4.

Figure 10. Case IV.T; four keyholes and a common weld pool, no buttonhole; (a) Cam2, (b) Cam1.

During the first 500 ms (figure 3), it was well observed that the right beam attempts (enlarged, penetrating) to form a buttonhole, as highlighted in figure 10. While the other three beams partially use their energy for the thicker sheet, this beam has excess power for the thinner sheet, though apparently not enough in common with the other beams to form a buttonhole. The more beams are generated, the less power per sub-beam, for same total power. Later the conditions have stabilized such that these attempts stopped throughout.

4. Concluding remarks

- For multi-spot laser beam patterns, except for Case IV.T, a stable buttonhole always formed for tailored blank laser welding, probably because of the asymmetric conditions, while a buttonhole was never observed for the symmetry by same thickness plates.
- The appearance of the buttonholes differs from earlier Catenoid-like shapes, which can be caused by the here observed fast melt jets and correspondingly different fluidmechanic melt pool conditions.
- Multi-spot patterns can generate buttonholes with variants of open boiling fronts, instead of keyholes.
- The length of the buttonhole can vary; disturbances, like oxide flakes at the surface, can switch a stable buttonhole into a different stable melt shape-mode, even keeping two melt streams separated.
- Although the multiple beams and the generated buttonhole cannot yet be assessed through pros and cons, the concept offers different mechanisms for welding, particularly for the asymmetric complexity of tailored blanks with different thickness; some results indicate possible advantages.

Acknowledgements
The research was funded by the European Commission, Research Foundation for Coal and Steel, project *Steel S4EV*, no. 800726 and by Sweden's Innovation Agency, VINNOVA, project *ACCEL*, 2019-00781.

A Kaplan https://orcid.org/0000-0002-3569-6795
S Robertson https://orcid.org/0000-0002-8298-292X
J Volpp https://orcid.org/0000-0003-0194-9018
J Frostevarg https://orcid.org/0000-0003-4265-1541

References
[1] Haglund P, I Eriksson, J Powell and A F H Kaplan 2013 *J Laser Appl* **25** 031501
[2] Schultz V 2019 *J Manuf Mat Proc* **3** 17
[3] Cho W and P Woizeschke 2021 *Intl J Heat Mass Trans* **164** 120623
[4] Cho W and P Woizeschke 2020 *Intl J Heat Mass Trans* **152** 119528
[5] Aalderink BJ, RGKM Aarts, DF de Lange and J Meijer 2007 *J Laser Appl* **19** 245
[6] Aalderink BJ, DF de Lange, RGKM Arts and J Meijer 2007 *J Phys D: Appl Phys* **40** 5388
[7] Zhang LJ, JX Zhang, A Gumenyuk, M Rethmeier and SJ Na 2014 *J Mat Proc Techn* **214** 1710
[8] Otto A, RG Vazquez, U Hartel and S Mosbah 2018 *Procedia CIRP* **74** 691
[9] Weberpals J and F Dausinger 2008 *Proceedings ICALEO 27* (LIA, Orlando, FL, USA) **27** 704
[10] Eriksson I, J Powell, AFH Kaplan 2011 *Sci Techn Weld Join* **16** 636
[11] Fabbro R, M Hamadou, F Coste 2004 *J Laser Appl* **16** 16
[12] Kaplan AFH 2015 *Appl Surf Sci* **328** 229
[13] Merklein M, M Johannes, M Lechner and A Kuppert 2014 *J Mat Proc Techn* **214** 151
[14] Kratzsch C, P Abels, S Kaierle, R Poprawe and W Schulz 2000 *Proceedings SPIE* **3888** 472
[15] Zhang X, L Li, Y Chen, X Zhu and S Ji 2019 *Metals* **9** 135
[16] Pang S, W Chen, J Zhou and D Liao 2015 *J Mat Proc Tech* **217** 131
[17] Li L, H Xia, G Ma and G Peng 2018 *J Mat Proc Tech* **255** 841
[18] Chen X, X Zhang, S Pang, R Hu and J Xiao 2018 *Opt Lasers Eng* **100** 239
[19] Hansen K S 2015 *Multispot fiber laser welding* (PhD thesis, Aalborg University, Denmark)
[20] Sundqvist J, AFH Kaplan, L Shachaf and C Kong 2017 *J Mat Proc Techn* **247** 48

Topographical evaluation of laser generated surfaces using statistical analysis of surface-normal vector distributions

Rickard Olsson[1,2], John Powell[1], Jan Frostevarg[1], Alexander F.H. Kaplan[1]
[1]Department of Engineering Sciences and Mathematics, Luleå University of Technology, 971 87 Luleå, Sweden.
[2]Laser Nova AB, SE-831 48 Östersund, Sweden
rickard.olsson@lasernova.se

Abstract. Surfaces generated by Additive Manufacturing or laser texturing can involve the solidification of droplets of liquid, which can give rise to overhanging features on the solidified surface. Overhanging features add a layer of complexity to the surface topography and are undetectable by standard surface roughness measurement techniques such as profilometry. Such features are important because they can have a considerable effect on surface properties such as wettability. New techniques and algorithms are therefore required to analyse and quantify convoluted surfaces with overhanging (re-entrant) features. Earlier work by the authors introduced the concept of using X-ray micro-computed tomography (Micro-CT) to identify the directions of vectors normal to the surface at any point and thus indicate the presence or absence of overhanging features. This paper divides overhanging features into two types; simple and compound, and introduces new, size independent, analysis techniques which measure what proportion of each type is on the surface. Another extension of the analysis is the comparison of surface profiles taken in different directions in order to identify any surface roughness anisotropies.

1. Introduction

Additive manufacturing (AM) [1], laser surface texturing [2–4] and thermal spray coating, often involve the attachment of droplets to a solid substrate. These techniques can generate new types of technical surface which, as Godi et al. have pointed out, often cannot be characterised according to the current ISO standards [5]. For example, some AM surfaces include overhanging features (also referred to as re-entrant features) like those described in figure 1. Overhanging features of this sort cannot be measured by mechanical or optical surface roughness measuring tools.

One example of the useful application of overhanging features on technical surfaces is the field of medical implants.[6] The classification of implant surface structures is a field where e.g. Wennerberg and Albrektsson[7] have pointed out the need for a more uniform terminology and have created a set of guidelines for topographic evaluation of surface defining parameters such as form, waviness and roughness. Compared to a smooth metal surface, a nanoscale oxidized surface layer improves the deposition of the bone matrix hydroxylapatite, the adsorption of proteins, the adhesion of cells, and the overall rate of osseointegration. [8][9] Integration can be improved and accelerated even further if this type of surface is produced on a micro-rough surface involving overhangs. This is because the overhangs provide areas for excellent mechanical locking [10], which is reinforced by apatite formation in the surface irregularities.[11] Laser pulses can be used to create "splashy" surfaces involving suitable overhanging features[12,13] including attached droplets of 1-5 μm diameter.[6,14,15] These types of laser-treated surfaces provide a better topography for cell growth[16] than those produced by acid-etching or sandblasting.[17] Laser texturing can also optimise surface energy, which improves protein adsorption and cell attachment.[18] Wennerberg et. al. [19] in their study of bone response to different surface roughness, also pointed out the limitations of R_a and S_a as measures when comparing different textures.

X-ray computed tomography (X-ray CT) can be used for stress analysis of Additive Manufacturing (AM) surfaces which are covered with attached droplets.[20] In this work Kantzos et al. identified overhanging features and surface porosity. A thorough discussion about the classification of complex surfaces (periodic, non-periodic, etc.) can be found in "A Taxonomy for Texture Description and Identification" by Rao[21]. An investigation about how traditional roughness metrics (R_a,R_z etc.) for surfaces generated by AM are affected by factors like laser power, slice size, scanning speed etc. has been carried out by Calignano [22]. Senin et al.[23] highlight the difficulties of reliably quantifying the topography of AM generated surfaces and comment that results can even be 'heavily dependent on instrument make and model'.

Pagani et al. presented in 2017 [24] a powerful technique for defining surface texture parameters for free-form surfaces using a triangular mesh with parametrization of the surfaces and B-spline reconstruction using the software platform SurfStand. The dominant factors were identified using Principal Component Analysis (PCA). Their ideas were developed further in [25] which describes complex freeform surfaces (also involving normal vectors) as a tool for classifying complex overhangs. (In their paper complex overhangs are named "re-entrant features"). Newton et al. [26] have presented developments in the field of feature-based surface characterisation which are particularly suited to AM generated surfaces.

Earlier work by the present authors [3,27-28] introduced a new approach for the characterization of complex surfaces which enables a quantitative description of overhanging features. The technique uses computed micro-tomography to map the surface. From the tomography data cross sectional traces are generated, which include all the overhanging features (see figure 2).

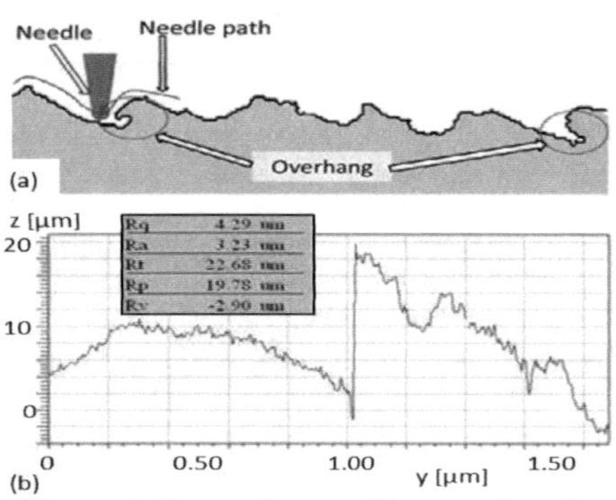

Figure 1. For surface profiles, overhanging features are invisible to traditional roughness measuring devices, such as (a) mechanical stylus profiling or (b) White Light Interferometry (WLI), measured here by a Wyko instrument.

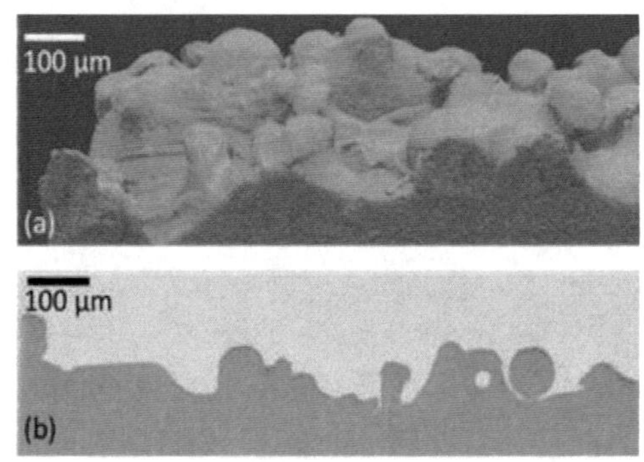

Figure 2. (a) Typical micro-tomography view of a rough surface created by additive manufacturing - note overhanging features, (b) a cross section taken from the tomography data.

These cross-sectional surface traces can have equally spaced surface-normal vectors allocated to them as shown in figure 3 (and for a simplified, artificial surface in figure 4). Statistical tools can then be employed to analyse the variation in surface-normal vector inclination for different types of surfaces. Zhang et al have investigated the usefulness of histogram representations of surface features [29] and Tang et al have explored the use of normal surface vector histograms as a method of surface classification [30]. The work involved recognition of objects and object classification in computer vision applications using information from a depth sensor. Similar techniques, using information from histograms of normal vectors for object classification in computer vision applications have been used by e.g. Watanabe [31] and Dalal [32]. However, this approach has not been used for the classification of engineering surfaces.

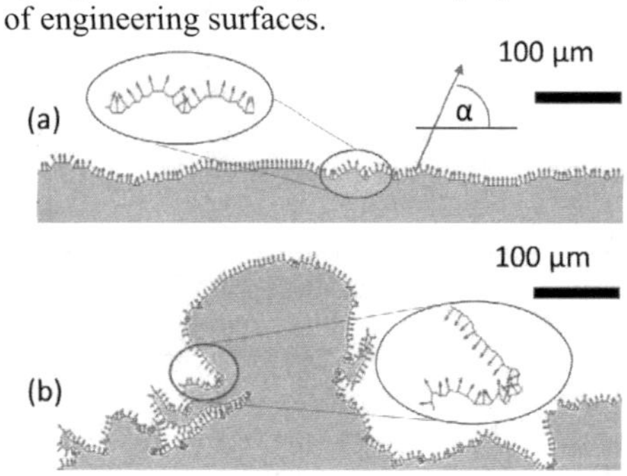

Figure 3. Surface profiles with equi-spaced surface-normal vectors: (a) surface normal vectors for mechanically roughened surfaces would have an inclination range of less than 180 degrees; (b) overhanging features introduce a much wider range of surface normal vectors.

For technical surfaces created by the accumulation of droplets, the distribution of the angles of surface-normal vectors can give quantitative and qualitative information about the presence or absence of overhanging features. This paper builds on the earlier work and presents, for the first time, analytical techniques for identifying two types of overhangs: simple and compound. Both types of overhangs can have a negative effect on wettability, but compound overhangs have a greater ability to trap air and thereby prevent local wetting. This work also presents a statistical tool which can be used for identifying cyclic patterns within the roughness of convoluted surfaces.

2. Methodology

Four different types of surfaces were created for this investigation. Titanium samples were prepared as follows:

- A grit blasted surface; grit blasted with Zirconium dioxide, Cerablast K125 (0-125) µm.
- A Laser BioHelix™ surface; a proprietary technique used for structuring the surface of medical implants.
- An Additive Manufactured (AM) surface; produced by additive manufacturing using 45-150 µm powder.
- An etched AM surface; The raw AM surface (as above) subsequently etched with a zinc chloride solution comprising ethyl alcohol (700 ml/l), isopropyl alcohol (300 ml/L), aluminum chloride (60 g/l), and zinc chloride (250 g/l) as described by [33] .

The AM-samples were manufactured in an ARCAM-A2 EBM machine using standard firmware and with a 50 µm layer thickness. The powder material was Ti6Al4V- Grade 5 with the particle distribution 45-150 µm supplied by Arcam EBM. Standard Arcam-A2 MultiBeam™ technology was used with two contours. The beam settings for the outer contour were: 340 mm/s and 4 mA, 40 spots, and for the inner contour 80 mm/s, 10 mA and 15 spots.

The BioHelix surface is a proprietary laser treated implant surface patented by Integrum AB, Gothenburg, Sweden. It is generated by exposing a Titanium surface to an intense bombardment of Nd:YAG Q-pulses of roughly 200 ns length and 1.3 W average power. (See Olsson [27,28]). All four samples were analysed with a Micro-CT instrument (Zeiss XRADIA 510 Versa) with a resolution of 0.5 µm in the X, Y and Z directions. Slices from the Micro-CT 3D topographies of the samples were analysed in a MATLAB-code using the steps described in in Table 1, which are illustrated (for a simplified artificial topography) in figure 4 (a). This analysis produced histograms of the distribution of the surface-normal vectors for each cross-section.

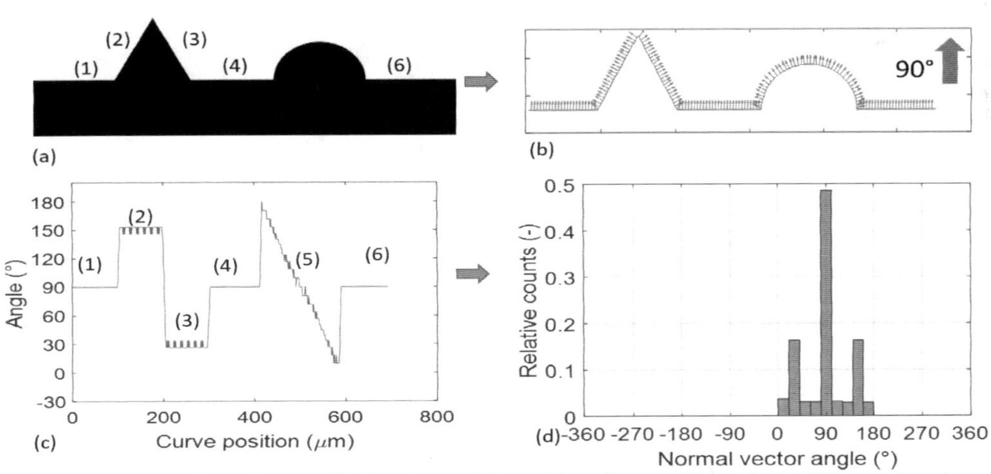

Figure 4. Example surface profile for the workflow: (a) surface profile with six different domains numbered (1)-(6), (b) calculation of equidistant surface vectors, (c) normal vector angle as a function of the curve position along the profile, (d) histogram of normal vector angles.

Table 1. The workflow applied in generating the surface-normal vector histograms.

Step	Task
a.	Roughly isolate the region of interest (ROI) from a Micro-CT image.
b.	Isolate the ROI in detail.
c.	Generate contour.
d.	At evenly spaced distances calculate normal vectors and normalize them to unit length.
e.	Calculate and plot the results and their distribution as a histogram (see figure 4).

In order to generate the contours (step c in table 1) an algorithm called "Snakes", or "Active Contour Tracking" was used [34]. This algorithm tracks contours in an image. The algorithm minimizes the distance between an object edge in an image and the generated curve, minimizes bending radii and keeps the contour and its first and second derivatives continuous.

The normal vectors were calculated as below:

The contour coordinates are stored in a 2D-vector $X = \begin{pmatrix} x_1 & .. & .. & x_N \\ y_1 & .. & .. & y_N \end{pmatrix}$,

and the tangent vectors TV are calculated as: $TV_i = X_{i+3} - X_{i-3}, i = 3..N-3.$ (1)

running in the positive X-direction, i.e., $$TV = \begin{pmatrix} tvx_1 & .. & .. & tvx_{N-6} \\ tvy_1 & .. & .. & tvy_{N-6} \end{pmatrix} \quad (2)$$

and the normal vectors are calculated as: $$NV = \begin{pmatrix} -tvy_1 & .. & .. & -tvy_{N-6} \\ tvx_1 & .. & .. & tvx_{N-6} \end{pmatrix} \quad (3)$$

A low pass filtering effect was introduced by the indices i-3, i+3 in equation (1) The low pass filter reduces edge effects and reduces an otherwise low angular resolution. Seven pixels of filter length gives an angular resolution of ~8 degrees and will at the same time reduce contour noise and was chosen as a good balance between noise sensitivity and angular resolution.

It should be noted that when a 2D slice is taken from a 3D surface, it will sometimes result in isolated features that are not connected to any other features. In these experiments such isolated features have been excluded. An obvious extension of this technique is to generate contour tracking in 3D and present the surface normal vectors in a spherical format, but this is outside the scope of the present paper.

Figs. 5(a) and (b) demonstrate two types of overhanging feature which can be created by additive manufacturing and other droplet deposition techniques: Simple and Compound overhangs. An overhanging feature can be generally described as any object which would cast a shadow if exposed to a light shining directly down upon it. Within this category there are sub-categories which could be important to surface phenomena such as wetting. For example, figure 5(a) shows a cross section of a simple overhanging feature which might be generated by a single molten droplet attaching itself to a substrate. Depending on the geometry of this type of overhang, and the viscosity/surface tension of the liquid involved, it might be possible to fully wet all of this surface. figure 5(b) on the other hand, shows a cross section of a more complex 'compound' overhang. In this case it would not be possible to entirely wet this surface because any encroaching liquid would find certain areas of the surface blocked by air bubbles trapped by the geometry of the compound overhang.

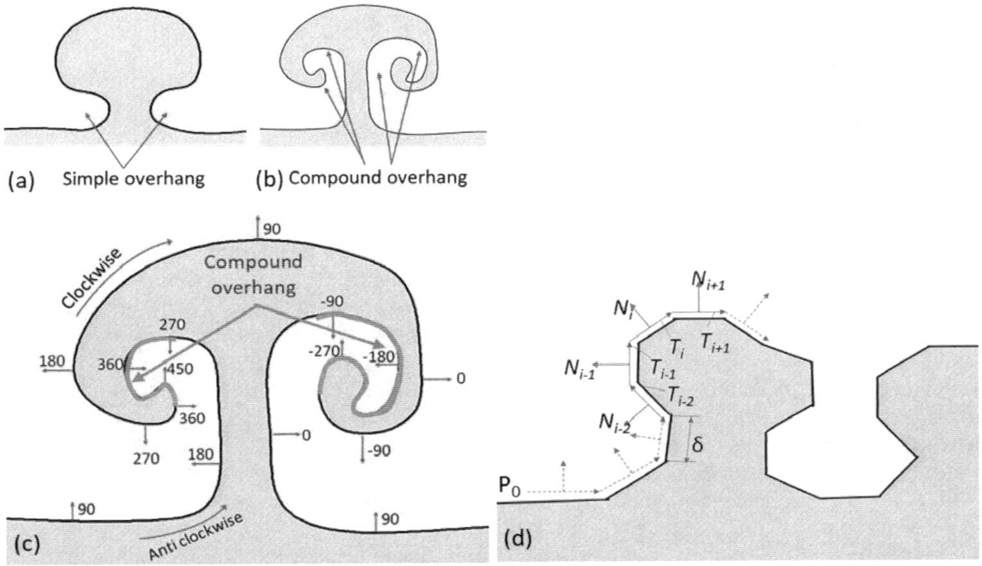

Figure 5. (a) A simple overhang surface profile, which might allow full surface wetting; (b) a 'compound' overhang surface profile, which would prevent full surface wetting due to trapped air bubbles; (c) surface normal vectors along the compound overhang surface profile. The presence of surface-normal vector angles greater than 270° degrees of or lower than -90° is indicative of a compound overhang. Angles in the ranges 0° to -90° (but no lower) or 180° to 270° (but no higher) indicate simple overhangs; (d) illustration of expressing the curve as a sum of linear sections T_i, each of equal length δ, with corresponding surface normal vectors N_i (starting at position P_0).

Compound overhangs cannot be adequately described or identified using a 0-360 degree scale for the direction of normal vectors which was employed in earlier work [28]. It is clear from figure 5(b) that, using a 360-degree scale, the shape of this compound overhang would, for example, give rise to two different types of area where the surface normal vector would be 270 degrees (i.e., vertically downwards). For this reason, we have developed the surface-normal vector angle analysis to include a range of angles greater than 360 degrees. figure 5(c) helps to provide an explanation of this approach. The assignment of angles to the surface normal vectors assumes that the surface contour is a line which is followed from its left-hand end to its right-hand end (i.e., in the positive x-direction). Surface normal vectors are calculated at equally spaced distances along this line. The angular value given to any vector is incrementally increased or decreased in comparison to the previous value depending on whether the normal vectors are progressing in an anticlockwise or clockwise direction. In this way two different numerical angular values will be given for the same surface-normal vector direction depending on what the previous vector value was. This allows us to differentiate between simple and compound overhangs.

Figure 5(c) contains only the ordinal vectors (which are not equally spaced) to illustrate this point. In this example the initial change in vector angle is from 90 (vertically upwards) to 450 (also vertically upwards). Between these two points the angle continuously increases to 450 because the progression of vector angle values was anticlockwise. At the point at which the value of 450 degrees is reached the angle progression changes to clockwise and all subsequent angular values are continuously reducing until we reach the next vertically upwards vector (at the top of the object) which now has an angular value of 90 degrees. Comparison between figure 5(a) and figure 5(c) makes it clear that the limits on surface normal vector achievable by a simple (non-compound) overhanging feature using this system would be a highest value of 270 degrees and a lowest value of -90 degrees. Therefore, angles outside this range are indicative of the presence of compound overhangs. Table 2 summarises the angular ranges of surface-normal vectors for surfaces with no overhangs, simple overhangs, and compound overhangs.

Table 2. Angular ranges of surface-normal vectors for different surface features.

Surface feature type	Surface-normal vector inclination range
No overhangs	0° to 180°
Simple overhangs	-90° to 0° and/or 180° to 270°
Compound overhangs	Below -90° and/or above 270°

Figure 6 Shows a typical result from the raw Additive Manufacturing sample investigated in this work as an example of how to interpret the surface-normal vector histograms which will be presented in the results section. The thresholds for simple and compound overhangs, see Table 2, are both exceeded in this example showing that both types of overhang are present on the surface. This is confirmed by observation of the black and white image of the cross section at the top of the figure.

One of the aims of this work is to develop tools which could be used to identify repeating patterns in the surface roughness which might be correlated to process parameters. For eample the Biohelix process produces a regular pattern of laser surface melts. If we consider the surface profile to be a signal then frequency analysis should reveal this repeating pattern. However, this simple approach is not possible in this case because, when overhangs are involved, the 'curve' of the surface trace sometimes includes two or more y values for the same x value, i.e. it has ambiguous coordinates. For a standard frequency (Fourier or Laplace) analysis it is important that the curve in question is single valued, i.e. unambiguous.

One way to get around this problem is not to express the profile curve not as y=f(x) but to divide it up into a succession of linear segments – as shown in figure 5(d).

If the profile curve is divided up into short straight lines each of equal length δ, pointing in the tangential direction (see figure 5(d)), the position at a point P_i on the curve, can then be expressed as:

$$\vec{P_i} = \vec{P_0} + \sum_{k=0}^{i} \delta \vec{T_k} \tag{4}$$

Calculating the normal vectors from the tangents and plotting normal vector angle vs. position now gives a single-valued set of data which is suitable for frequency analysis.

3. Results and Discussion

3.1 General comparison of sample surfaces

Figure 7 presents Micro-CT images of the four sample surfaces. Certain features of the surfaces are clear;

- The grit blasted surface has the smoothest surface.
- The BioHelix surface was produced as a pattern of parallel overlapping lines (like a ploughed field) Olsson.[27]

The AM surface (c) and its etched counterpart (d) have larger surface features than the BioHelix sample and the grit blasted sample.

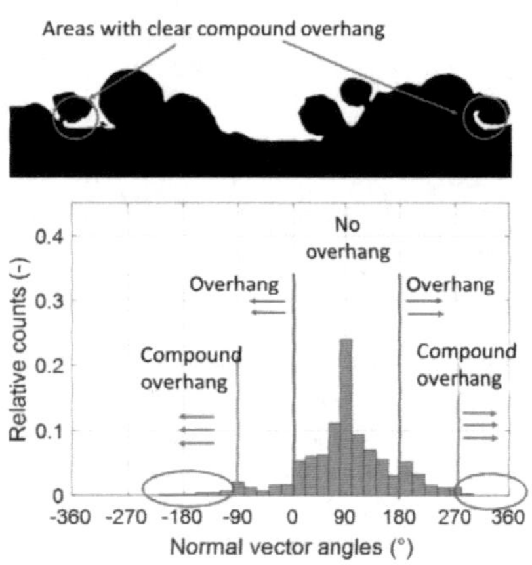

Figure 6. Surface profile (raw AM) with compound overhangs (upper) and its histogram (lower) with thresholds of the angle ranges for no, simple and compound overhang, respectively.

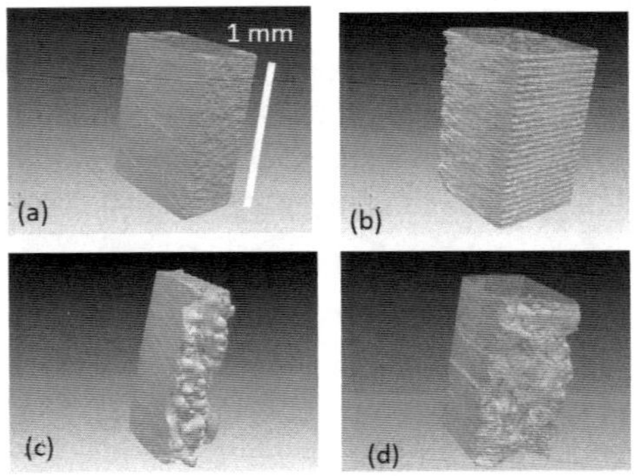

Figure 7. Micro-CT images of the samples: (a) grit blasted, (b) laser BioHelix, (c) raw AM, (d) etched AM.

Slices through the Micro-CT data can now be taken, giving cross sections from which the surface profile and its normal vector at any point can be identified.
Following the procedure laid out in table 1 and figure 4, figure 8 presents typical profile cross sections of the four samples together with the histograms of the surface-normal vector distributions.

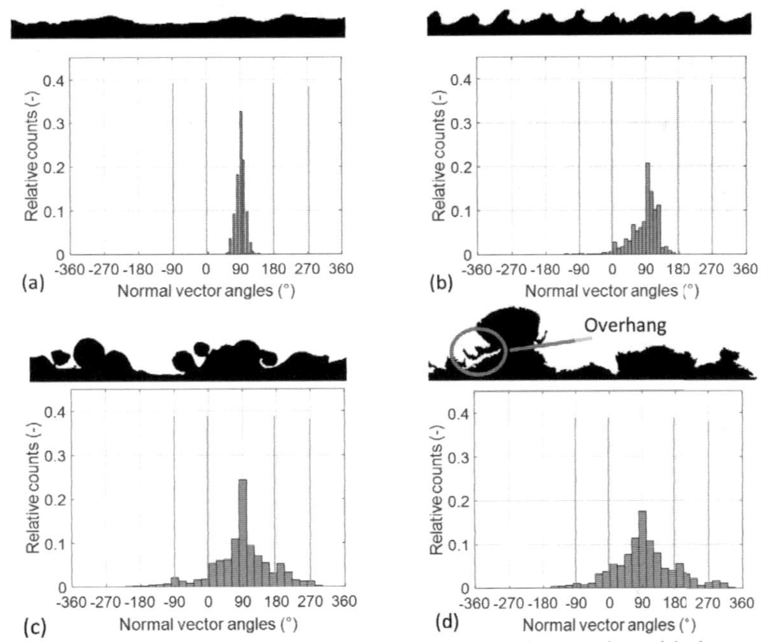

Figure 8. Typical surface profiles of the four samples together with the histograms of the surface-normal vector distributions: (a) grit blasted sample,(b) laser BioHelix sample, (c) raw AM sample, (d) etched AM sample.

The span of surface-normal vector inclinations and their standard deviations are presented in Table 3 together with the relevant R_a roughness measurements.
Figure 8(a) confirms the assumption that grit blasting does not create overhanging features in a surface. In this case the range of surface-normal vector inclinations is only 85 degrees; from 50 to 135, with a standard deviation of only 16 degrees. In the case of the BioHelix sample, figure 8(b), the range of vector angles is much wider; from -122 degrees to 160, with a standard deviation of 59 degrees. The existence of angles below -90 clearly indicates that there are simple and compound overhanging features on this surface. The presence of such features is obvious in the SEM image of this surface presented in figure 9. The BioHelix process produces a micro-rough surface which is optimised for osseointegration. The distinctive surface texture is achieved by producing a 'splashy' surface of parallel, overlapping rows of shallow melts that experience melt disruption during processing [27,28]. This gives a complex surface containing overhangs.

Table 3. Statistical surface properties (surface-normal angle range and roughness) of the four surfaces.

Surface	Standard deviation of the angles (°)	Angular range (°) of surface-normal vectors.	R_a (μm)
Grit-blasted	16	50 to 135	1.5
Laser BioHelix	59	-122 to 160	2
Raw AM	52	-190 to 270	24
Etched AM	79	-153 to 321	16

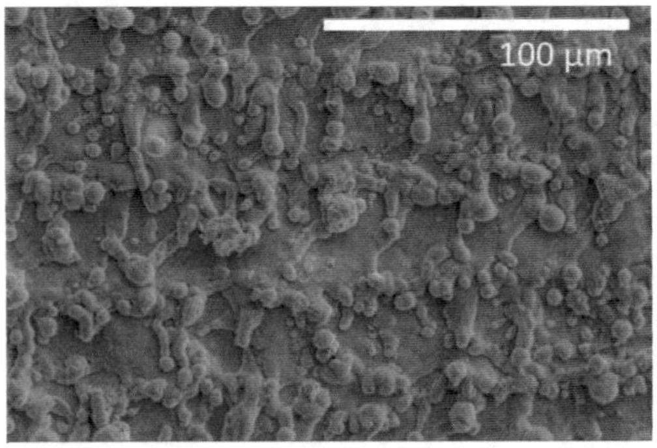

Figure 9. This higher magnification SEM image of the laser BioHelix sample surface reveals the presence of overhanging features.

As might be expected the Raw Additive Manufacturing surface figure 8(c) also has a considerable number of overhanging features, surface-normal vector angles from -190 to 270 degrees. The overhanging features of the AM surface are created by the impingement of overlapping laser-melted droplets. It is important to note that the results indicate that the level of overhanging features on a surface is not related to R_a roughness (see Table 3).

Figure 8(d) shows the strong evidence of overhanging features for the etched AM surface, with an inclination angle range of between -153 and 321 degrees. Further investigation of this surface (compare Figs. 8(c) and (d)) indicates that the zinc chloride etchant has corroded and undermined the overhanging features created by the AM droplet deposition process.

3.2 Directional roughness analysis

The next stage in our analysis of these surfaces was to investigate the differences between tomography cross sections taken in different directions across the sample surface. Slices from each sample were taken using the ParaView-software [35] at four different angles; 0°, 45°, 90° and 135°, as shown in figures 10(a)-(d).

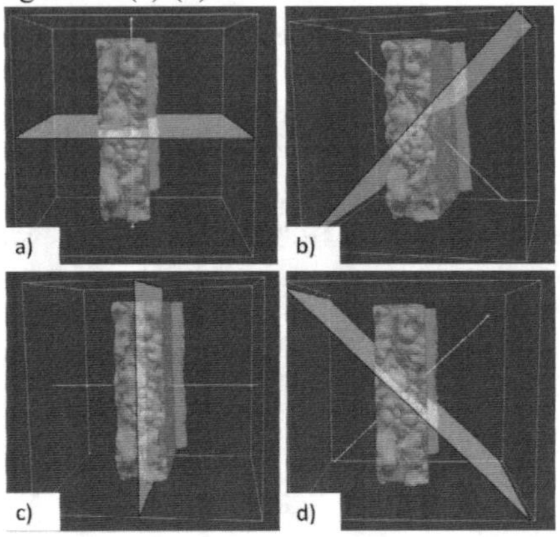

Figure 10. The four chosen slicing orientations to extract surface profiles from 3D-data measured by Micro CT: (a) 0°, (b) 45°, (c) 90°, (d) 135°.

IOP Conf. Series: Materials Science and Engineering 1135 (2021) 012023 doi:10.1088/1757-899X/1135/1/012023

The sliced results for the four directions in each of the four samples are presented in Figs 11-14.

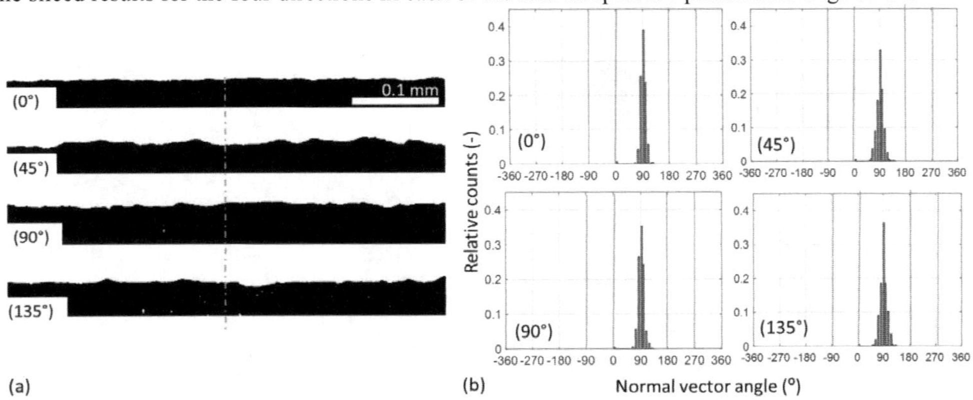

Figure 11. Grit blasted sample; (a) surface profiles and (b) normal vector angle distribution for the four slicing angle orientations; the angle limits for simple and compound overhang are highlighted.

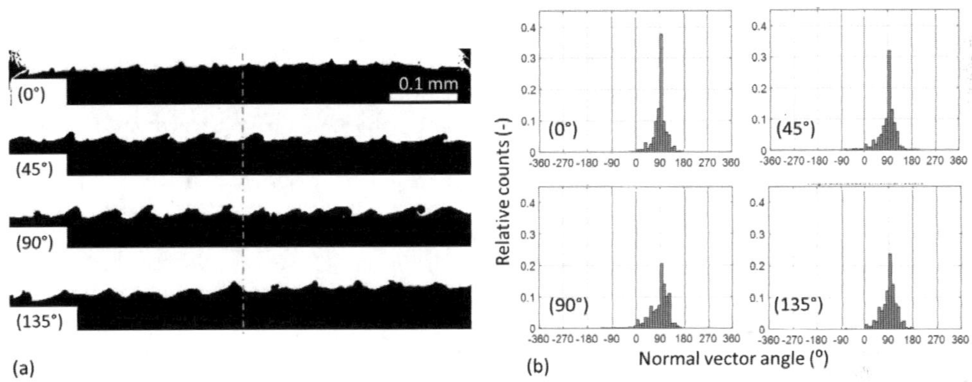

Figure 12. BioHelix sample; (a) surface profiles and (b) normal vector angle distribution for the four slicing angle orientations; the angle limits for simple and compound overhang are highlighted.

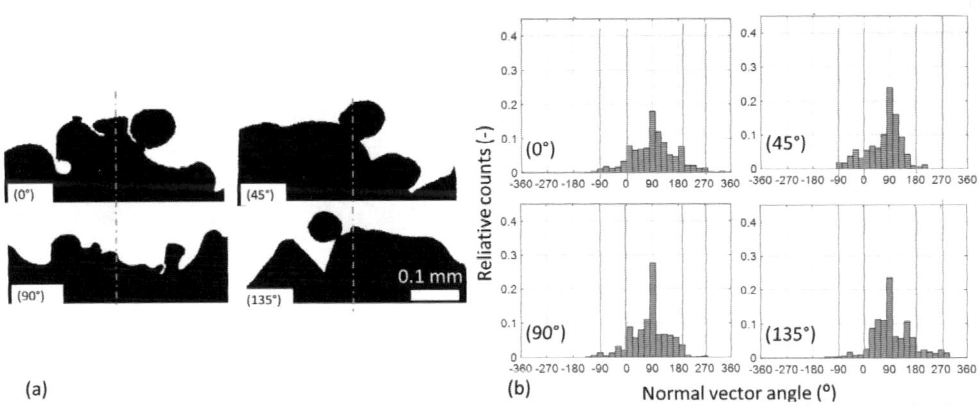

Figure 13. Raw AM sample; (a) surface profiles and (b) normal vector angle distribution for the four slicing angle orientations; the angle limits for simple and compound overhang are highlighted.

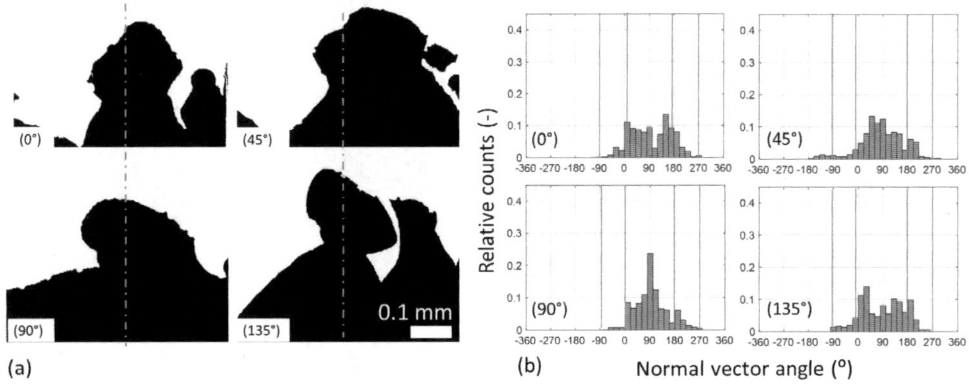

Figure 14. Etched AM sample; (a) surface profiles and (b) normal vector angle distribution for the four slicing angle orientations; the angle limits for simple and compound overhang are highlighted.

Figure 11 shows no noticable trends between the cross section directions, with a low level of roughness and low levels of inclination of the surface normal vectors. There are no vector angles below zero or above 180 and so no overhanging features of any sort are present. This is the expected result from a random roughness grit blasted surface. Figs. 13 and 14 also show minimal evidence of differences in profile or normal vector distribution between the four different sampling directions. For these samples there is evidence of simple and compound overhangs in both the cross section images and the vector histograms. Figure 13 generally has a cluster of vector direction at 90 degrees which is not as distinct in the case of figure 14. This indicates that the AM surface has a flatter surface than the etched AM sample.

Figure 12 is the only sample which shows clear directional differences between the traces, i.e. anisotropy. The BioHelix sample shows evidence of simple and occassional compound overhangs but also appears to have a repeating pattern within the general roughness in certain cross sectional directions. This makes sense because the surface is generated from parralell rows of surface 'splashes' created at a pitch of 59 μm between laser pulses.

3.3 Quantification of overhang levels

The final stage in this work was to estimate how much of the surface would be shadowed by overhanging features if illuminated from directly above (i.e., how much of the surface is involved in overhanging features). This was calculated by establishing the proportion of surface-normal vectors which were above the inclination thresholds for simple and compound overhangs.

The results are presented in Table 4.

Table 4. The proportion of the sample surface covered by simple and compound overhangs expressed as a percentage of the width of the sample, for the four slicing angles.

Trace angles (°) (0, 45, 90, 135)	Proportion of simple overhang (%)	Proportion of compound overhang (%)
(a) Grit-blasted	0	0
	0	0
	0	0
	0	0
(b) Laser - BioHelix	0.4	0
	2.5	0
	2.4	0.4
	0.8	0
(c) Raw AM	28	0
	36	1
	18	1.6
	14	2.7
(d) Etched AM	41	0.4
	43	7.0
	19	0
	49	0

Table 4 reveals the following points; Once again we can see that there are no overhanging features on the grit blasted surface. The laser-Biohelix surface has evidence of simple overhanging features with a small amount of compound overhangs. Approximately 25% of the Raw AM surface is covered by simple overhangs and, on average, 1.25% of the surface involves compound overhangs. In the case of the etched AM surface approximately 40% of the surface is covered in simple overhangs with 2% of the surface having compound overhangs. This could be expected because of etchant corrosion undermining the overhangs created by the AM process.

3.4 Advantages and limitations of the technique

The surface analysis technique presented here has the advantages that features which are hidden from optical and mechanical methods can be included in qualitative and quantitative assessments. Deeply complex surfaces can be described, and the method is essentially size-independent. The technique could be expanded into 3D and Standard Tessellation Language (STL) models can be used.

Limitations of the technique include the fact that it involves expensive equipment and sampling which is not non-destructive. Also, it is not appropriate in cases where the features under investigation are of a similar, or larger, scale to the specimen size used for Micro-CT.

3.5 Future work

In future work the algorithms developed here to study overhanging features need to be tested for the limits of their applicability and generalisation. This would lead on to full three-dimensional spatial analysis involving even more complex overhangs and surface features.

The rapid growth in additive manufacturing and laser surface texturing means that technical surfaces with overhanging features are likely to become more common. Future work on the characteristics of such surfaces will depend on the type of analytical tools presented here for the quantitative assessment of overhanging features.

4. Conclusions

- Additive manufacturing and laser surface texturing can create surfaces covered in overhanging features; this can have substantial effects on surface properties such as wettability.
- Overhanging features can usefully be divided into two types: simple and compound.
- Micro-CT can reveal the topological details of convoluted surfaces containing both simple and compound overhangs.
- From micro-CT results surface profiles can be extracted; surface-normal vectors can be applied to these profiles along with an angle-based algorithm to quantitatively identify the presence and proportion of simple and compound overhangs; different characteristics were demonstrated for four surfaces.
- Studying surface profiles in different directions facilitates the identification of topological anisotropies.
- Convoluted surface profiles containing overhanging features cannot be subjected to standard Fourier analysis to search for repeating patterns because they have ambiguous vertical coordinates. However, frequency analysis can be used if the surface is divided up into a series of short, straight-line segments with surface-normal vectors.
- Analysis tools of this type will become increasingly useful as surfaces including overhangs generated by lasers and other technologies become more common.

Acknowledgments
The authors would like to thank the Mid Sweden University, Sports Tech Research Centre for kindly providing samples of AM Ti samples. The authors are also grateful for funding granted by EU-ERDF-InterregNord project C3TS, no. 20201279.

Orcid iDs
R. Olsson https://orcid.org/0000-0002-7706-6994
J. Powell https://orcid.org/0000-0001-6390-4494
J. Frostevarg https://orcid.org/0000-0003-4265-1541
A. Kaplan https://orcid.org/0000-0002-3569-6795

References

[1] Shin Y C, Wu B, Lei S, Cheng G J and Lawrence Yao Y 2020 *Journal of Manufacturing Science and Engineering* **142** 110818
[2] Lei S, Zhao X, Yu X, Hu A, Vukelic S, Jun M B G, Joe H E, Lawrence Yao Y and Shin Y C 2020 *Journal of Manufacturing Science and Engineering, Transactions of the ASME* **142** 031005
[3] Olsson R, Powell J, Palmquist A, Brånemark R, Frostevarg J and Kaplan A F H 2018 *Journal of Laser Applications* **30** 042009
[4] Frostevarg J, Olsson R, Powell J, Palmquist A and Brånemark R 2019 *Applied Surface Science* **485** 158-69
[5] Godi A, Kühle A and De Chiffre L 2013 *Measurement Science and Technology* **24** 085009
[6] Rupp F, Liang L, Geis-Gerstorfer J, Scheideler L and Hüttig F 2018 *Dental Materials* **34** 40–57
[7] Wennerberg A and Albrektsson T 2000 *The International journal of oral & maxillofacial implants* **15** 331–44
[8] Grandfield K, Gustafsson S and Palmquist A 2013 *Nanoscale* **5** 4302
[9] Shah F A, Nilson B, Brånemark R, Thomsen P and Palmquist A 2014 *Nanomedicine: Nanotechnology, Biology, and Medicine* **10** 1729–37
[10] Brånemark R, Emanuelsson L, Palmquist A and Thomsen P 2011 *Nanomedicine:Nanotechnology, Biology and Medicine* **7** 220–7
[11] Wang X, Shah F A, Palmquist A and Grandfield K 2017 *ACS Biomaterials Science & Engineering* **3** 49–55
[12] Sisti K E, de Rossi R, Antoniolli A M B, Aydos R D, Guastaldi A C, Queiroz T P, Garcia I R, Piattelli A and Tavares H S 2012 *Journal of Oral Implantology* **38** 231–7

[13] Liang C, Wang H, Yang J, Cai Y, Hu X, Yang Y, Li B, Li H, Li H, Li C and Yang X 2013 *ACS Applied Materials and Interfaces* **5** 8179–86

[14] Palmquist A, Johansson A, Suska F, Brånemark R and Thomsen P 2013 *Clinical Implant Dentistry and Related Research* **15** 96–104

[15] Palmquist A, Emanuelsson L and Sjövall P 2012 *Applied Surface Science* **258** 6485–94

[16] Mirhosseini N, Crouse P L, Schmidth M J J, Li L and Garrod D 2007 *Applied Surface Science* **253** 7738–43

[17] Nayar S, Bhuminathan, Mahadevan, Santhosh S and Vignesh 2015 *Journal of Pharmacy and Bioallied Sciences* **7** 89

[18] Pfleging W, Kumari R, Besser H, Scharnweber T and Majumdar J D 2015 *Applied Surface Science* **355** 104–11

[19] Wennerberg A and Albrektsson T 2009 *Clinical Oral Implants Research* **20** 172–84

[20] Kantzos C A, Cunningham R W, Tari V and Rollett A D 2018 *Computational Mechanics* **61** 575–80

[21] Rao A R 2012, *A Taxonomy for Texture Description and Identification,* (New York, NY: Springer US)

[22] Calignano F 2018 *Virtual and Physical Prototyping* **13** 97–104

[23] Senin N, Thompson A and Leach R K 2017 *Measurement Science and Technology* **28** 095003

[24] Pagani L, Qi Q, Jiang X and Scott P J 2017 *Measurement: Journal of the International Measurement Confederation* **109** 281–91

[25] Pagani L, Townsend A, Zeng W, Lou S, Blunt L, Jiang X Q and Scott P J 2019 *Measurement: Journal of the International Measurement Confederation* **141** 442–59

[26] Newton L, Senin N, Smith B, Chatzivagiannis E and Leach R 2019 *Surface Topography: Metrology and Properties* **7** 045020

[27] Olsson R, Powell J, Palmquist A, Brånemark R, Frostevarg J and Kaplan A F H 2019 *Journal of Laser Applications* **31** 022508

[28] Olsson R, Powell J, Frostevarg J and Kaplan A F H 2020 *Journal of Laser Applications* **32** 022030

[29] Zhang M, Levina E, Djurdjanovic D and Ni J 2008 Journal of Manufacturing Science and Engineering, Transactions of the ASME **130** 0310101–9

[30] Tang S, Wang X, Lv X, Han T X, Keller J, He Z, Skubic M and Lao S 2013 *Lecture Notes in Computer Science (including subseries Lecture Notes in Artificial Intelligence and Lecture Notes in Bioinformatics)* vol 7725 LNCS pp 525–38

[31] Watanabe T, Ito S and Yokoi K 2009 *Lecture Notes in Computer Science (including subseries Lecture Notes in Artificial Intelligence and Lecture Notes in Bioinformatics)*

[32] Dalal N and Triggs W 2004 2005 *IEEE Computer Society Conference on Computer Vision and Pattern Recognition CVPR05*

[33] Tajima K, Hironaka M, Chen K-K, Nagamatsu Y and and Kozona Y 2008 *Dental Materials Journal* **27** 258-65

[34] Kass M, Witkin A and Terzopoulos D 1988 *International Journal of Computer Vision **1** 321-32*

[35] Ahrens, James, Geveci, Berk, Law C 2005 *ParaView: An End-User Tool for Large Data Visualization, Visualization Handbook* (Elsevier)

IOP Conf. Series: Materials Science and Engineering 1135 (2021) 012024 doi:10.1088/1757-899X/1135/1/012024

Influence of laser hardening on laser induced periodic surface structures on steel substrates

Lewin Rathmann[1], Tim Radel[1]

[1]Bremer Institut für angewandte Strahltechnik GmbH, Klagenfurter Str. 5, 28359 Bremen, Germany

rathmann@bias.de

Abstract. Laser-induced periodic surface structures (LIPSS) are used to structure surfaces for functionalization. Thus, hydrophilic states are generated using LIPSS. However, these nanostructures do not withstand mechanical loads and therefore cannot be used for most tribological applications. Within this work the approach of laser hardening of LIPSS is investigated. It is shown that laser hardening leads to an alteration of prior structured surfaces. That effects the wetting behaviour. The higher the laser power during hardening, the more increases the contact angle of a single droplet on the surface and the more the surface lacks in terms of wetting behaviour. This phenomenon is attributed to changes in LIPSS' aspect ratio. A high ratio leads to low contact angles and is shifted to low values when the laser power increases resulting in high contact angles. Hence, it is concluded that the thermal load during laser hardening, and it's influence on the wettability must be taken into account when LIPSS are subjected to laser hardening.

1. Introduction

Laser-induced periodic surface structures (LIPSS) are used to structure surfaces for their functionalization in mechanical applications [1]. These structures enable for example the influence of wetting properties on the one hand and of friction and wear behavior of contact partners on the other and are therefore subjected to many investigations [2-4]. Changing wetting properties becomes evident if conventional lubricants should be substituted with water. And alternatives to conventional lubricants are of great interest from ecological and economical points of view. But the low viscosity and the resulting insufficient film thickness due to poor wettability represent a major obstacle to use water as an effective lubricant [5]. Structuring a surface with laser-induced periodic surface structures which create a hydrophilic surface and therefore increase the wettability, it was shown that this obstacle could be overcome, and water be used as lubricant in a lubricated sliding application [6].

Nevertheless, LIPSS wear out quickly in direct tribological contact even when loads of 1 N (corresponds to 467 MPa) [7] down to 25 mN (corresponds to 218 MPa) [8] are applied. Consequently, for many technical applications, their wear resistance must be significantly increased while simultaneously preserving their functionality. Due to the local applicability, the realizable velocities, the generally achievable high hardness [9-14] and the feature that no quenching medium is needed, which could influence the nanostructures, laser hardening seems to be the tool of choice for processing LIPSS while minimizing substrate distortion [15]. El-Khoury et al. [16] showed the feasibility of laser hardening to increase the hardness of direct laser interference patterning (DLIP) micro-textures.

The current state of the art shows the beneficial use of LIPSS in mechanical contact situations between two steel substrates. However, since LIPSS are subjected to wear, a treatment is needed to improve their resistance in terms of mechanical contact. Consequently, this work deals with the influence of laser hardening on geometrical and functional properties of laser induced periodic surface structures on X153CrMoV12 steel substrates. Besides contact angle measurements to determine the change in the wetting behavior of LIPSS, confocal laser scanning microscopy (CLSM) and scanning electron microscopy (SEM) are additionally used to characterize the geometrical changes.

2. Methodology

2.1. Substrate Material

The cold work steel X153CrMoV12 is used as substrate material. It has an initial hardness of 250 HV1 and is delivered as flat material. It is cut in substrates of 30 mm x 10 mm x 10 mm in length x width x height. Primary to structuring and hardening, the substrates are polished to an initial surface roughness Sa of 100 nm and then cleaned in an ultrasonic bath with isopropanol for five minutes. The chemical composition of X153CrMoV12 is given in Table 1.

Table 1. Nominal chemical composition of the material 1.2379 (X153CrMoV12).

material number	material designation	C in mass-%	Cr in mass-%	Mo in mass-%	V in mass-%
1.2379	X153CrMoV12	1.55	12.00	0.70	0.80

2.2. Laser Structuring

The applied structures are low spatial frequency LIPSS first order (LSFL-I). These are applied via pico-second laser TruMicro 5050 made by Trumpf using the listed parameters in Table 2. The LIPSS are generated scanning parallel lines with a certain line spacing. The parameters are listed in Table 3.

Table 2. Parameter of the laser TruMicro 5050 made by Trumpf.

wavelength in nm	1030
pulse duration in ps	6
repetition rate in kHz	50
spot diameter in µm	46

Table 3. Parameter to create LSFL-I.

pulse energy in µJ	200
scan velocity in mm/s	1840
line spacing in µm	36.8
number of cycles	1

2.3. Laser Hardening

A continuous wave (cw) disk laser TruDisk 12002 made by Trumpf was used for laser hardening. The laser parameter to harden the substrates with LIPSS on the surface are given in **Table 4**. The laser spot is rectangular. The energy distribution along the short axis is quasi-Gaussian while it is homogeneous along the long axis. Using a rectangular spot counteracts the increased heat influence of track overlap that would occur with a single, scanning laser spot. Therefore, the substrate is hardened with one cycle.

IOP Conf. Series: Materials Science and Engineering 1135 (2021) 012024 doi:10.1088/1757-899X/1135/1/012024

Table 4. Laser parameter to harden the substrates.

wavelength in nm	1030
spot size in mm x mm	10 x1.6
laser power in W	250 to 1100
scan velocity in mm/s	10
shielding gas	Argon
flow rate in L/min	20

2.4. Sessile-Drop Tests

Alterations in the wetting properties of the LIPSS were determined using contact angle measurements based on the "Sessile-Drop"-method. A detailed description of the basic experimental setup is given in [14]. In the present work a highspeed camera Phantom VEO 410 L was aligned perpendicular to the structured surface to record the development, fall and spreading of a single droplet of deionized water. The droplet is produced by a dosing device model 900 E made by GLT and develops at the tip of a cannula with a diameter of 300 µm which is placed 5 mm above the substrate. The scenery is illuminated using an illumination laser CAVILUX HF system made by CAVITAR with a wavelength of 810 nm. The emitted light cone is directed onto a screen behind the sample and thus created a high contrast shadow image of the droplet. Actuating the dosing unit started the high-speed camera to record enabling an evaluation after a certain period. This was chosen to 80 ms since no state changing dynamic was to be detected after this. The contact angle was derived from the certain image using a self-written software program. An overview of the experimental setup is given in Figure 1.

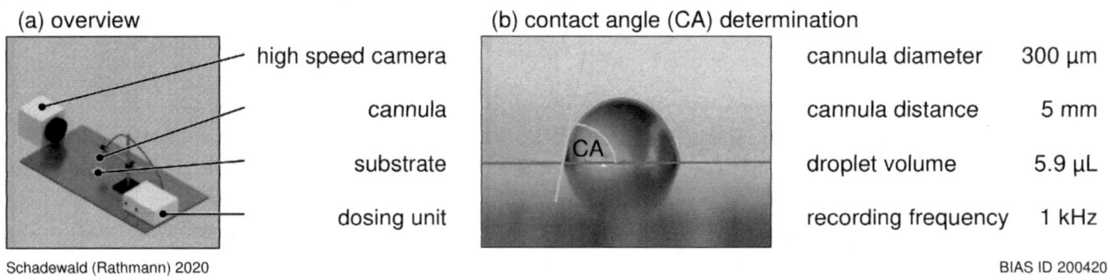

Figure 1. Overview of the experimental setup regarding the "Sessile-Drop"-method according to [18].

2.5. Confocal Laser Scanning Microscopy and Scanning Electron Microscopy

The results of structuring and hardening were on the one hand analysed with a confocal laser-scanning microscope (CLSM) of type VK-9700 made by KEYENCE. Its fifty-fold magnification was used to record an overview of the altered surfaces.

On the other hand, the Auriga 40 scanning electron microscope (SEM) built by Zeiss was used for a more detailed investigation of the influence of laser radiation during hardening on the LIPSS in the sub-micrometre range. The parameters used to generate the images can be found in **Table 5**.

Table 5. SEM parameters for imaging the surface and structure alterations due to laser hardening.

magnification	1.19k-fold	6.29k-fold
working distance in mm	11.2	11.2
EHT in kV	10	10
signal	SE	SE
aperture size in µm	30	30

3. Results

The alteration of the surface and thus of the nanostructures due to laser hardening at different laser powers is shown in **Figure 2** There, for the laser powers 500 W and 1100 W, the resulting changes in the nano structured surfaces compared to the reference without hardening are clearly visible. As the power increases, the structures disappear over a wide area, see in **Figure 2** the red marked areas. It is additionally to observe that both the number of these areas and their size increase with increasing laser power.

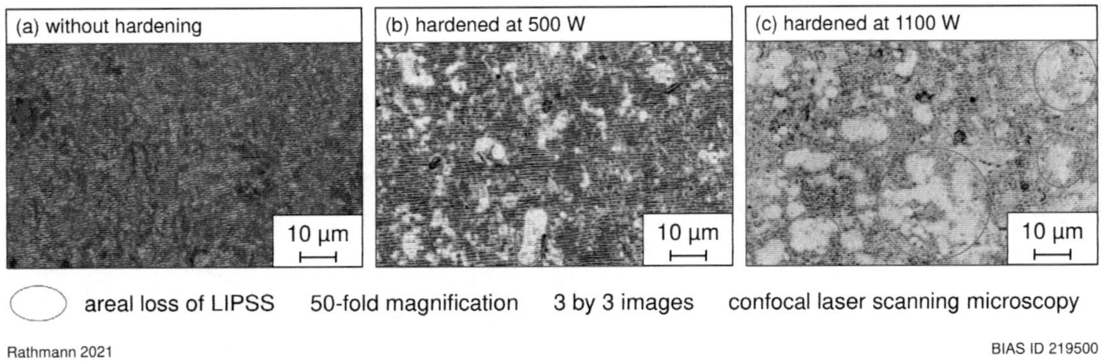

Figure 2. CLSM images of the substrate depicting the surface alterations due to the laser hardening.

The qualitatively structural reduction of the LIPSS is reflected in the change in their heights. Thus, **Figure 3** shows that the width of the nanostructures remains unchanged, while there is a decrease in the structure height with increasing laser power.

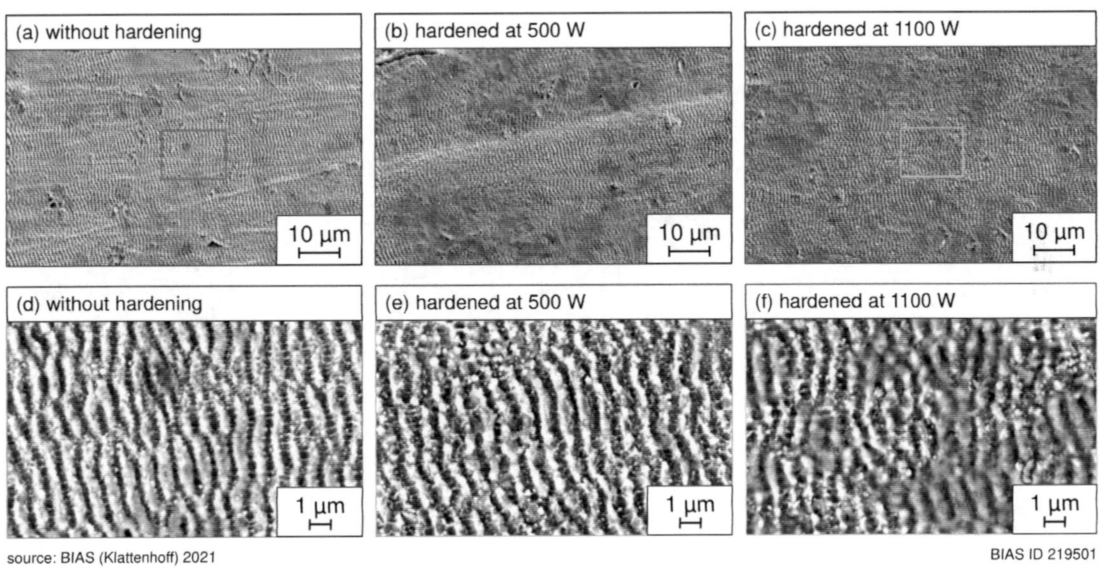

Figure 3. Alteration of the LIPSS after being exposed to different laser powers during laser hardening at 1.19k-fold ((a) to (c)) and 6.29k-fold ((d) to (f)) magnification.

The quantitative change in nanostructures is shown in **Figure 4**. There, the height, and width of the structures are plotted against the laser power used. For comparison, the height and width of the unhardened nanostructures are additionally indicated. While the width shows no change on average, the height shows a decrease on average.

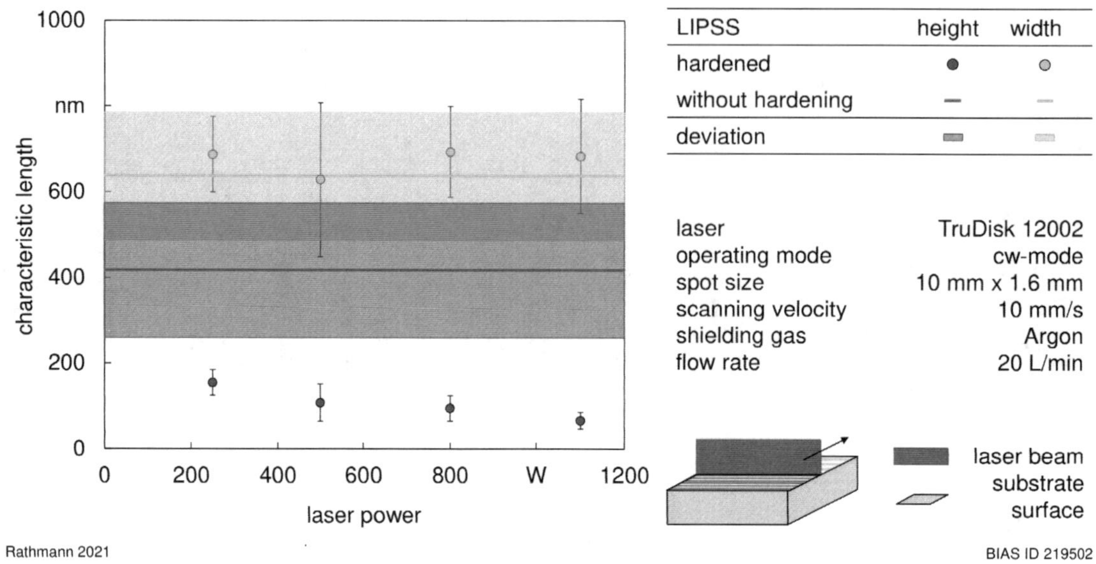

Figure 4. Height and width of LIPSS alteration in dependency of the laser power used.

Contact angle determinations carried out with these geometrical prerequisites led to the result shown in **Figure 5**. While laser hardening has no effect on unstructured substrates and their wettability, the nanostructures hardened by laser show an increase in the contact angle from a laser power of 700 W. The contact angle converges to the contact angle of a drop on a polished and hardened substrate.

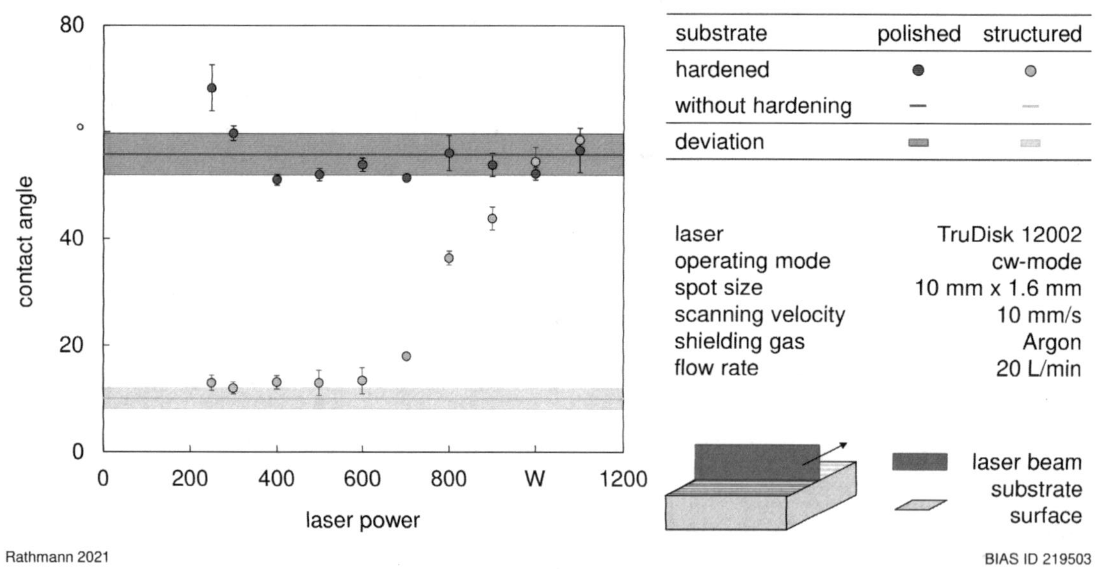

Figure 5. Contact angle in dependency of the operated laser power on structured and hardened steel substrates.

4. Discussion

The literature shows that LIPSS are suitable to influence and even change the surface properties. Thus, surfaces that initially exhibit hydrophobic properties change to hydrophilic properties after structuring with LIPSS. However, the literature also shows that these nanostructures do not withstand mechanical wear. As a result, the surfaces provided with them lose the properties previously set via the LIPSS. To increase the wear resistance of these structures they must be protected. Therefore, they can be put in lubrication pockets as shown by Rathmann et al. [18], or laser hardening can be used to increase the wear resistance, as El-Khoury et al. [16] demonstrated when they hardened DLIP microstructures via laser. Hence, LIPSS were hardened in this work using laser hardening and at different laser powers. Thereby the wettability decreases with increasing laser power. There is, to our best knowledge, no literature dealing with short term heat treatment of LIPSS. Nevertheless, geometrical changes and chemical changes can impair the wettability. It is known that LIPSS change their wettability over time based on changes in the chemical composition of the structured surface, as Kietzig et al. [19] presented. They generated LIPSS at a fluence of 5.16 J/cm² on substrates of 304L and stored them in different atmospheres (air, nitrogen, water, and carbon dioxide). Within five days, the previously hydrophilic behaviour of the nanostructures changed to hydrophobic. Even if the laser hardening was carried out under inert gas atmosphere, a chemical influence e. g. oxidation of the surface cannot be completely excluded. Further the hight of the LIPSS and therefore the aspect ratio changes. This distinguishes the results presented here from those by El-Khoury et al. [16] where there was no effect on geometry, the DLIP structures present there were in the micrometre range and the LIPSS present here are in the nanometre range. Regarding the shape changes, according to Wenzel [20], the surface enhancement affects the wetting behaviour. For the given results, the LIPSS hight decreases when laser hardened with 500 W without a reduction of wettability. The wettability is unaffected up to a laser power of 600 W. Hence, in this range, the geometric effect seems not to have a dominant effect. The shape changes may be a result of local melting, as they appear like those shown in Gregorcic et al. [21]. Nevertheless, this must be clarified further high-resolution images like transmission electron microscopy. Further investigations must clarify which mechanism if geometrical or the chemical effects are the dominating one for the decreasing wettability.

5. Conclusion

In this work, the influence of laser radiation during laser hardening on laser-induced periodic surface structures was presented. These nanostructures initially have a super-hydrophilic behaviour, which, however, changes to a hydrophobic one, like those of polished specimens, with increasing laser power during hardening. Thereby, the geometrical shape of the LIPSS changes. Hence, it is concluded that the thermal load during laser hardening, and it's influence on the wettability must be considered when LIPSS are subjected to laser hardening.

Acknowledgements

Funding by the Deutsche Forschungsgemeinschaft DFG (project 416530419 "Wasser als Schmierstoff für die Hochgeschwindigkeitsumformung mithilfe von LIPSS") is gratefully acknowledged.

ORCID iDs

T Radel https://orcid.org/0000-0002-6681-2111

References

[1] Yu J J and Lu Y F Laser-induced ripple structures on Ni-P substrates 1999 *Appl. Surf. Sci.* **148** 248-252
[2] Duarte M, Lasagni A, Giovanelli R, Narciso J, Louis E and Mücklich F Increasing lubricant film lifetime by grooving peridocial patterns using laser interference metallurgy 2008 *Adv. Eng. Mater.* **10** 554-558
[3] Whang Z, Thao Q and Wang C Reduction of friction of metals musing laser-induced periodic surface nanostructures 2015 *Mircomachines* **6** 1606-1616
[4] Bonse J, Höhm S, Kirner S V, Rosenfeld A and Krüger J Laser-induced periodic surface structures – A scientific evergreen 2017 *IEEE J. Select. Topics Quantum Electron* **23** 9000615
[5] Kim H-J and Kim D-E Water lubrication of stainless steel using reduced graphene oxide coating 2015 *Sci. Rep.* **5** 17034

[6] Chen C-Y, Wu B-H, Chung C-J, Li W-L, Chien C-W Wu, P-H and Cheng C-W Low-friction characterisitcs of nanostructured surfaces on silicon carbide of water lubricated seals 2013 *Tribol. Lett.* **51** 127-133
[7] Camilo F, Kirner S V, Krüger J and Bonse J Surface functionalization by laser-induced periodic surface structures 2020 *J. Laser Appl.* **32** 022063
[8] Bonse J, Koter R, Hartelt M, Spaltmann D, Pentzien S, Höhm S, Rosenfeld A and Krüger J Femtosecond laser-induced periodic surface strucutres on steel and titanium alloy for tribological applications 2014 *Appl. Phys.* A **117** 103-110
[9] Eichstädt J, Römer G R B E and Huis in't Veld A J Towards friction control using laser-induced periodic surface structures 2011 *Phys. Proc.* **12** 7-15
[10] Babu P D, Balasubramanian K R and Buvanashekaran G Laser surface hardening: a review 2011 *Int. J. Surf. Sci. Eng.* **5(2/3)** 131-151.
[11] Khorram A, Davoodi Jamaloei A, Jafari A and Moradi M Nd:YAG laser surface hardening of AISI 431 stainless steel; mechanical and metallurgical investigation 2019 *Opt. Laser Technol.* **119** 105617.
[12] Buling A, Sändker H, Stollenwerk J, Krupp U and Hamann-Steinmeier A Laser surface pretreatment of 100Cr6 bearing steel – Hardening effects and white etching zones 2016 *Appl. Surf. Sci.* **378** 564-571.
[13] Sorgente D, Corizzo O, Ancona A, Scintilla L D, Palumbo G and Tricarico L Laser hardening of AISI 52100 bearing steel with a discrete fiber laser spot 2014 *Proc. SPIE Int. Soc. Opt. Eng.* **8963** 896308.
[14] Klocke F, Schulz M and Gräfe S Optimization of the laser hardening process by adapting the intensity distribution to generate a top-hat temperature distribution using freeform optics 2017 *Coatings* **7** 77.
[15] Vollertsen F, Partes K and Meijer J State of the art of laser hardening and caldding 2005 Proc. 3rd Int. WLT-Conf. LiM.
[16] El-Khoury M, Seifert M, Bretschneider S, Zawischa M, Steege T, Alamri S, Lasagni A F and Kunze T Hybrid processing of bearing steel by comining Direct Laser Intereference Patterning and laser hardening for wear restistance applications 2021 *Mat. Let.* **303** 130284
[17] Zorba V, Stratakis E, Barberoglou M, Spanakis P, Tzanetakis P and Fotakis C Tailoring the wetting response of silicon surfaces via fs laser structruing 2008 *Appl. Phys. A: Mat. Sci. Proc.* **93** 819-825.
[18] Rathmann L, Beste L-H and Radel T Laser based process chain to use LIPSS on forming tools 2021 *Surf. Coat. Tec.* **426** 127761.
[19] Kietzig A-M, Hatzikiriakos S G and Englezos P Patterned superhydrophobic metallic surfaces 2009 *Langmuir* **25(8)** 4821-4827
[20] Gregorcic P, Sdlacek M, Podgornik B and Reif J Formation of laser-induced periodic surface structures (LIPSS) on tool steel by multiple picosecond laser pulses of different polarizations 2016 *Appl. Surf. Sci.* **387** 698-706
[21] Wenzel R N Resistance of solid surfaces to wetting by water 1936 *Ind. Eng. Chem.* **28** 988-994

IOP Conf. Series: Materials Science and Engineering 1135 (2021) 012025 doi:10.1088/1757-899X/1135/1/012025

An Approach to Predict the Residual Stress-Depth-Profile of Thin Sheet Metal Processed by Laser Shock Peening Without Coating

Tobias Valentino[1], Andreas Stephen[1], Tim Radel[1]

[1]BIAS - Bremer Institut für angewandte Strahltechnik GmbH, Klagenfurter Str. 5, 28359 Bremen, Germany

valentino@bias.de

Abstract. For conventional laser shock peening, the positive influence of compressive residual stresses on fatigue strength is well understood. To protect the material's surface from ablation, a sacrificial layer is applied. This, however, leads to an additional process step, which deteriorates its economic efficiency. Thus, laser shock peening without coating (LPwC) is more frequently investigated for industrial applications. However, LPwC increases the thermal impact on the material, which may provoke tensile residual stresses in the surface region. In this regard, understanding the influence of LPwC on the residual stress state and deriving a suitable state, e.g., for subsequent applications or forming operations, result in a design of experiment with numerous residual stress measurements. Residual stress-depth-profiles obtained by X-ray diffraction are time-consuming and cost intensive. Hence, a model is proposed to predict the residual stress-depth-profile of LPwC-processed thin sheets. The analytical model is based on the source stress model and uses experimental results, namely hardness as well as shape change measurements. Sheets made of X5CrNi18-10 and with a thickness of 1 mm are LPwC-processed with a nanosecond fiber laser. In the thermally dominated area where tensile residual stresses are present, the model agrees well with the experimental measurements. Moreover, it is revealed that LPwC leads to a saturation of residual stress level maximum and depth in dependence of pulse energy, repetition rate and number of repetitions. Subsequently, the model is used for tailoring the stress profile of thin sheets by LPwC for subsequent bottom bending.

1. Introduction

Metallic micro components are found in many products of our daily life such as in laptops, smartphones, or tablets [1]. To produce such small components economically, they are often mass-produced in billions per year. Cold forming processes, such as bending or deep drawing, are suitable for mass production because they allow low costs and cycle times [2]. A major challenge in forming thin sheet metal with a thickness $t \leq 1$ mm is the increased dispersion of springback [3]. Springback is an elastic shape change of material, which appears upon removal of forces e. g., when releasing of constraints after forming [4]. When using the same forming tool and punch force, dispersion of springback is solely affected by material deviations, such as sheet thickness, grain size and residual stress state [5]. It is found that the latter has the strongest effect on dispersion of springback [6]. Residual stress state deviations in a material are the result of the production process: Locally different solidification and cooling conditions during casting or manufacturing result in varying residual stresses [7]. It is therefore be presumed that a targeted adjustment of the residual stress state prior to forming leads to lower dispersion of springback.

Processes known to modify the residual stress state of metals are shot peening and laser shock peening (LSP). LSP has a higher depth impact compared to shot peening [8]. Moreover, LSP is applied to enhance the fatigue strength of turbine blades [8]. Still, a major challenge is the prediction of residual stresses in the material to optimize the process for its application. There are only a limited number of models accessible for determining residual stresses in LSP-processed sheets [9]. The reason is that many parameters, such as high strain rates of more than 10^6 s^{-1}, stress wave attenuation and dispersion, influence the residual stress state. These increase the complexity of the models [9].

In LSP, a confining medium is applied on the target to increase the peak pressure. Peak pressures of more than 5 GPa can be achieved [10]. Water is best suited for most industrial applications because most common lasers cause wear on glass, which reduces the peak pressure over time [11]. However, water has a lower acoustic impedance in comparison to glasses, which results in lower peak pressures [12]. To protect the base material from ablation, a sacrificial layer is usually applied on the surface [13]. Although, a sacrificial layer increases the peak pressure by at least a factor of 1.4, it also doubles the dispersion of peak pressure [14]. Moreover, the application of a sacrificial layer implies an additional process step. Thus, current studies investigate LSP without sacrificial layers. In literature, this process is referred to as LPwC [15]. LPwC may also enhance the fatigue strength of metals, despite the increase in surface roughness and higher heat deposition [16]. The latter may cause the formation of tensile residual stresses in the surface region [17]. They are the result of elevated temperatures, surface softening and resolidification [18]. To reduce these detrimental effects, it is recommended to achieve intensities in the range of GW/cm^2 with lower pulse energies, shorter pulse durations and smaller focus diameters [19]. Pulse energies for LPwC are approximately in the range of 6 mJ [20] up to 200 mJ [21].

So far, it was demonstrated that a favorable residual stress state can be induced in the forming zone of a sheet metal prior to bending by means of LPwC [22]. In the experiments, the dispersion of bending angle was decreased by the factor 2 because of LPwC. Consequently, the residual stress state, present in the initial state, only played a subordinate role. However, it has not been fully understood yet, how it is possible to obtain a suitable residual stress state of thin sheet metal for subsequent forming processes. Deriving a suitable residual stress state from experiments would result either in numerous residual stress measurements or in Finite element modelling (FEM). Residual stress-depth-profiles obtained by X-ray diffraction are cost and time-intensive. FEM becomes complex when considering all the mechanisms and need suitable data and validation. Hence, a simple model is proposed to predict the residual stress-depth-profile of LPwC-processed sheets. The analytical model is based on the source stress model [23] and uses results from hardness-depth as well as shape change measurements. Here, solution annealed sheets made of X5CrNi18-10 with a thickness of 1 mm are LPwC-processed with a nanosecond fiber laser. Subsequently, bottom bending experiments with a nominal bending angle of 90° are performed on these sheets. The dispersion of bending angle is compared to those obtained with the initial state.

2. Experimental Setup and Methods

2.1. Laser peening without coating

LPwC experiments are conducted on solution annealed sheets made of X5CrNi18 10 (initial state). The dimensions of the sheets are shown in **Fig. 1a**. The sheets were processed single-sided with a single-mode fiber laser (YLP-HP) from IPG. The laser has a beam quality factor of $M^2 < 1.1$, and a wavelength of 1064 nm. All the investigated laser parameters are listed in **Tab. 1**.

The laser beam was deflected via a scan head intelliSCAN 20 from Scanlab and focused on the sheet surface. The scan head has a F-Theta lens and a focal length of 160 mm resulting in a spot diameter of 35 µm. The intensity distribution of the laser beam is a Gaussian distribution. The laser was focused on the sheet metal, which was positioned in a sheet holder (see **Fig. 1b**). Spot diameter, scan speed and repetition rate result in a pulse overlap of 67 %. Parallel scanning is used to process the sheets. The hatching distance is set to 12 µm to enable a similar overlap compared to the pulse overlap. The scan direction is transverse to the rolling orientation and the sheets are processed only in the center region, which is affected by the subsequent bending process (see **Fig. 1c**). The sheet metal surface was continuously supplied with laminar water flow. The water layer had a thickness of 4 mm.

Shape changes of the sheets and ablation depth prior to bending were measured with a Keyence confocal microscope (model VK 9700). The deflection of each sheet was measured with a 10x objective. Here, thermal as well as mechanic induced shape changes are characterized by the deflection of the sheets (see **Fig. 2**). The ablation depth describes the maximum depth that was removed on the sheet after LPwC.

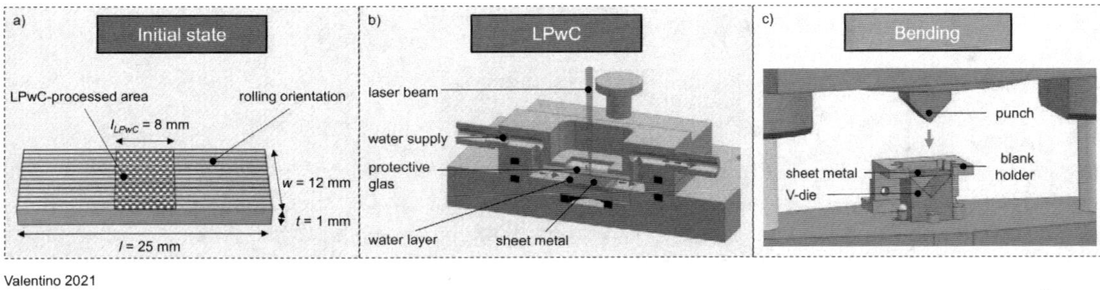

Figure 1: Process chain of LPwC-processed and bended sheets: a) dimensions of solution annealed sheets made of X5CrNi18 10, b) setup for LPwC with continuous water supply in sheet holder unit, c) setup for bottom bending with a nominal bending angle of 90°.

Table 1: Experimental design of investigated parameters.

Pulse duration t_p in ns	Repetition rate f in kHz	Pulse energy E_p in mJ	Scan speed v_s in m/s	Scan repetitions n	Average laser power P_L in W
120	200	1.00	2.3	1, 4, 8, 12, 15, 30, 60 and 100	200
120	200	0.20	2.3	15	40
120	100	1.00	1.2	15	100
120	100	0.20	1.2	15	20
120	20	1.00	0.2	1, 4, 8, 12, 15 and 30	20
120	20	0.20	0.2	15	4
30	200	0.20	2.3	15	40
30	200	0.15	2.3	15	30
30	150	0.20	1.7	15	30
30	150	0.15	1.7	15	23
30	100	0.20	1.2	15	20
30	100	0.15	1.2	15	15

2.2. Bottom bending

Bottom bending experiments were performed on a testing machine AllroundLine Z250 from Zwick. The positioning accuracy of the testing machine is ±2 µm. The crosshead speed was set to 1 mm/s for bending experiments. The sheet was placed on top of a V-die in a blank holder unit (see **Fig. 1c**). The V-die hard a nominal bending angle of 90°. The LPwC-processed surface faced toward the punch.

The bending angle was measured with a coordinate measuring machine Crysta-Apex C from Mitutoyo. A touch probe stylus from Renishaw was used to access the bending angle. The coordinate measuring machine has a length measurement deviation below 1.7 µm and a resolution of 0.1 µm. On the bent sheets, 6 measuring points were recorded on both sides of the inner leg. The distance between

IOP Conf. Series: Materials Science and Engineering 1135 (2021) 012025 doi:10.1088/1757-899X/1135/1/012025

each measuring point was set to 0.5 mm. A measurement line was derived for each side from the measuring points. The bending angle was derived from the two measurement lines.

2.3. Metallography

Vickers micro hardness measurements at 0.01 kgf (HV0.01) were performed with a DuraScan hardness tester from Struers. A hardness matrix was created in the cross sections of the solution annealed and LPwC-processed sheets.

An X-ray diffractometer with Cr-kα radiation and a beam diameter of 1 mm was used to detect the sheets' residual stress states. To determine residual stresses in dependence of the depth of the sheet, a defined amount of material was removed by electrochemical etching.

3. Analytical Model

3.1. Approach

The goal of the model is to predict the residual stress state of sheet metal processed by LPwC. Instead of directly measuring the residual stress state in dependence of the sheet thickness with an X-ray diffractometer, the residual stress state is derived from hardness-depth as well as shape change measurements. The approach is based on the theoretical description of plastically deformed layers by the stress source model [23]. Moreover, Hooke's law is applied to determine the residual stresses [24].

First, the source stress is determined. It is a measure of the bending moment, which acts on the sheet leading to its deflection (**Fig. 2**). A bending moment M acting on the sheet with a length l and a second moment of area I causes a maximal deflection u in dependence of the Young's modulus E [25]. All the constant material parameters are listed in **Tab. 2**.

$$M = -\frac{12 \cdot u \cdot E \cdot I}{l^2} \tag{1}$$

$$I = \frac{w \cdot t^3}{12} \tag{2}$$

w is the width of the sheet along which the residual stresses are assumed to be constant, and t is the sheet thickness. The bending moment acting on a sheet after LPwC is the result of the plastically deformed surface region and can be calculated by integrating the source stresses σ_{source} along the transition depth z_0 [25].

$$\sigma_{source} = \frac{M}{w \cdot \int_0^{z_0} \left(z - \frac{t}{2}\right) \cdot dz} = \frac{2 \cdot M}{w \cdot z_0 \cdot (z_0 - t)} \tag{3}$$

Table 2: Constant material parameter and values used for model.

Material parameter	Value
Sheet length l	25 mm
Sheet thickness t	1 mm
Sheet width w	12 mm
Second moment of area I	1 mm^4
Young's modulus E	195,000 N/mm² [26]

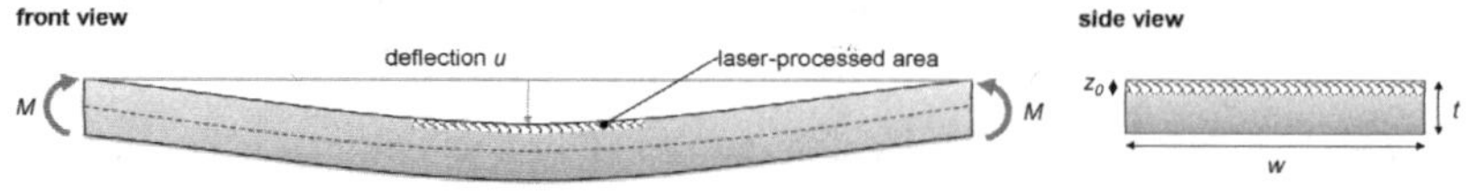

Figure 2: bending moment which leads to characteristic deflection of sheets.

IOP Conf. Series: Materials Science and Engineering 1135 (2021) 012025 doi:10.1088/1757-899X/1135/1/012025

Further, the sheet's cross section is divided into a laser-affected surface region and a core region to estimate the sheet's residual stress-depth profile (see **Fig. 3**). The transition depth from a harder surface to softer core region is determined according to [27] via the hardness-depth-profile. A "u-shaped" hardness-depth-profile is typically found after LPwC. Thus, a 2nd degree polynomial function f is derived for each hardness-depth profile. Moreover, the maximum hardness value in the surface region ($z < 0.1$ mm) and the mean hardness value from the core region (0.4 mm $< z < 0.6$ mm) are estimated. A mean hardness value is derived from the surface and core region. The transition depth z is determined via the 2nd degree polynomial function. A transition depth of $z_0 = 0.165$ mm is obtained from **Fig. 3**.

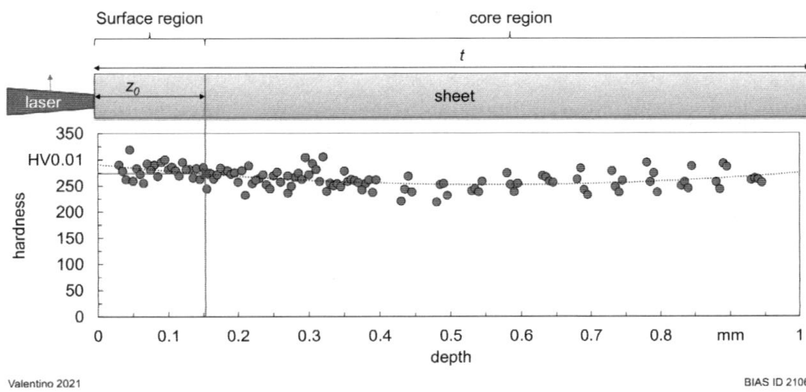

Figure 3: Determination of transition depth from surface to core region via hardness-depth-profile.

As one condition, the net sum of all residual stresses over a cross-section must be equal to zero. By assuming a step function, the mean stress for surface $\bar{\sigma}_{sur}$ and core $\bar{\sigma}_{cor}$ region is described accordingly:

$$0 = z_0 \cdot \bar{\sigma}_{sur} + (t - z) \cdot \bar{\sigma}_{cor} \tag{4}$$

As second condition, there must be a force equilibrium between the source force F_{source} and the forces of the surface F_{sur} and core region F_{cor}.

$$F_{source} = F_{sur} + F_{cor} \tag{5}$$

where $F = A \cdot \sigma$ and A is assumed to be constant. From Eq. (4) and Eq. (5) following step functions are derived:

$$\sigma_{rs}(z) = \begin{cases} z < z_0 \Rightarrow \bar{\sigma}_{sur} = \frac{\sigma_{source} \cdot (t - z_0)}{(t - z_0) - z_0} \\ z \geq z_0 \Rightarrow \bar{\sigma}_{cor} = \sigma_{source} - \frac{\sigma_{source} \cdot (t - z_0)}{(t - z_0) - z_0} \end{cases} \tag{6}$$

The core region is assumed to be constant in the following. However, it is expected that the residual stress profile is continuous over the entire depth of the sheet. Furthermore, it is assumed that the residual stress is maximal on the surface. The value then approaches the core region's one. A simple function that satisfies both conditions for the surface region is a 2nd degree polynomial function:

$$\sigma_{rs,sur}(z) = a_1 \cdot z^2 + a_2 \cdot z + a_3 \tag{7}$$

Following, three conditions must be met:

I) Surface function and core have an intersection at $(z;\bar{\sigma}_{cor})$.

$$\bar{\sigma}_{cor} = a_1 \cdot z^2 + a_2 \cdot z + a_3 \tag{8}$$

II) The slope in the core region is 0 since it is assumed to be constant. Thus, the slope at the intersection must also be 0.

$$0 = \frac{d}{dz_0}(a_1 \cdot z_0{}^2 + a_2 \cdot z_0 + a_3) = 2 \cdot a_1 \cdot z_0 + a_2 \tag{9}$$

III) The integral of the surface function must remain constant

$$\bar{\sigma}_{sur} \cdot z_0 = \int_0^{z_0}(a_1 \cdot z^2 + a_2 \cdot z + a_3)dz = \frac{1}{3} \cdot a_1 \cdot z_0{}^3 + \frac{1}{2} \cdot a_2 \cdot z_0{}^2 + a_3 \cdot z_0 \tag{10}$$

Accordingly, by inserting the three conditions in Eq. (7), the surface function can be written as follows:

$$\sigma_{rs,sur}(z) = \frac{3 \cdot (\bar{\sigma}_{sur} - \bar{\sigma}_{cor})}{z_0{}^2} \cdot z^2 + \frac{6 \cdot (\bar{\sigma}_{cor} - \bar{\sigma}_{sur})}{z_0} \cdot z + 3 \cdot \bar{\sigma}_{sur} - 2 \cdot \bar{\sigma}_{cor} \tag{11}$$

By inserting Eqs. (1), (2), and (3) in Eq. (11), the following expression for residual stresses at depths $z < z_0$ is derived:

$$\sigma_{rs,sur}(z) = \frac{2 \cdot E \cdot t^3 [3 \cdot t(z^2 - 2 \cdot z_0 \cdot z + z_0{}^2) - z_0{}^3] \cdot u}{l^2 \cdot z_0{}^3 \cdot (t - 2 \cdot z_0) \cdot (t - z_0)} \tag{12}$$

4. Results

Fig. 4 shows the base material's residual stress depth profile and the comparison between model and experiment of residual stress depth profiles of LPwC-processed sheets. The residual stress depth profile of the base material shows that the residual stresses fluctuate around 0. In contrast, the results for the LPwC-processed sheets reveal a strong increase in tensile residual stress in the surface region in comparison to the base material. Although the pulse energy is the same for both illustrated profiles, the increase in tensile residual stress on the surface is stronger for the material processed with a repetition rate of $f = 200$ kHz in comparison to $f = 20$ kHz.

The model agrees well with the experimental results. When considering the standard deviations of the residual stress measurements, the relative deviation between calculated and measured residual stresses is on average 5 %.

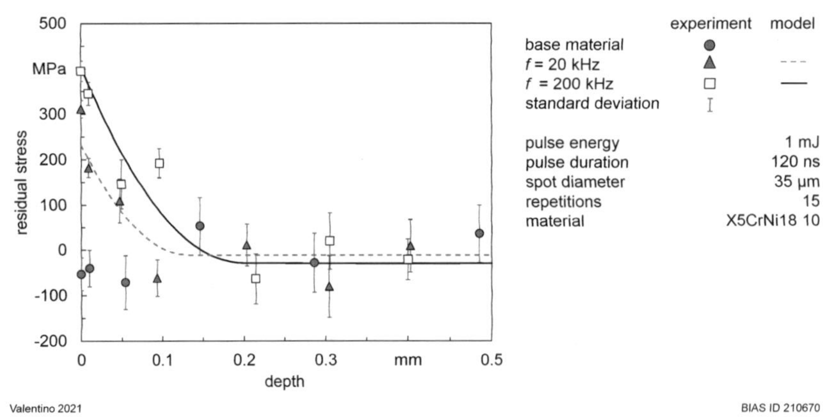

Figure 4: Residual stress depth profile of base material and comparison between model and experiment of residual stress depth profiles of LPwC-processed sheets.

IOP Conf. Series: Materials Science and Engineering 1135 (2021) 012025 doi:10.1088/1757-899X/1135/1/012025

In **Fig. 5**, the measured and predicted residual stress on the surface are compared. The dotted line shows the expected trend. It is to be expected that the measured residual stresses correspond equally to the determined ones. The dispersion of the predicted residual stresses is derived from the determined dispersion of deflection (shown in **Fig. 9**). The predicted values agree relatively well with the measured ones. Significant increases in deviation between predicted and measured stresses are observed for smaller stresses. The residual stresses are underestimated for lower deflections.

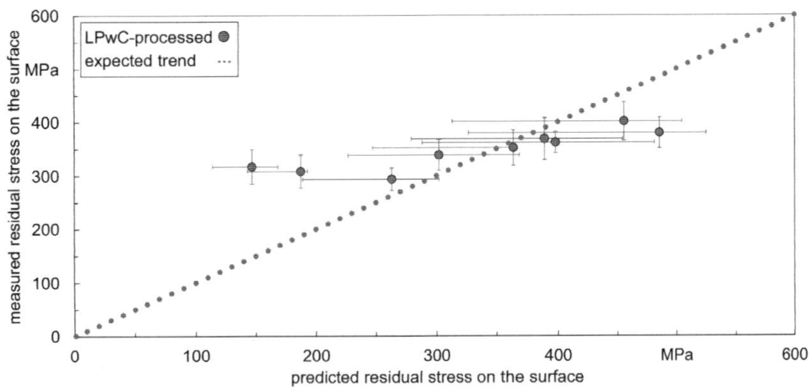

Figure 5: Comparison of measured and predicted residual stress on the surface of LPwC-processed sheets.

Induced stresses can cause a shape change of sheets. Almost no deflection is detected for the solution-annealed sheets before LPwC treatment. Their average deflection is $u = 0.53$ µm ± 1.50 µm. **Fig. 6** shows the deflection and ablation depth in dependence of scan repetitions for two different repetition rates. For both, the deflection increases until 8 scan repetitions. After 8 scan repetitions, the deflection seems to decline. In contrast, the ablation depth increases constantly for both repetition rates. Deflection and ablation rate are higher for larger repetition rate.

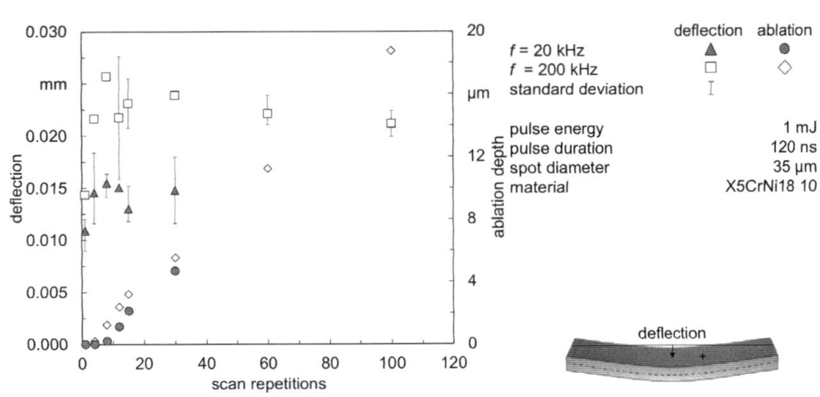

Figure 6: Deflection of sheets and ablation depth in dependence of scan repetitions.

Fig. 7 shows the derived transition depths in dependence of the scan repetitions. In these and all the other experiments (as listed in **Tab. 1**), no trend can be identified. An average transition depth of 0.14 mm ± 0.04 mm is derived from all the experiments.

Fig. 8 shows the deflection in dependence of average laser power for 15 repetitions. The deflection seems to increase when higher average laser power is used. In **Fig. 9** the maximum residual stresses obtained from the model are compared with the bending angle dispersion. Dispersion decreases with increasing tensile residual stress. 0 MPa residual stress is taken for the base material. A linear trend is observed between bending angle dispersion and maximum residual stress with a coefficient of determination of $R^2 = 0.31$. The largest bending angle dispersion is found for the initial base material.

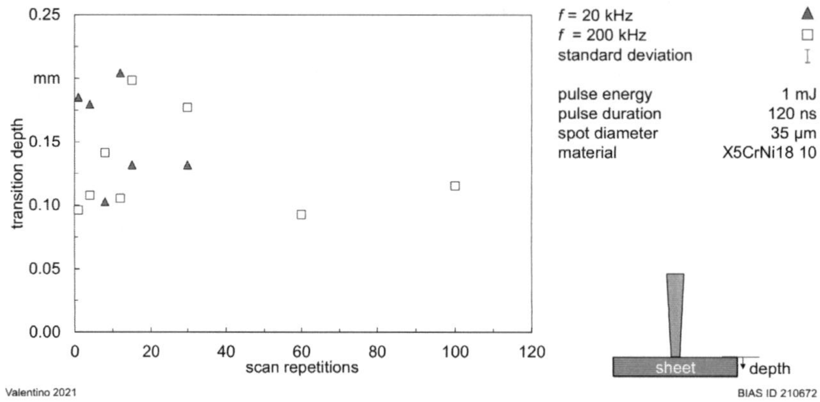

Figure 7: From hardness-depth profiles determined transition depth z in dependence of the total deposited energy.

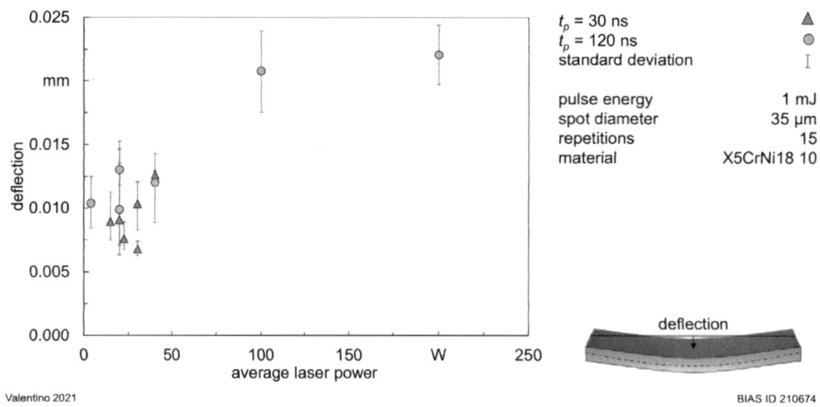

Figure 8: Deflection in dependence of the average laser power.

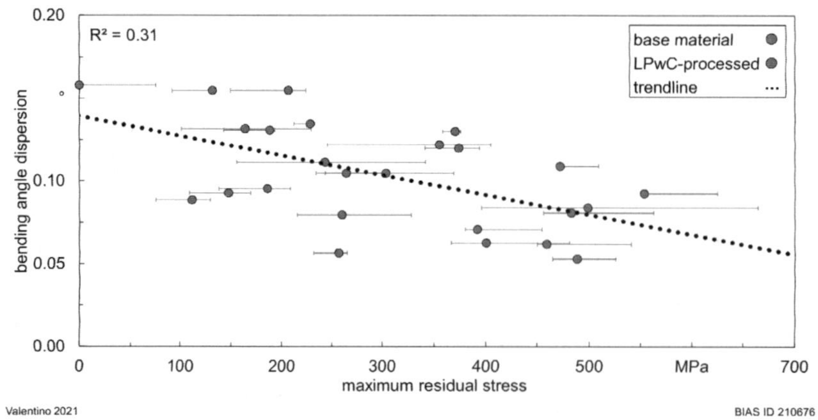

Figure 9: Bending angle dispersion in dependence of maximum residual stress

Fig. 10 compares the bending angle dispersion with the residual stress integral. The residual stress integral describes the sum of tensile residual stresses up to the depth where the residual stresses change from tensile to compressive residual stresses. The bending angle dispersion declines with increasing residual stress integral. Here, a linear trend can also be observed between bending angle dispersion and residual stress integral, but with a stronger coefficient of determination of $R^2 = 0.61$.

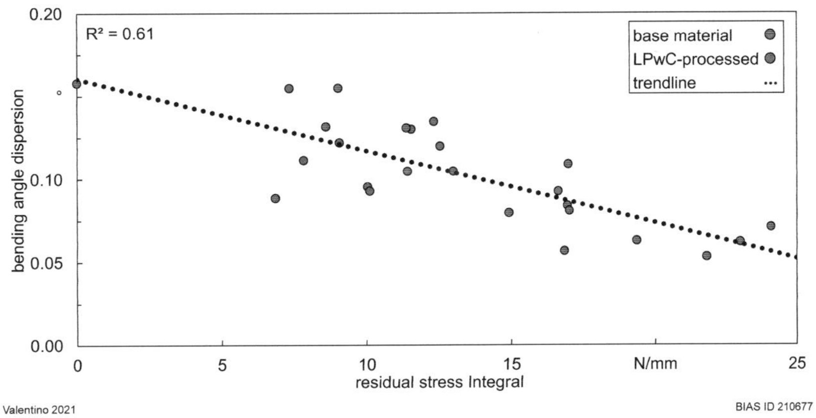

Figure 10: Bending angle dispersion in dependence of residual stress integral

5. Discussion

A pragmatic model is proposed to predict the residual stress-state caused by LPwC. The idea is to derive the change in residual stress depth profile of LPwC-processed sheets from experimentally determined shape deviations and hardness depth profiles. The advantage of this model is that numerous residual stress measurements by X-ray diffractometry can be avoided. The comparison between model and experiments shows in **Fig. 4** that the derived residual stress depth profiles correlate well with the measured profiles. Tenile residual stresses are undesirable in many applications, as they e.g., reduce the fatigue strength. Still, they can be estimated with the model based on hardness and deflection. The relative deviation between calculated and measured residual stresses is on average 5 %. Tensile residual stresses are also described by Peyre et al. in the surface region when using LPwC [17].

However, the relative deviation increases on the surface and for smaller sheet deflections (see **Fig. 5**). This increase can be explained by the formation of martensite. LPwC leads to material ablation because no sacrificial layer was applied. Material ablation with a nanosecond-pulsed laser can cause melting and rapid cooling [28] resulting in the formation of a martensitic layer on the surface [22]. Although, the martensitic content on the surface differs in dependence of the laser parameters, the martensitic layer thickness stays below 10 µm independent of the investigated laser parameters. The martensitic layer counteracts thermal-induced sheet deflections. Accordingly, the measured sheet deflection is the result of both mechanisms. However, thermal-induced tensile residual stresses are dominant for all the investigated laser parameters. Thus, it is suggested that the influence of counteracting forces by the martensitic layer become stronger relative to the thermal-induced forces when sheet deflections are below 10 µm (see **Fig. 5**).

Deflections and accordingly, tensile residual stresses are higher on the surface with increasing repetition rate although the transmitted energy per unit length is the same (see **Fig. 4**). In contrast to the measured deflection, the transition depth is not affected within the observed parameter spectrum (see **Fig. 7**). When increasing the repetition rate, the heat may not be removed fast enough by heat conduction into the sheet leading to higher thermal impact and therefore, to higher tensile residual stresses [29]. Moreover, when heat accumulation occurs in the material [30], the ablation depth may increase because the energy required to vaporize the material is reduced. The latter is observed in **Fig. 6** for higher repetition rates.

When considering a larger deflection created with many scan repetitions, the ablation depth may also deteriorate the predictability of the model. The sheet thickness decreases by LPwC, which may overestimate the residual stress, as can be derived from Eq. (8).

Despite material ablation (see **Fig. 6**) and the deflection prior to bending (see **Fig. 8**), subsequent bending experiments reveal a positive impact of LPwC on bending angle dispersion (see **Fig. 9**). It is demonstrated that sheets with tensile residual stresses on the surface have smaller bending angle dispersion. For the observations to be consistent with the descriptions by Reissner et al. concerning the influence of deviations in material state on springback [5], this means that the residual stress should saturate at a maximum level in dependence of the laser parameters. This also means that LPwC minimizes the influence of the initial sheet state. A quantity that better describes the observations by Reissner et al. is the residual stress integral. The model reveals that a larger residual stress integral (see **Fig. 10**) has a positive effect on bending angle dispersion. Consequently, two main parameters are found to achieve low bending angle dispersion with LPwC. Firstly, the residual stress must be maximal at the surface. Secondly, the transition depth from tensile to compressive residual stress must be as deep as possible in the sheet. Moreover, it is expected that the method is neither limited to LPwC nor to thin sheets. This effect may be the key to derive favorable stress-profile for subsequent forming operations. It is suggested that a residual stress-depth-profile can be created by LPwC, which locally converges to the one induced by bending operations.

6. Conclusion

A simple model based on hardness-depth profiles and deflection was successfully derived to determine the residual stress-depth profile of LPwC-processed sheets. From the model, it is concluded that two parameters are essential for low bending angle dispersion, which increases the robustness of subsequent bending operations:

- Higher tensile residual stresses on the surface are beneficial
- The transition depth from tensile to compressive residual stress must be deep in the sheet

Since the transition depth is nearly constant at about 0.14 mm ± 0.04 mm in the observed parameter spectrum, the model can be used for tailoring the stress profile of thin sheets by LPwC for subsequent bending processes. To achieve higher tensile residual stresses on the surface, it is revealed that higher repetition rates are advantageous for low bending angle dispersion because of heat accumulation effects.

Acknowledgments

Financial support of the project "Increasing reproducibility of the springback angle of thin metal sheets by inducing residual compressive stress by means of laser shock peening", which is funded by the Deutsche Forschungsgemeinschaft (DFG, German Research Foundation) – Project No. 399930874 – is gratefully acknowledged.

ORCID iDs

T Valentino https://orcid.org/0000-0002-1643-7819
A Stephen https://orcid.org/0000-0002-1207-1346
T Radel https://orcid.org/0000-0002-6681-2111

References

[1] Geiger M, Kleiner M, Eckstein R, Tiesler N and Engel U 2001 CIRP Ann. – Manuf. Technol.
[2] Diehl A (2010) Größeneffekte bei Biegeprozessen - Entwicklung einer Methodik zur Identifizierung und Quantifizierung, Dissertation, Universität Erlangen-Nürnberg, Germany.
[3] v Ravensvaaij R, v Tijum R, Hora P, v d Boogaard T and Engel, U 2013 Proc. IDDRG 2013, Zurich, Switzerland.
[4] Xia Z C, Cao J 2014 CIRP Encyclopedia Prod. Eng.
[5] Reissner J, Müller-Duysing M, Dannemann E and Ladwig J 1990 Umformtechnik

[6] Tekkaya A E, Allwood J M, Bariani P F, Bruschi S, Cao J, Gramlich S, Groche P, Hirt G, Ishikawa T, Lübbe C, Lueg-Althoff J, Merklein M, Misiolek W Z, Pietrzyk M, Shivpuri R and Yanagimoto J 2015 CIRP Ann. – Manuf. Technol.
[7] Tekkaya A E, Gerhardt J and Burgdorf M 1985 CIRP Ann. – Manuf. Technol.
[8] Clauer A H 1999 Surface Performance of Titanium, TMS.
[9] Ding K and Ye L 2006 Laser Shock Peening.
[10] Peyre P and Fabbro R 1995 Opt. and Quantum Electron.
[11] Fox J A 1974 Appl. Phys. Lett.
[12] Fabbro R, Fournier J, Ballard P, Devaux D and Virmont J 1990 J. Appl. Phys.
[13] Fairand B P, Wilcox B A, Gallagher W J and Williams D N 1972 J. Appl. Phys.
[14] Fabbro R, Peyre P, Berthe L, Sollier A and Bartnicki E 2000 Proc. SPIE 3888, Osaka, Japan.
[15] Clauer A H 2019 Metals.
[16] Sano Y, Altenberger I, Scholtes B, Masaki K, Ochi Y and Akita K 2006 Proc. ASME Pressure Vessels and Piping Division Conference, Vancouver, Canada.
[17] Peyre P, Berthe L, Scherpereel X and Fabbro R 1998 J. Mater. Sci.
[18] Kulkarni A, Prabhakaran S, Chettri S and Kalainathan S 2018 Int. J. Eng. Sci. Technol.
[19] Ivetic G, Meneghin I, Troiani E, Molinari G, Ocaña J, Morales M, Porro J, Lanciotti A, Ristori V, Polese C, Plaisier J and Lausi A 2012 Mater. Sci. Eng.
[20] Sakino Y and Sano Y 2018 Yosetsu Gakkai Ronbunshu.
[21] Tanaka H, Akita K, Sano Y, Mukai N and Ohya S 2006 J. Soc. Mater. Sci. Jpn.
[22] Valentino T, Stephen A and Radel T 2021 J. Laser Appl.
[23] Brinksmeier E, Cammett J T, König W, Leskovar P, Peters J and Tönshoff H K 1982 CIRP Ann. – Manuf. Technol.
[24] Tönshoff H K 1966 Eigenspannungen und plastische Verformungen im Werkstück durch spanende Bearbeitung, Dissertation, Technische Universität Hannover, Germany.
[25] Brinksmeier E and Sölter J 2009 CIRP Ann. – Manuf. Technol.
[26] Meschede D 2004 Gerthesen Physik, Springer-Verlag, Berlin.
[27] Beer O 2019 J. Heat Treat. Mater.
[28] Aramcharoen A, Mativenga P T and Manufacturing & Laser Processing Group 2007 Int. J. Adv. Manuf. Technol.
[29] Weber R, Graf T, Berger P, Onuseit V, Wiedenmann M, Freitag C, and Feuer A 2014 Opt. Express.
[30] Momma C, Chichkov B N, Nolte S, v Alvensleben F, Tünnermann A, Welling H and Wellegehausen B 1996 Opt. Commun.

IOP Conf. Series: Materials Science and Engineering 1135 (2021) 012026 doi:10.1088/1757-899X/1135/1/012026

Ultrashort-pulsed laser processing with spatial and temporal beam shaping using a spatial light modulator and burst modes

Christian Lutz[1], Cemal Esen[2], Ralf Hellmann[1]

[1]Applied laser and photonics group, University of Applied Sciences Aschaffenburg, Wuerzburger Strasse 45, 63743 Aschaffenburg, Germany

[2]Applied Laser Technologies, Ruhr University Bochum, Universitaetsstrasse 150, 44801 Bochum, Germany

Christian.Lutz@th-ab.de

Abstract. We report on the effect of simultaneous spatial and temporal beam shaping on the ablation rate, ablation efficiency and the resulting surface characteristics of micromachined stainless steel using ultrashort-pulsed lasers. Beam shaping and the use of pulse bursts are promising methods to allocate the over the last decades increasing laser power of ultrashort-pulsed lasers in ablation processes. While the individual effects of beam shaping and pulse bursts on the ablation characteristics have recently been examined, the combination of both has not yet been adequately investigated. Using a spatial light modulator to generate different spot distributions with up to six spots and different separations it is possible to spatially distribute the available laser power. In combination with temporal beam shaping using a 200 kHz repetition rate and pulse bursts with a 40 MHz intra-burst rate, we investigate the influences in a scanning-based process and find an increasing ablation rate and efficiency for higher fluences. Subsequently using bursts in combination with a multi-spot beam profile, we found a distinctive emergence of cone like protrusions and a smoothing effect for fluences between 1.5 J/cm² and 3 J/cm² with six spot beam profile.

1. Introduction

The use of ultrashort-pulsed lasers has been proven to exhibit advantages for high quality requirements in laser micromachining. Especially processing stainless steel has been investigated for several years to optimize removal rates and the processed surface quality by using different laser parameters [1–3], pulse durations [4; 5],beam profiles [6–8] or machining strategies [9]. However, for industrial use often the throughput of ultrashort-pulsed laser based machining approaches is not high enough. And since for steel the preferable fluence level for efficient machining is determined to be within the range of 0.2-0.4 J/cm², a simple increase of the available average to several hundred Watts does not advance micromachining [1; 2]. Quite the contrary, a higher fluence results in negative effects like heat accumulation [10; 11], plasma shielding [12] and as a consequence in a lower ablation efficiency [11].

An approach to distribute the available laser power is by using spatial beam shaping technology [13; 14]. Optical beam shaping can be achieved by different solutions like digital mirror devices, deformable mirrors, diffractive optical elements, acusto-optic deflectors and spatial light modulators (SLM) [15]. With these optical elements it is possible to shape the beam towards the desired spatial intensity distribution including multi-beam arrays, top-hat profiles and Bessel beams [16–18].

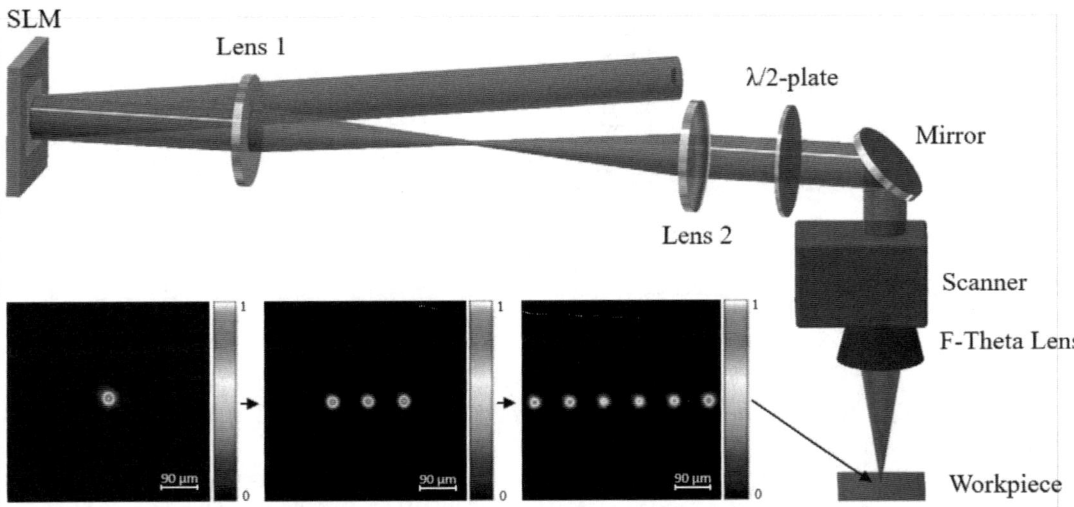

Figure 1. Schematic illustration of the setup for femtosecond laser ablation using a spatial light modulator (SLM) and a 4f-telescope with a galvo scanner including an F-Theta lens. Also shown are measured intensity profiles with one, three, and six spots with a separation of 90 µm.

In the time domain, also the burst mode emerges to be a promising method of increasing the throughput of ultrashort-pulsed laser processes due to an increasing removal rate [19–21]. With the possibility to distribute the energy of a single pulse to a certain number of sub-pulses, the potentially available fluence can be split over the sub-pulses to maintain efficient material removal per sub-pulse. Using a MHz burst, two intra-burst pulses have a distance of a few nanoseconds, whereas in GHz burst regime the delay of two pulses is in the picosecond range. [22] While for a delay of a few nanoseconds the residual heat of a preceding pulses can lead heat accumulation effects with an adverse effect on the removal rate [23; 24] a delay in the picosecond range may results in stronger shielding effects induced by plasma formation [20; 25].

While the individual effect of beam profiling and pulse bursts on the ablation characteristics have recently been examined by various studies, the combination of both has not yet been adequately investigated. In the context, here we investigate the combination of different beam profiles and pulse burst in the MHz regime in a scanning based process on stainless steel with particular focus on the ablation rate, ablation efficiency, and surface characteristics.

2. Experimental

We use an Yb:YAG laser (Amplitude Tangor) with a wavelength of 1030 nm and a maximum power of 100 W. The laser is integrated into a micromachining system (Pulsar Photonics, RDX-1000), which uses a galvo scanner (Scanlab, IntelliSCAN$_{se}$14) and a F-Theta lens (QiOptiq, LINOS F-Theta Ronar) with a focal length of 100 mm. The focal diameter d_o, measured by a high-resolution CCD camera (IDS, UI149xLE) is 41µm $(1/e^2)$. The respective fluence is calculated by $\Phi = E_p/(\pi \cdot r_0^2)$, with E_p being the pulse energy and r_0 the radius of the focal beam.

Experiments are performed using a pulse repetition rate of 200 kHz with 800 fs pulse duration (FWHM, measured after laser aperture by an autocorrelator) and a scanning speed of 2968 mm/s, resulting in a pulse overlap of 64 %. While the intra-burst repetition rate is given by the oscillator frequency at a constant value of 40 MHz, the laser can emit burst packages with a size of 200 maximum, dependent to the used repetition rate. The SLM based beam shaping module (cf. Fig.1) allows dynamic beam shaping to generate different spot distributions with a maximum frame rate of 60 Hz. To protect the LCOS-SLM (Hamamatsu X15223), an effective active liquid cooling system is integrated. Required computer generated (CGH) holograms are calculated using an iterative flourier

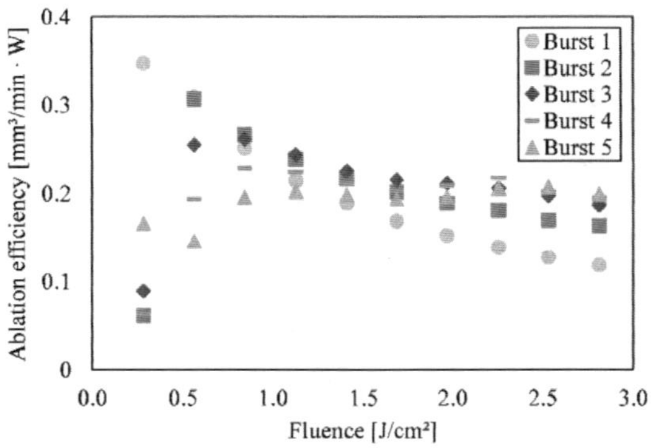

Figure 2. Ablation efficiency for fluencies between 0.28 and 2.8 J/cm² for different bursts. The ablation efficiencies are calculated of the basic surface level and the average depth of the ablated geometry, the calculated process time and the used laser power.

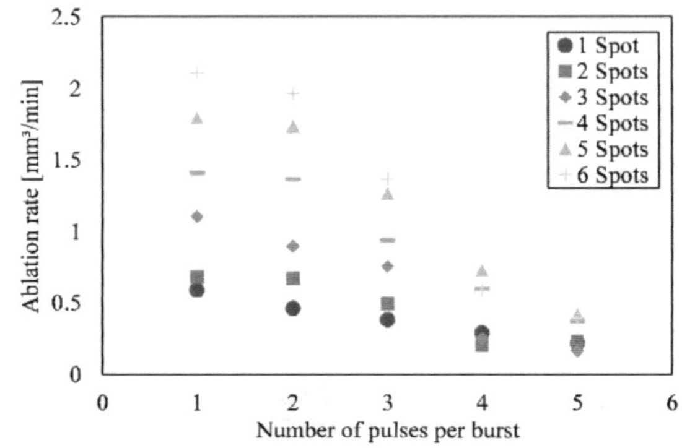

Figure 3. Ablation rate for a fluence of 0.6 J/cm² for different bursts and number of spots. The ablation rate is calculated of the basic surface level and the average depth of the ablated geometry after 50 scanning layers.

transform algorithm. After the initial calculation of the CGH, the uniformity of the single sub-beams in the focal plane is further improved by a feedback loop based weighted Gerchberg-Saxton algorithm [26]. As the diffraction efficiency for each CGH differs, we control the total reflective power from the SLM using a power meter (Gentec EO, UP25N 100H H9) to ensure equal power impingement for each beam profile. The ultrashort-pulse laser is used to ablate a defined geometry, i.e. a cavity in a precise layer-by-layer ablation process by scanning the laser spot over the sample. The bidirectional hatched area has a size of 2 x 2 mm². By adding an additional rotation of $\alpha = 11°$ after each hatch, we obtained a homogenous irradiated surface after 50 scans. The ablated stainless steel (X5CrNi18-10) structures are analyzed by a laser-scanning microscope (Keyence, VK-X200series) and a scanning electron microscope (Tescan, Maia3 TriglarTM).

3. Results and discussion

3.1. Ablation rate and efficiency

Ablation characteristics shown in Fig. 2-5 are studied by employing different laser powers and burst package sizes in a scanning based process with 50 layers in a fluence range between 0.28 J/cm² and 2.8 J/cm². For the differently used beam profiles with a constant spot separation of 90 µm, the laser power range is customized on the spot number to retain the same fluence per spot.

The results shown in Fig. 2 reveal a higher ablation efficiency for a single pulse and a lower number of bursts for fluences below 1.5 J/cm², where the graphs intersect. This is in accordance to previous studies by Jaeggi et al., showing a similar devolution in a fluence zone around 1.5 J/cm² for a spot radius of 16 µm with a wavelength of 1064 nm on stainless steel [27]. Above this level, the ablation efficiency decreases with a low gradient for the single pulse and burst two, while for higher bursts the ablation efficiency increases slightly to a maximum of 0.2 mm²/min·W. However, the optimum fluence is shifted to higher values using of a higher number of bursts, concurring with investigations of Kramer et al. [19]. As shown in Fig. 3, for a fluence of 0.6 J/cm² all beam profiles show a similar trend of the ablation rate with a decreasing character for higher burst numbers. The comparatively low fluence of 0.6 J/cm² results in a high ablation rate for a single pulse and lower rates for a higher number of bursts, because the used energy is equal or below the optimal fluence of 0.3 J/cm² [6] for the machined material, agreeing with studies using a single spot process [22]. This behaviour is supported by the increasing negative slope for a higher number of spots and thus the distribution of the energy used over several sub pulses.

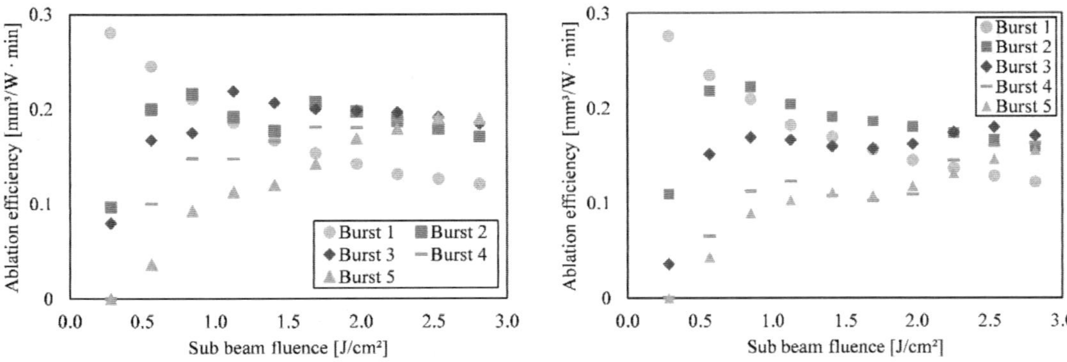

Figure 4. Ablation efficiency for a three (left) and six (right) spot profile and fluencies between 0.28 and 2.8 J/cm² for different bursts. The ablation efficiencies are calculated of the basic surface level and the average depth of the ablated geometry, the calculated process time and the used laser power.

Based on these results, we following study the influence of a different number of sub beams in the same fluence range to detect any influence of the combination of temporal and spatial beam shaping on the ablation efficiency and ablation rate. Due to an intensity uniformity of minimum 0.95 regarding the multi spot profiles, the fluence in each spot is almost the same.

As depicted in Fig. 4 both, the three and the six spot beam profile exhibit the same characteristics. A low fluence in combination with bursts tends to result in a decreasing ablation efficiency, while for a higher fluence bursts from two to five induce an ablation efficiency higher than a single pulse, a typically behaviour for using a single spot [25]. As depicted in both charts, the disparities for the tested bursts are very small for fluences between 2.2 and 2.8 J/cm² also observed for a fluence of 2.5 J/cm² for two to 10 bursts form Lickschat et al. [22].

In Fig. 5 we illustrate the achievable ablation rates for a spot number from one to six and bursts from one to five for a fluence of 2.8 J/cm². Using the six spot profile and a burst number of three, we achieve a maximum of 7.7 mm³/min, which corresponds with a factor of 1.4 as compared no burst usage. For almost all investigated beam profiles a burst number of tree results in the highest ablation rate, which can be concluded from an optimum in terms of process efficiency. The lower ablation rate of burst 2 could be due to the absorption of the second impinging pulse by the still existing plasma formation, so

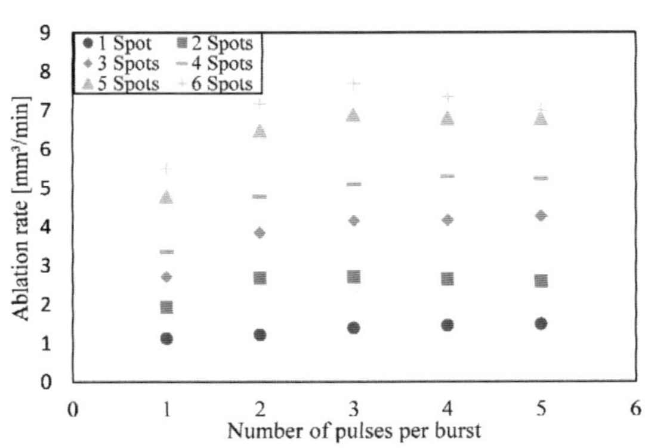

Figure 5. Ablation rate for a fluence of 2.8 J/cm² for different bursts and number of spots. The ablation rate is calculated of the basic surface level and the average depth of the ablated geometry after 50 scanning layers.

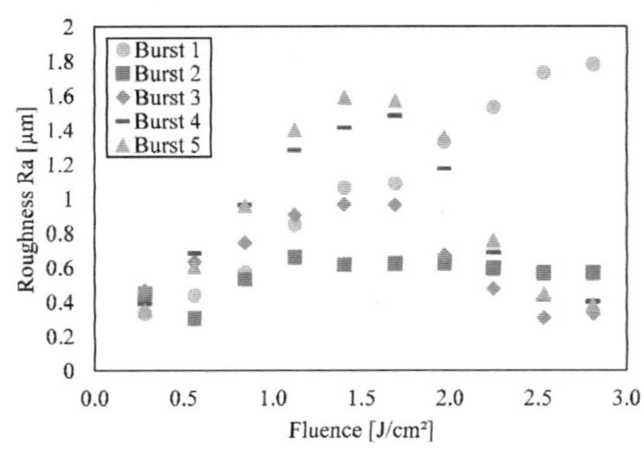

Figure 6. Effect on the roughness of different bursts, a single spot and fluencies between 0.28 and 2.8 J/cm². The measured surface is a rectangle of 80x60 µm on the bottom with 50 layers ablated volume, recorded with a 150x enlargement.

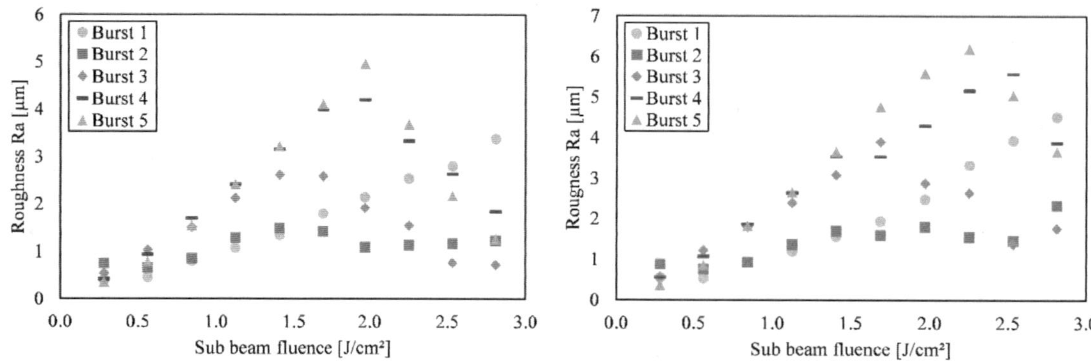

Figure 7. Effect on the roughness of different bursts, for four (left) and six (right) spots, fluencies between 0.28 and 2.8 J/cm². The measured surface is a rectangle of 80x60 µm on the bottom with 50 layers ablated volume, recorded with a 150x enlargement.

that it hardly produces any ablation. The use of burst 4 and thus the division of the pulse energy into four sub-pulses leads to a lower ablation efficiency (Fig. 4) and thus also to a lower ablation rate.

However, we found no negative influence using multi spot beam profiles with a separation of 90 µm and bursts with a repetition rate of 40 MHz, but a possibility to distribute a laser power of 45 W for six spots in an efficient micromachining process with a maximum ablation rate of 7.7 mm³/min, which is a great value compared to those reported in literature.

3.2. *Surface characteristics*

Equal to the prior documented parameters ablation rate and –efficiency, another important aspect for micromachining processes for industrial applications is the surface quality. In particular, ultrashort-pulsed lasers are distinguished for low surface roughness, but with an increasing available laser power and resulting high process fluences this advantage is competing with machining time.

Figure 6 shows the influence of bursts one to five for a fluence between 0.28 and 2.8 J/cm² using a single spot. While the increasing roughness of the single pulse shows an almost linear increase in a typical range [28; 29], burst two and three commute at a constant roughness around 0.6 µm above a fluence of 1 J/cm². Using burst four and five, the roughness increases to the highest value of 1.7 µm, leveling down for fluences above 1.5 J/cm to the measured minimum of 0.3 µm. Lickschat et al. measured a surface roughness in the same range for a burst of four and five with a pulse duration of 1 ps on stainless steel [22].

As illustrated in Figure 7, obviously, higher fluence results in larger roughness for multi spot ablation processes using no burst mode, similar to previous studies using the same laser setup but different beam distributions [6]. For the tested bursts, the results show a corresponding behavior with a slightly drift of the maximum roughness to higher fluence, increasing the number of spots. However, expecting the high ablation rate above 7 mm³/min for a six spot profile, a burst of three and a fluence of 2.5 J/cm², we reached a comparatively low roughness of around 1.5 µm. This low roughness might be attributed to the removal of the previously generated CLP structures (Fig. 8) and possibly to a melting of the material due to higher heat accumulation, which resembles a smoothing effect. In combination, the high removal rates achieved in conjunction with a low roughness demonstrates the high potential of the combined burst and multi spot ablation.

To investigate the different slopes regarding the influence of bursts, fluence and the spot number, in Figure 8 we depict different surfaces, taken with a scanning electron microscope for the single spot and the 4 spot beam profile. Apparently, the variation of roughness can be identified by structure formations influenced by the fluence, the number of burst and the spot number. Using a single spot and higher bursts, the formation of so called cone like protrusions (CLP) is reinforced for a fluence of 1.4 J/cm². After exceeding this fluence, the CLPs are ablated and a smooth surface results, corresponding to the

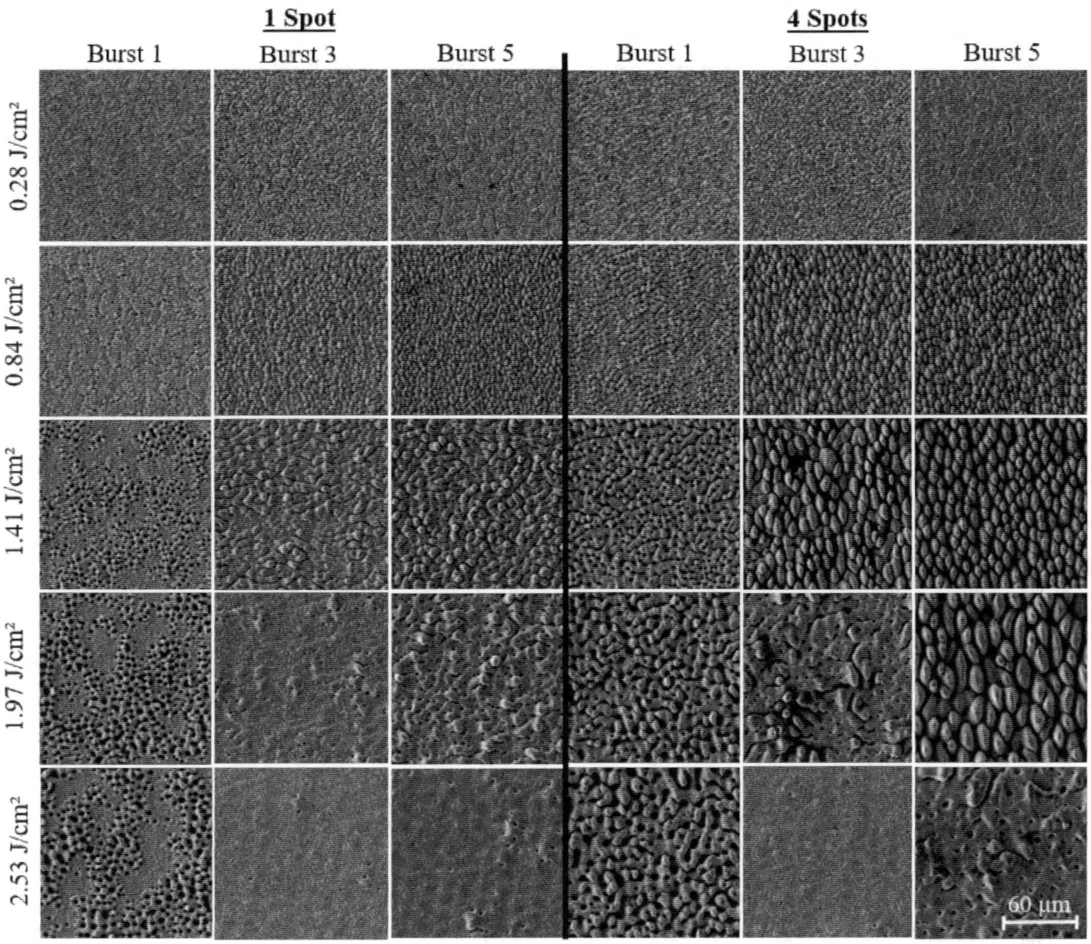

Figure 8. Illustrations of the different surfaces after 50 layers ablation, taken with a scanning electron microscope. Resulting surfaces for various fluences and numbers of bursts are shown using a single spot (left part) or a four spot profile (right part).

results shown in Fig. 6 and previous studies [22; 30]. The ablation of material at too high peak fluencies causes a pronounced formation of cavities, craters and CLPs. This behaviour can be deferred to higher fluences using bursts [31]. As the surfaces ablated with the four spot profile show in Fig. 8, these effect can be reinforced with the consequence of distinct CLP formations until a fluence of 0.8 J/cm². Comparing the first and fourth column in Figure 8 shows that the use of a higher average power distributed over four spots results in a more pronounced emergence of CLPs even without the use of bursts. This effect increases with the use of burst three and five, which leads to the emergence of CLPs already in the range of lower fluences and generally larger CLPs in the higher fluence range. Furthermore, using SLM based beam shaping, higher diffraction orders occur around the obtained beam profile, or scattered radiation between the spots. Although these do not reach high fluences, so that no material is ablated, they can still heat it before the ablation process through the spots. These assessments are consistent with studies by other authors that identify heat accumulation as one of the driving factors in the formation of CLPs [32].

4. Conclusion

We have demonstrate the combination of spatial and temporal beam shaping and their effects on ablation efficiency, ablation rates and surface characteristics in a scanning based ultra-short pulsed laser ablation micromachining process on stainless steel. To increase the maximum ablation rate and for investigation of effects induced of a 40 MHz burst and different spot distributions, we combine bursts from one to 5 with a maximum of six spots, finding ablation rates up to 7.7 mm^3/min with an ablation efficiency of 0.2 mm^3/min·W. Due to the high ablation efficiency and a suitable pulse interaction behaviour, the highest ablation rates are achieved with Burst 3.We found no negative effects regarding the ablation rate and efficiency using spot numbers from one to six but a correlation between the charts of the corresponding bursts and an efficient way to distribute a laser power of 45 W with six spots in a micromachining process. In addition, we show different roughness characteristics between a single pulse and the usage of bursts, with only a minor drift to higher fluences, increasing the number of spots. Compared to a single spot process with the highest used fluence, combining bursts of two or three and a multi-spot results in a lower surface roughness during a higher removal rate. Furthermore, we note reinforced formation and size of cone like protrusions in some fluence zones when we use the multi-spot profiles and bursts, which could be related to increased heat accumulation resulting from higher average power, higher orders or scattered radiation due to the SLM based beam shaping or the heat influence of the bursts in MHz range. Finally, both a higher removal rate and a low roughness can be achieved by the right combination of temporal and spatial beam shaping, using bursts and multi-spot profiles.

Acknowledgement

This work has been financially support by the German Federal Ministry of Education and Research (BMBF grant FKZ 13FH150IN6).

ORCiD iDs

C Lutz https://orcid.org/0000-0001-8698-9902
C Esen https://orcid.org/0000-0001-6518-9914
R Hellmann https://orcid.org/0000-0001-5592-3388

References

[1] Hodgson N, Heming S, Steinkopff A, Haloui H and Lee T Ultrafast Laser Ablation at 1035nm, 517nm and 345nm as a Function of Pulse Duration and Fluence *Lasers in Manufacturing Conference 2019* 151

[2] Neuenschwander B, Jaeggi B, Schmid M and Hennig G 2014 Surface Structuring with Ultra-short Laser Pulses: Basics, Limitations and Needs for High Throughput *Phys. Procedia* **56** 1047

[3] Raciukaitis G 2009 Use of High Repetition Rate and High Power Lasers in Microfabrication: How to Keep the Efficiency High? *J. Laser Micro Nanoeng.* **4** 186

[4] Schille J, Schneider L and Loeschner U 2015 Process optimization in high-average-power ultrashort pulse laser microfabrication: how laser process parameters influence efficiency, throughput and quality *Appl. Phys. A* **120** 847

[5] Neuenschwander B, Jaeggi B, Schmid M, Rouffiange V and Martin P-E 2012 Optimization of the volume ablation rate for metals at different laser pulse-durations from ps to fs *Proceeding of SPIE* **8243** 824307

[6] Lutz C, Roth G, Rung S, Esen C, and Hellmann R 2021 Efficient Ultrashort Pulsed Laser Processing by Dynamic Spatial Light Modulator Beam Shaping for Industrial Use *J. Laser Micro Nanoeng.* **16** 62-67

[7] Gillner A, Finger J, Gretzki P, Niessen M, Bartels T and Reininghaus M 2019 High Power Laser Processing with Ultrafast and Multi-Parallel Beams *J. Laser Micro Nanoeng.* **14** 129
[8] Bruening S, Du K, Jarczynski M, Jenke G and Gillner A 2018 Ultra-fast laser micro processing by multiple laser spots *Procedia CIRP* **74** 573
[9] Finger J and Hesker M 2021 High power ultrashort pulse laser processing using a flexible multibeam approach *J. Phys. Photonics* **3** 21004
[10] Bauer F, Michalowski A, Kiedrowski T and Nolte S 2015 Heat accumulation in ultra-short pulsed scanning laser ablation of metals *Opt. Express* **23** 1035
[11] Finger J and Reininghaus M 2014 Effect of pulse to pulse interactions on ultra-short pulse laser drilling of steel with repetition rates up to 10 MHz *Opt. Express* **22** 18790
[12] König J, Nolte S and Tünnermann A 2005 Plasma evolution during metal ablation with ultrashort laser pulses *Opt. Express* **13** 10597
[13] Bernard O, Audouard E, Schöps B, Delaigue M, Dalla-Barba G, Mishchik K, Hönninger C and Mottay E 2018 Efficient micro processing with high power femtosecond lasers by beam engineering and modelling *Procedia CIRP* **74** 310
[14] Schille J, Schneider L, Mauersberger S, Szokup S, Höhn S, Pötschke J, Reiß F, Leidich E and Löschner U 2020 High-Rate Laser Surface Texturing for Advanced Tribological Functionality *Lubricants* **8** 33
[15] Salter P S and Booth M J 2019 Adaptive optics in laser processing *Light, science & applications* **8** 110
[16] Hendriks A, Naidoo D, Roux F S, López-Mariscal C and Forbes A 2012 The generation of flat-top beams by complex amplitude modulation with a phase-only spatial light modulator *Proceeding of SPIE* **8490** 849006
[17] Orlov S, Vosylius V, Gotovski P, Grabusovas A, Baltrukonis J and Gertus T 2018 Utilizing Fundamental Beam-Mode Shaping Technique for Top-Hat Laser Intensities in Direct Laser Interference Patterning *J. Laser Micro Nanoeng.* **13** 268-272
[18] Schwarz S, Rung S, Esen C and Hellmann R 2018 Homogeneous Low Spatial Frequency LIPSS on Dielectric Materials Generated by Beam-Shaped Femtosecond Pulsed Laser Irradiation *J. Laser Micro Nanoeng.* **13** 90-94
[19] Kramer T 2017 Increasing the Specific Removal Rate for Ultra Short Pulsed Laser-Micromachining by Using Pulse Bursts *J. Laser Micro Nanoeng.* **12** 107
[20] Žemaitis A, Gečys P, Barkauskas M, Račiukaitis G and Gedvilas M 2019 Highly-efficient laser ablation of copper by bursts of ultrashort tuneable (fs-ps) pulses *Scientific reports* **9** 12280
[21] Hu W, Shin Y C and King G 2010 Modeling of multi-burst mode pico-second laser ablation for improved material removal rate *Appl. Phys. A* **98** 407
[22] Lickschat P, Metzner D and Weißmantel S 2021 Burst mode ablation of stainless steel with tunable ultrashort laser pulses *Journal of Laser Applications* **33** 22005
[23] Metzner D, Lickschat P and Weißmantel S 2019 Laser micromachining of silicon and cemented tungsten carbide using picosecond laser pulses in burst mode: ablation mechanisms and heat accumulation *Appl. Phys. A* **125** 3439
[24] Jäggi B, Förster D J, Weber R and Neuenschwander B 2018 Residual heat during laser ablation of metals with bursts of ultra-short pulses *Advanced Optical Technologies* **7** 175
[25] Neuenschwander B, Kramer T, Lauer B and Jaeggi B 2015 Burst mode with ps- and fs-pulses: Influence on the removal rate, surface quality, and heat accumulation **9350** 93500U
[26] Memmolo P, Miccio L, Merola F, Paciello A, Embrione V, Fusco S, Ferraro P and Antonio Netti P 2014 Investigation on specific solutions of Gerchberg–Saxton algorithm *Opt. Lasers Eng.* **52** 206
[27] Jaeggi B, Remund S, Zhang Y, Kramer T and Neuenschwander B 2017 Optimizing the Specific Removal Rate with the Burst Mode Under Varying Conditions *J. Laser Micro Nanoeng.* **12** 258-266

[28] Žemaitis A, Gaidys M, Gečys P, Račiukaitis G and Gedvilas M 2019 Rapid high-quality 3D micro-machining by optimised efficient ultrashort laser ablation *Opt. Lasers Eng.* **114** 83
[29] Cheng J, Perrie W, Edwardson S P, Fearon E, Dearden G and Watkins K G 2009 Effects of laser operating parameters on metals micromachining with ultrafast lasers *Appl. Surf. Sci.* **256** 1514
[30] Metzner D, Lickschat P and Weißmantel S 2021 Optimization of the ablation process using ultrashort pulsed laser radiation in different burst modes *Journal of Laser Applications* **33** 12057
[31] Sailer M, Bauer F, Kleiner J and Kaiser M Scaling of ablation rates. Ablation efficiency and quality aspects of burst-mode micromachining of metals *Lasers in Manufacturing Conference 2015* 265
[32] Kam D H, Bhattacharya S and Mazumder J 2012 Control of the wetting properties of an AISI 316L stainless steel surface by femtosecond laser-induced surface modification *J. Micromech. Microeng.* **22** 105019

IOP Conf. Series: Materials Science and Engineering 1135 (2021) 012027 doi:10.1088/1757-899X/1135/1/012027

Thermal imaging of high power ultrashort pulse laser ablation of alumina towards temperature optimized micro machining strategies

Stefan Rung[1], Niklas Häcker[1], Ralf Hellmann[1]

[1]Applied laser and photonics group, University of Applied Sciences Aschaffenburg, Wuerzburger Strasse 45, 63743 Aschaffenburg, Germany

stefan.rung@th-ab.de

Abstract. The application of pulsed laser systems with pulse durations in the pico- and femtosecond regime for material processing is commonly associated with a cold ablation. Due to the minimized interaction-time between the ultrashort laser pulses and the material, this statement is almost valid as long as no heat accumulation effect appears. With the increasing demand of high productivity processes, the average power of ultrashort pulsed laser systems increases above 100 W, which leads, however, to increased thermal effects during laser processing. This is especially important for laser processing of technical ceramics like alumina. Large temperatures gradients, which locally occur during laser processing using high average power could lead to thermal modifications and cracks in the material. In this study, we present a process-optimization method for high power laser ablation of alumina based on thermal imaging. The use of a 2D IR camera enables the estimation of the temperature distribution during the laser processing. We investigate the influence of laser power up to 80 W, pulse duration between 900 fs and 10 ps and processing duration on the resulting material temperature. Beside the material temperature we evaluate the material removal rate and the resulting surface quality.

1. Introduction

Nowadays technical ceramic materials are an essential building block in different important fields of modern products. Materials like alumina (Al_2O_3), zirconia (ZrO_2) or alumina nitride (AlN) provide outstanding mechanical, thermal, electrical and chemical properties for high technical demands [1; 2]. The probably most well-known oxide ceramic material is alumina and it combines very good electrical insulation (10^14 – 10^16 Ohm m), high mechanical strength (< 600 MPa), high corrosion and wear resistance, a low density (< 4 g/cm^3) and operating temperatures above 1000°C [3]. These features enable the application as substrates and heat sink for power electronics e.g. for engine control unit or charging unit for e-mobility, or heavy duty forming tools and wear protection. However, the processing of this brittle material is a challenge for conventional manufacturing processes like milling or Electrical Discharge Machining (EDM) [4; 5]. Limiting factors for a suitable production technology are the material removal rate and quality factors like cracking and the resulting surface roughness.

The application of laser irradiation for ceramic processing enables versatile material processing approaches without mechanical induced stress of the material. For ceramics processing, CO_2 lasers are primarily used for processes such as cutting, drilling and perforation [6; 7], For micro structuring, however, Q-switched solid state [8; 9] and Excimer-lasers [10; 11] have been applied. While Excimer-laser processing is linked to small ablation rates, the application of solid state lasers in the nanosecond regime enables moderate ablation rates of 1.5 mm^3/min using 20 W average laser power [12]. However, micro structuring using pulse durations in the nanosecond regime still introduces a high thermal load leading to melting effects which can cause micro cracks and stress break [8; 11; 13].

To overcome this thermal induced damages generated by the processing technology, the application of laser systems having ultrashort pulses is highly interesting [14–16]. By the reduction of energy-material interaction time using picosecond and femtosecond pulse durations, the thermal penetration depth of the individual laser pulse decreases and enables a so called "cold processing". However, with increasing pulse energy and pulse repetition rate, heat effects, e.g. by heat accumulation, cannot be avoided and lead to heating of the processed specimen. The heat accumulation is either a result of subsequent laser pulses by the pulse repetition rate or by a successive laser path during layer-wise processing. With upcoming high power industrial grade pico- and femtosecond lasers, there is a growing demand for machining strategies to deploy high average laser power and to overcome the limiting factors of the material removal rate while maintaining high quality surfaces and temperature minimized processes.

Typically, process parameter optimization is based on the resulting ablation characteristics, e.g. optimizing the material removal rate and/or the surface quality. Beyond this, in this study we focus on the thermal stress, which is introduced by the laser micromachining of Al_2O_3. Therefore the temperature of the processing region is analysed during laser micromachining and the effect of the resulting temperature as a function of the applied laser parameters is investigated. Information about the thermal load can either be gathered by a calorimetric investigation [17] or by the emission of thermal radiation using a pyrometer [18; 19] or a thermographic camera [20; 21]. The benefit of a thermographic camera is the possibility of a spatially resolved determination of the surface temperature and therefore a simultaneous measurement at different location on the processed specimen.

In this study we use a 2D-thermographic camera to evaluate the surface temperature as a function of the applied laser power up to 80 W, a pulse duration between 900 fs and 10 ps and the number of scans of a layer-wise processing, respectively. By a scanning electron microscope analysis we determine the effect of the accumulated temperature on the material in the direct vicinity of the processed area.

2. Experimental

In this study, alumina (Ceramtec) with a purity of 96%, specified grain size of 3 to 5 µm and an initial substrate thickness of 1 mm is used for laser ablation experiments. The initial surface roughness is Ra 1.5 µm, as measured using laser scanning microscope (Keyence VH-X).

For laser surface processing, we use a micro-machining station (WSMH, Optec) equipped with a high power ultrashort pulsed laser (Amphos 200) having a variable pulse duration between 900 fs and 10 ps (FWHM) and a repetition rate of up to 40 MHz. The maximum average power is 200 W with a maximum pulse energy of 1 mJ. For the presented ablation studies, the fundamental emission wavelength of 1030 nm is used. Figure 1 shows the experimental setup for the surface treatment.

The energy of the laser is adjusted by an external attenuator based on a rotating wave plate and a polarizer. Using an adjustable beam expander telescope, the raw beam diameter is adjusted to a diameter of 5.5 mm ($1/e^2$). A galvo scanner (Excelliscan 14, Scanlab) is used in combination with a telecentric lens (f = 163 mm) to focus the beam onto the sample with a spot diameter of 50 µm ($1/e^2$). All given values of the pulse energy and the resulting laser fluence are derived by the laser power measured directly behind the processing optic. The pitch p between two consequent pulses is chosen to be 10 µm, i.e. 80% spatial overlap, for both, scanning direction and perpendicular thereto. Using a pulse repetition rate of 200 kHz for all shown experiments, the corresponding scanning speeds is 2020 mm/s.

The temperature profile of the Al_2O_3 substrate during the laser processing is captured using a thermal imaging camera (TIM-640-VGA ,Micro Epsilon) with a measuring range of -20°C to 900°C and a frame rate of 32 Hz. The optical configuration of the temperature camera enables a spatial resolution of 100 µm x 100 µm per pixel with a total image size of 640 x 480 pixel. To avoid undesired IR reflection on the surface and to optimize the radiation of the thermal emission, we apply a thin layer of a black paint on the Al_2O_3 substrate. A single layer ablation test performed on the substrate with and without the thin coating reveals, that within the observed parameter range the coating does not influence the ablation behaviour of the Al_2O_3, yet ensures an emissivity between 0.9 and 0.95.

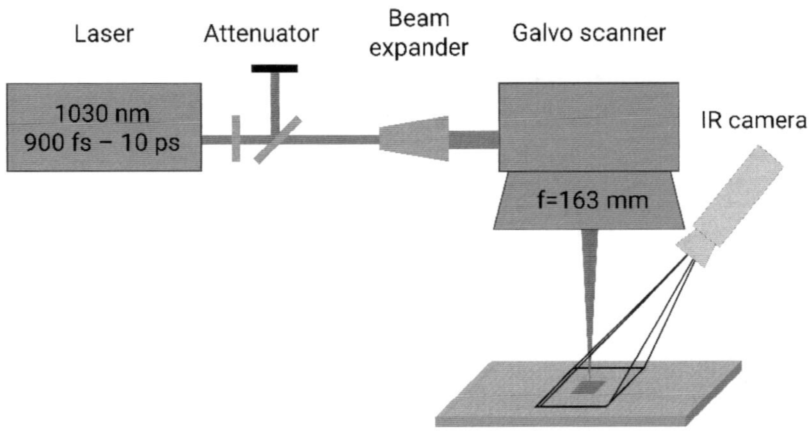

Figure1. Experimental setup for thermal imaging during laser ablation of Al2O3.

To evaluate the ablation characteristics, square cavities with an edge length of 2 mm are ablated using multifarious laser and processing strategy parameters. The measurement of the resulted cavity depth d is performed using an optical 3D profilometer (VR 3200, Keyence). For a detailed visual inspection of the cavity surface we apply a 15 nm thin gold layer on the material and use a scanning electron microscope (SEM, MAIA3, TESCAN).

The ablation efficiency AE (ablated volume per time and power) is calculated by the following equation with d being the depth d of the ablated cavities, p the pitch, n the number of ablated layers, f_R the repetition rate of the laser and P the average power of the laser [22], respectively.

$$AE = \frac{d * p^2 * f_R}{n * P}$$

3. Results

3.1. Evaluation of the region of interest

In a first step of our study, we define areas of interest for a substantial analysis of thermal load and select calculation approaches towards representative statements for the thermal load of the processed area. Therefore, we define several areas and calculation approaches of temperature data selection. Figure 2 a shows a schematic illustration of the evaluated regions. In general, two different categories are used, areas including the ablation area and measuring sites excluding the ablation area (heat affected zone, HAZ). The top edge of the HAZ measure area is 200 µm apart of the scan path, i.e. the distance between the ablation area and the measure area HAZ is 200 µm, to avoid an interaction of the scan path for both, HAZ small and HAZ large. Beside the choice the region of interest (ROI), the calculation method has to be defined. In this study we use either the Peak-mode, where the highest value inside the region of interest is used, or the Average-mode, where the unweighted mean value of all included pixel is calculated. Figure 2b shows the result of the temperature measurement for different ROI and calculation methods using an average laser power of 40 W, a pulse duration of 5 ps and 20 layers, leading to a total processing time of approximately 4 s, i.e. a processing time of 0.2 s per layer.

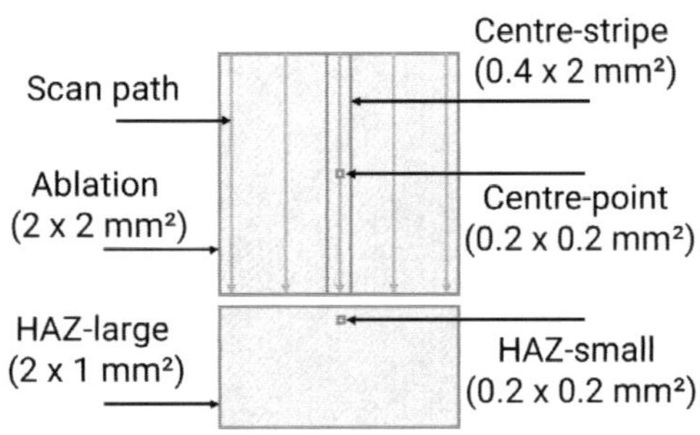

(a) Illustration of the measure areas

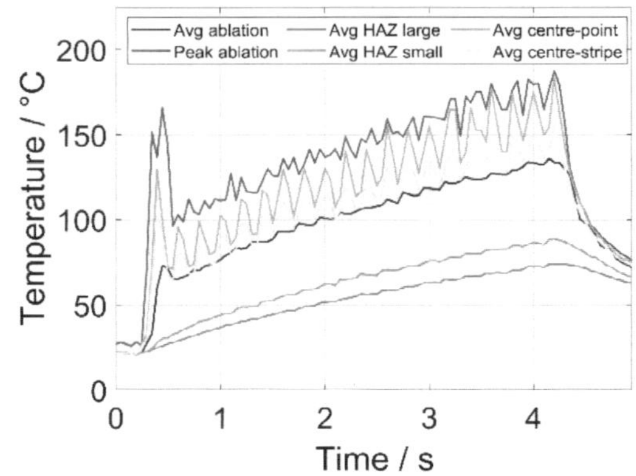

(b) Temperature date of different areas

Figure 2. Determination of suitable measuring areas with resulting temperature data using an average laser power of 40 W, a pulse duration of 5 ps and 20 layers

The comparison of the different temperate courses in figure 2b reveals a strong influence of the ROI and the calculation method. However, all measuring data show, by trend, an increasing temperature with time, reflecting the heating of the entire region. The temperatures within the ablation area (Avg ablation, Peak ablation, Avg centre-point and Avg centre-stripe) are approximately twice as high as the values in the conjunct heat affected zone (Avg HAZ large and Avg HAZ small). In detail, the temporal curves of the confined center areas point and stripe exhibit a strong modulation. This reflects the heating and cooling phase during ablation of a layer. The frame rate of the camera enables approximately 6 measuring values within this duty cycle and makes it possible to detect the thermal irradiation within the ablation area without the overlaid laser scan path. The excessive first peak of the temporal curves within the ablation area leads to the assumption that the emissivity changes after the ablation of the first layer, i.e. the removal of the black coating. The uncovered pure Al_2O_3 has an emissivity below 0.7 [10], which explains the temperature drop approximately of a third from the first to the second layer.

In the area of the HAZ, the black coating is unimpaired for all investigated parameters, therefore the emissivity in this region is assumed as being unaffected. The temporal curves of the HAZ show a smooth increase without a temperature modulation. The heat capacity and the inertial heat transfer lead to a homogenous temperature increase even in the direct vicinity of the ablation area (HAZ-small) 200 µm apart of the ablation area. Hence, we choose the average value of the HAZ-small area for further depiction of the temperature values. This value is a suitable indicator for the thermal load generated by the laser micromachining to the surrounding material.

3.2. Influence of laser power and pulse duration

Based on these findings, we evaluate the temperature of the HAZ-small area as a function for different laser processing parameters. Again, the temporal curves in Figure 3a-3c represent the layerwise ablation with a total amount of 20 scans. We applied three different laser powers to represent a fine, intermediate and rough processing, defined by the ablated depth per layer. Using a spatial pulse overlap of 80%, a laser power of 8 W, 40 W and 80 W lead to an ablated depth per layer in the region of 5 µm, 25 µm and 50 µm, respectively. The exact achieved ablation depth per layer, which is influenced by the applied laser pulse duration, was measured and later considered for the calculation of the ablation efficiency (c.f. Figure 3d).

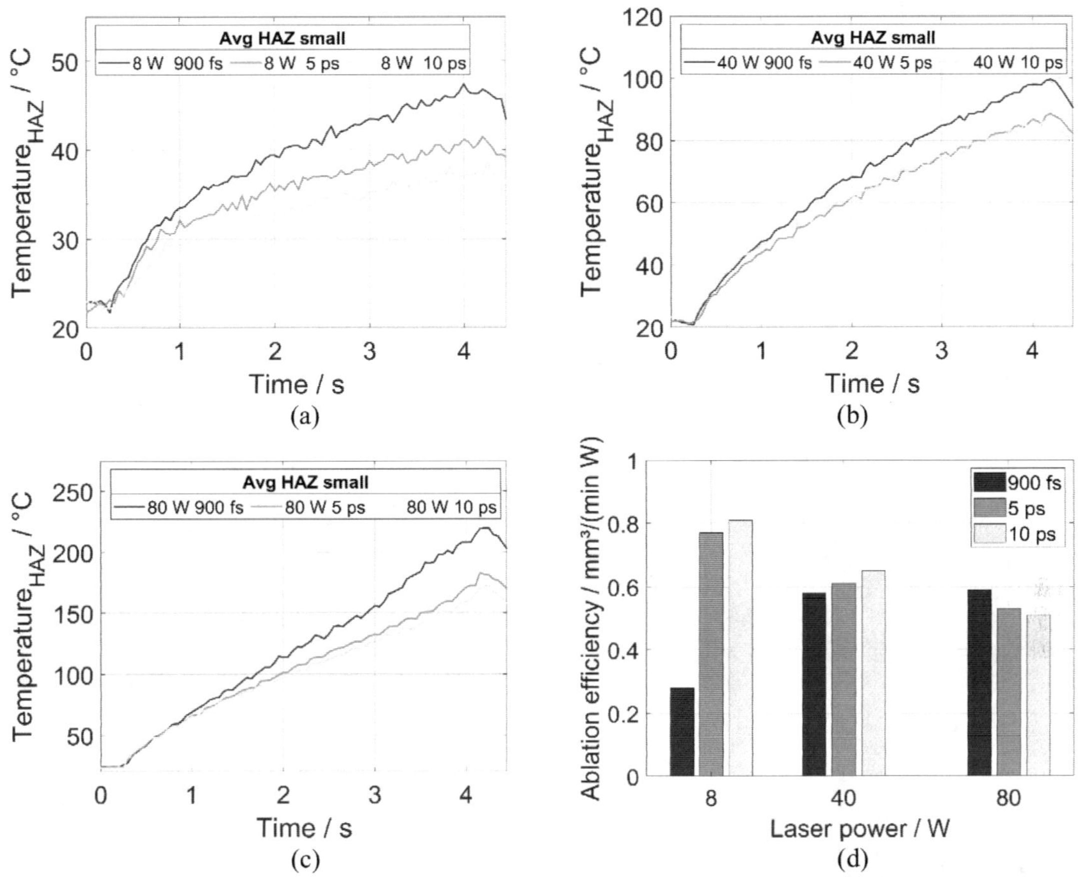

Figure 3. Temperature measurement of HAZ-small for 900 fs, 5 ps and 10 ps using a laser power of (a) 8 W, (b) 40 W and (c) 80 W with the resulting values of the ablation efficiency (d)

Based on the common presumptions of pulsed laser material processing, shorter pulses lead to less material heating. However, the temperature measurements reveal, that a decrease of the laser pulse duration leads, quite the contrary, to higher temperatures in the HAZ. This behaviour is valid for all investigated laser powers and also for all analysed measuring areas. Within the considered parameter range, the temperature of the heat affected zone is approximately 20% higher using 900 fs compared to 5 ps and 10 ps at the end of the ablation test. For laser processing using 8 W and 40 W average laser power, the decreased thermal load of the substrate is also associated with an increased ablation efficiency. Especially for 8 W, the ablation efficiency for 5 ps and 10 ps is more than 2.5-times higher as compared to 900 fs, which implies a distinct higher ratio of ablation converted energy of the laser process. While the ablation efficiency of 5 ps and 10 ps processing decreases with increasing laser power, i.e. increasing laser pulse energy, the ablation efficiency using 900 fs increases with increasing laser power. The ceramic processing with 80 W reveals that the efficiency using 900 fs is slightly higher as compared to the picosecond regime. However, the higher efficiency does not reflect in a lowered temperature increase. During laser processing, several thermal processes can be responsible for the measured temperature rise. On the one hand, the absorbed laser energy cannot be fully converted for the ablation process and leads to material heating. On the other hand, the generated plasma plume emits thermal radiation that is capable to increase the substrate temperature. In addition, the optical behaviour like the reflectivity and optical penetration depth changes with altering pulse duration and laser fluence

IOP Conf. Series: Materials Science and Engineering 1135 (2021) 012027 doi:10.1088/1757-899X/1135/1/012027

[23]. The complex interaction of the laser parameters leading to the measured heating effects are beyond the scope of this paper and has to be studied in further research. The high efficiency at 80 W average laser powers shows that Al_2O_3 is well suitable for high power laser ablation processes using ultrashort laser pulses. The application of 80 W laser power can be implemented in an ablation rate of approximately 45 mm³/min with a comparable small surface roughness of 2.5 µm (Ra).

The SEM analysis of the surrounding material does neither reveal any surface modification nor visible micro cracks, in accordance to the comparative low temperature observed. However, according to Figure 2b, the temperature in the ablation area is significant higher during the laser process and therefore requires a separate consideration. The in-depth modification of the Al_2O_3 material is analysed by a cross section of a fabricated cavity. Therefore a cavity is ablated using a laser power of 75 W, a pulse duration of 10 ps, a pulse repetition rate of 200 kHz, an overlap of 80% and with 7 passes. Figure 4 shows the 65° tilted overview and the detailed SEM image of this cross section. The melting effect is clearly limited to the surface and no visible in-depth modification is generated by the laser ablation process. The unimpaired grain structure of the Al_2O_3 is directly under the molten surface, which is less than 0.5 µm thick. The inspection of the cross section also reveals, that the high power laser processing using ultrashort laser pulses produces no cracks or other visual defects to the underlying material.

Figure 4. SEM analysis of the material cross section (65° tilt) revealing no noteworthy in-depth material modification of the initial grain structure.

3.3. Temperature depending ablation

A further investigation is performed to reveal the influence of the substrate temperature onto the ablation depth. Therefore we use an average power of 80 W and a pulse duration of 10 ps for cavity ablation with successive increment of the number of scans from n=1 to 11. Figure 5a shows the temperature curves for this layer-wise ablation.

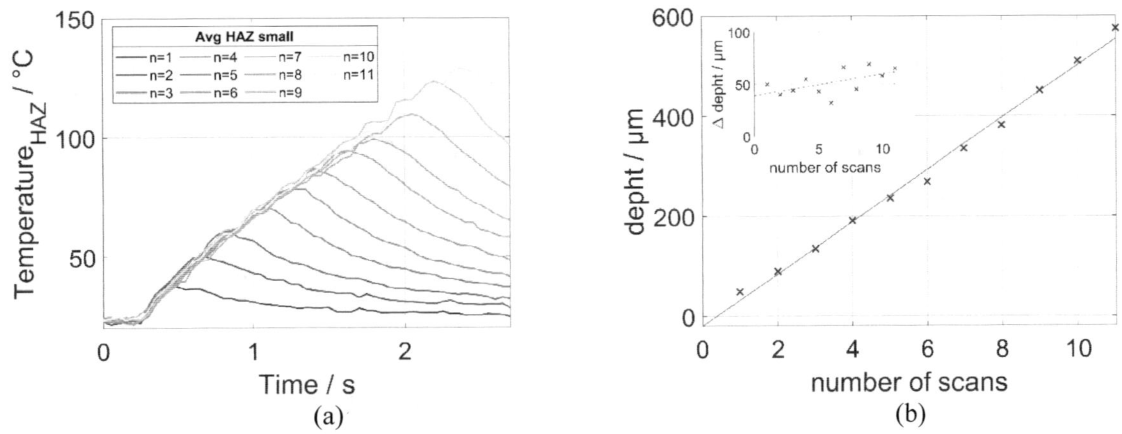

Figure 5. Layerwise laser ablation shows linear temperature rise (a) and cavity depth development (b).

The measured temperature increment of approximately 10°C per layer is almost constant over the observed test duration. The corresponding evolution of the cavity depth in shown in Figure 5b, with increasing number of scans, the depth increases with a mean slope of 52µm per layer. The detailed analysis of the depth increment reveals a small increased depth by approximately 2 µm per layer, i.e. per 10°C. For a precise depth ablation towards a desired ablation depth this temperature depending layer depth has to be considered especially for a high number of scans.

4. Conclusion

This paper shows the possibility of high power laser micromachining of Al_2O_3 using ultrashort laser pulses with simultaneously low thermal loads of the processed material. A thermal imaging camera was used to evaluate the appearing temperature during the laser process. By the analysis of the heat affected zone in the direct vicinity of the ablation area, a maximum temperature of 220°C can be measured. The comparison of three different structuring regimes for fine (8 W), intermediate (40 W) and rough (80 W) processing reveals, that a pulse duration of 900 fs lead to higher temperatures than the application of 5 ps and 10 ps. The observed unintuitive effect of increased heating with smaller pulse duration needs further investigations to reveal the underlying heating process. High ablation rates of up to 45 mm³/min does not lead to critical temperatures and therefore does not generate structural defects or micro cracks. Overall, it turn out that the thermographic camera is suitable for online monitoring of laser processing and helps to identify critical process parameters towards a temperature optimizes laser processing.

Acknowledgments
This work has been supported within the DFG project ultrashort pulsed lasers (grant 671/39-1).

ORCID iDs
Stefan Rung https://orcid.org/0000-0003-4029-9185
Niklas Häcker https://orcid.org/0000-0003-1278-399X
Ralf Hellmann https://orcid.org/0000-0001-5592-3388

References

[1] Samant, A N. and Dahotre, N B. 2009 Laser machining of structural ceramics—A review *Journal of the European Ceramic Society* **29** 969
[2] Bilal, A; Jahan, M Pervej; Talamona, D and Perveen, A 2018 Electro-Discharge Machining of Ceramics: A Review *Micromachines* **10** 10

[3] Auerkari, P 1996 Mechanical and physical properties of engineering alumina ceramics *VTT Technical Research Center of Finland* **1792**
[4] Sabur, A; Ali, M Yeakub; Maleque, M Abdul and Khan, A Ali 2013 Investigation of Material Removal Characteristics in EDM of Nonconductive ZrO2 Ceramic *Procedia Engineering* **56** 696
[5] Calignano, F; Denti, L; Bassoli, E; Gatto, A and Iuliano, L 2012 Studies on electrodischarge drilling of an Al2O3–TiC composite *Int J Adv Manuf Technol* **161** 182
[6] Molian, R; Shrotriya, P and Molian, P 2008 Thermal stress fracture mode of CO2 laser cutting of aluminum nitride *Int J Adv Manuf Technol* **39** 725
[7] Hirayama, Y; Yabe, H and Obara, M 2001 Selective ablation of AlN ceramic using femtosecond, nanosecond, and microsecond pulsed laser *Journal of Applied Physics* **89** 2943
[8] Wang, X; Shephard, J D.; Dear, F C. and Hand, D P. 2008 Optimized Nanosecond Pulsed Laser Micromachining of Y-TZP Ceramics *J American Ceramic Society* **91** 391
[9] Adelmann, B 2014 Investigation on Flexural Strength Changes of Alumina Caused by Cutting using Fiber Laser *JLMN* **9** 153
[10] Shaheen, M E.; Gagnon, J E. and Fryer, B J. 2021 Experimental studies on ablation characteristics of alumina after irradiation with a 193-nm ArF excimer laser *Eur. Phys. J. Plus* **136**
[11] Sciti, D; Melandri, C and Bellosi, A 2000 Excimer laser-induced microstructural changes of alumina and silicon carbide *J Mater Sci* **35** 3799
[12] Preusch, F; Adelmann, B and Hellmann, R 2014 Micromachining of AlN and Al2O3 Using Fiber Laser *Micromachines* **5** 1051
[13] Ohmura, E 2018 Effect of Circular Arc Radius on Laser Scribing Width and Depth of Al2O3 Ceramics: Experiment and Numerical Simulation *JLMN* **13** 10
[14] Gärtner, E; Polise, V; Tagliaferri, F and Palumbo, B 2018 Laser Micro Machining of Alumina by a Picosecond Laser *JLMN* **13** 76
[15] Friedrich, M; Seiler, M; Waechter, S; Bliedtner, J and Bergmann, J P. 2018 Precision structuring and functionalization of ceramics with ultra-short laser pulses *Journal of Laser Applications* **30** 32501
[16] Friedrich, M; Waechter, S; Giesecke, J; Bliedtner, J and Bergmann, J Pierre 2017 Investigations of surface processing of functional ceramics applying ultrashort laser pulses *Journal of Laser Applications* **29** 22502
[17] Jäggi, B; Förster, D J.; Weber, R and Neuenschwander, B 2018 Residual heat during laser ablation of metals with bursts of ultra-short pulses *Advanced Optical Technologies* **7** 175
[18] Canuisuis, M; Oberlander, M; Herzog, D; Schmidt-Lehr, M; Ploog, P; Albert, F and Emmelmann, C 2015 Correlation between Temperature Field and Heat Affected Zone during Laser Cutting of Carbon Fiber Reinforced Polymers *Lasers in Manufacturing Conference (München)*
[19] Maiwald, F; Englmaier, S and Hierl, S 2021 Online Pyrometry for Weld Seam Localization in Absorber-Free Laser Transmission Welding of Transparent Polymers *JLMN* **16** 8
[20] Safdar, S; Li, L; Sheikh, M A. and Schmidt, M J. 2006 Thermal history analysis of surface heating of mild steel with different laser beam geometries *Proceedings of the Institution of Mechanical Engineers, Part C: Journal of Mechanical Engineering Science* **220** 1549
[21] Dittmar, H; Herda, P; Jäschke, P; Suttmann, O and Overmeyer, L 2016 Optimising Process Efficiency During Remote Laser Cutting of CFRP by Utilisation of a Double-scan-head *Physics Procedia* **83** 1279
[22] Schwarz, S; Rung, S; Esen, C and Hellmann, R 2021 Enhanced ablation efficiency using GHz bursts in micromachining fused silica *Opt. Lett.* **46** 282
[23] Förster, D; Bui, D A; Onuseit, V; Weber, R and Graf, T 2015 Energy transfer mechanisms during laser pulsed processing of metals *Lasers in Manufacturing Conference (Munich)* 1

IOP Conf. Series: Materials Science and Engineering 1135 (2021) 012028 doi:10.1088/1757-899X/1135/1/012028

Comparative study of cell growth and cellular adhesion on Ti-6Al-4V surfaces made by Selective Laser Melting followed by different surface post processing steps

Benedikt Adelmann[1], Melanie Abb[1], Ralf Hellmann[1]

[1]Applied laser and photonics group, University of Applied Sciences Aschaffenburg, Wuerzburger Straße 45, 63739 Aschaffenburg, Germany

benedikt.adelmann@th-ab.de

Abstract. Selective laser melting is generally considered as to improve the design of medical implants, thus supporting medical treatment and maintaining mobility of invalid and older people. In particular, medical grade titanium alloys are in favour for spinal implants, as being nowadays manufactured by, e.g., milling. Selective laser melting offers the advantage of an adapted elasticity as to avoid stress shielding within the backbone by including complex lattice structures inside the individualized implant. For the integration into the backbone, surface properties, particularly surface roughness, are crucial with respect to biocompatibility and cell growth. Opposite to conventional milling, selective laser melting, however, results in an inferior surface roughness, leading to the necessity of downstream process steps.
We report on cell growth and cellular adhesion of human primary fibroblasts on medical grade Ti-6Al-4V fabricated by selective laser melting followed by combinations of milling, hot isostatic pressing, chemical surface treatment and steam-sterilization to generate different surface conditions for cell growth. For example, cell growth is studied for varying milling path spacing on SLM parts exhibiting different surface roughness. Our results reveal good cell growth for milling path spacing lower than 0.18 mm as compared to higher milling path spacing and not milled surfaces. Cell fluorescence images and SEM images show that the cell growth is additionally hampered by the edges of the milling path. Conveniently, process failures such as pores originating from the selective laser melting process do not hamper the cell growth.

1. Introduction

Due to the worldwide increasing life expectance, the number of elderly people rises strongly. This socio-demographic development effects numerous areas of life and results in both entrepreneurial and social challenges as well as high demands on medical technology. In addition to the treatment of degenerative diseases of the central nervous system, the main focus is on maintaining mobility. This increasing demand inevitably leads to growth in the medical technology sector. Selective laser melting (SLM) is generally considered as to improve the design of medical implants, thus supporting medical treatment and maintaining mobility of invalid and elder people. In particular, medical grade titanium alloys are in favor for spinal implants, as being nowadays manufactured by, e.g., milling or casting. Despite numerous efforts, however, milled implants may still exhibit strong deficits, such as, e.g., stress shielding, which occurs due to the high stiffness of implants as compared to natural bone tissue. It leads

to a capacity underload of the protected natural bone, which in turn may results in a reduction in bone density and weakening of the healthy bone tissue.
Stress shielding occurs when the insertion of the implant results in mechanical shielding of the remaining tissue, thus damaging the balance between formation and degradation in the bone tissue [1]. To avoid stress shielding in SLM produced implants, Fousová et al. produced complex lattice structures of 316L stainless steel to mimic mechanical properties (especially elasticity) similar to trabecular human bone [2]. With fully porous material produced by SLM, stress shielding was reduced by 75% compared to a fully solid implants [3]. With porous materials produced by SLM, the Young´s modulus can be varied by one order of magnitude to adapt it to the application and to avoid stress shielding [4]. SLM even allows a mixture of powders like hydroxyapatite (a matrix for bones) and 316L stainless steel to produce a material with tensile strengths close to human bones [5].

Since literature reveals the possibility to deliver implants with the required elasticity [6], biocompatibility is a second important requirement to implants. While in the past, various materials were used for implants such as different high, medical-grade steels and cobalt-chrome based alloys [7], nowadays usually titanium alloys are preferred because of its good biocompatibility [8]. Next to the Ti-6Al-4V alloy also Ti-35Zr-28Nb [9], Ti-30Zr-7Mo [10] or Ti-10Ta-2Nb-2Zr [11] reveal high biocompatibility. To evaluate biocompatibility, mainly cell growth behavior tests with fibroblasts or osteoblast cells [12] are performed, the results of such studies, however, do not only depend on the material but also on the specific surface properties. The high surface roughness after the SLM process [13] is generally hampering the cell growth and results in a low biocompatibility [8], thus post processes are required. To improve the cell growth behavior after SLM, exhibiting pores with a size of up to 250 µm, Matena et al. functionalized the SLM surface with a polycaprolactone (PCL) coating [14]. In Ref. [15], chemical etching was found to be the most effective surface post-processing treatment (vs. sandblasting, polishing) to improve cellular affinity and simultaneously increase the fatigue performance.

In order to further improve the functional accuracy of the implant, i.e. to provide a reliable interface for the insertion tool or to remove unwanted support structures, generally, a milling process after SLM is compulsory. Therefore, the objective of this contribution is to evaluate the cell growth behavior of SLM build Ti-6Al-4V alloy which is post processed by milling compared to other surface treatments.

2. Experimental

2.1. Process Chain

The complete process chain, employed in this study, is shown in figure 1, beginning with the building the of the titanium samples using SLM. Subsequently, the samples are optionally milled, sandblasted or untreated handled to the following hot isostatic pressing step. After pressing, the samples are removed from the base plate by wire EDM. Before the cell growth tests, the surface is etched and cleaned to remove nitrogen residues from the hot isostatic pressing process.

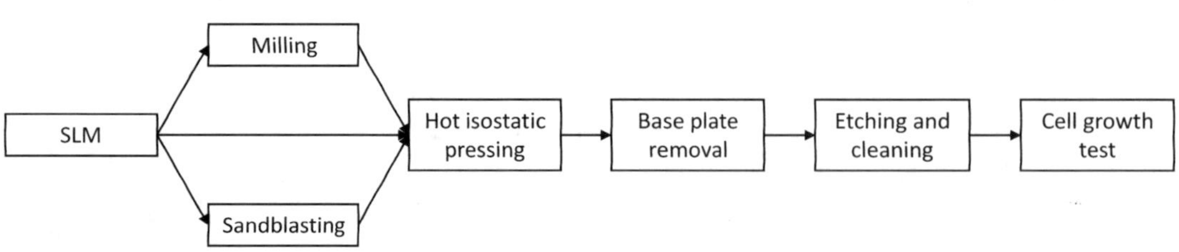

Figure 1: Process chain

2.2. *Selective Laser Melting*

For the SLM process, DMG Mori Lasertec 30 machine is used to process Ti-6Al-4V (material number 3.7165). Samples are cylinders with a diameter of 15 mm and a height of 7 mm to fit into the cell growth measurement vessel. The samples are directly build on the base plate without support structures to increase the accuracy and stiffness for the milling step. The layer height during SLM is 50 µm which requires 140 passes to produce the cylinders. As laser process parameters, a laser power of 220 W and a scan velocity of 0.74 m/s are used in combination with a hatch distance of 0.10 mm. This parameter combination results in high density and comparatively low surface roughness of the samples in the range of 10 µm.

2.3. *Milling and other mechanical post processing*

For the milling process, which is depicted in figure 2a, a ball cutter with a diameter d of 2 mm is used. The most influential parameters on surface roughness after milling are the milling path distance a_e, the feed per tooth f_z and cut velocity v_c, respectively. While the feed per tooth and the cut velocity are limited by the manufacturer's specification in order to reduce tool wear and are set to a cut velocity v_c of 190 m/min and a feed per tooth f_z to 0.03 mm, the milling path distance can be varied over a wide range. The required rotation speed n and feed rate v_f can be calculated with the equations 1 and 2.

$$n = \frac{v_c}{d \cdot \pi} \quad (1)$$

$$v_f = z \cdot f_z \cdot n \quad (2)$$

The calculation results in a rotation speed of 30 240 1/min and a feed rate of 1 815 mm/min. According to figure 2b, the surface shows a defined waviness with a distance between the valleys and the depth of the valleys defined by the milling path distance. Alternative to the milling process and for comparative reasons, further samples are also sandblasted manually or are brought untreated to the following process step (as build).

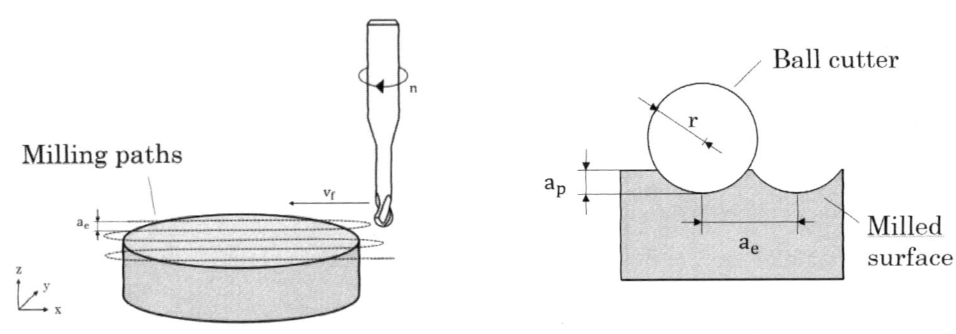

Figure 2: Milling path on the specimen (a) and cross section of the milled surface (b)

2.4. *Hot isostatic pressing and chemical post treatment*

For medical implants, a high service life and thus the reduction of residual stresses in the material are required, which means in particular defect-free microstructures and high material strength. However, since additively manufactured components do not adequately meet these requirements, post-treatment of the specimen is inevitable. For this purpose, hot isostatic pressing (HIP) has been established, which allows for the reduction of inner defects such as pores and cavities and for an increase of density [16]. Furthermore, hot isostatic pressing can significantly reduce the residual stress in the component. This leads to an increase in the fracture toughness and thus to a significantly stronger plastic area. As a result, the HIP process has also influence on the machinability [17] and therefore on the achieved roughness after milling.

The material Ti-6Al-4V used here is present as α-martensite after the SLM process, due to the high cooling rate. When the samples are hot isostatically pressed below the β-transus temperature (995 °C), it converts to a lamellar-shaped mixture of α and β fractions. In this state, fractions of pure α-martensite are still present. If, on the other hand, a temperature above the β-transus temperature is selected for the HIP cycle, large β-grains are formed, which convert to a lamellar α-β mixture during the cooling process [17, 18]. In detail for the hot isostatic pressing of our samples according to the norm ASTM F2924 – 14, a temperature of 925° C and a pressure of 100 MPa are used for a duration of 3 hours. This HIP process is necessary for the cell growth behaviour test because the ratio α and β phase influences the cell growth behaviour strongly [19] and so the parameters of the HIP process directly influence the biocompatibility.

To remove the samples from the base plate wire, EDM is used to ensure good height accuracy. Since the samples show undesired contamination of nitrogen at the surface after hot isostatic pressing, they are, additionally, chemically post treated. For this purpose, the samples are placed in boiling hydrogen peroxide for a total of 20 minutes and then cleaned with isopropanol and water.

2.5. Cell growth measurement

To make a statement about and rate the biocompatibility of SLM build parts, the cell growth behaviour of fibroblast cells is investigated which gives information of the cytotoxicity and adherence of the surface. For the cell growth test, the specimens are first steam sterilized at a temperature of 121° C for 20 minutes. The cells are seeded with a density of 5 000 cells/cm² together with a cell suspension on the specimen. After a period of 24 h, the nutrient solution is changed. After reaching confluency, the largely gapless coverage of the surface of a culture vessel with adherent cells, equilibration (reseeding of the cells) is performed in a period of 30 minutes. The seeding density corresponds to 10 000 cells/cm². After further 24 h in the incubator (37° C), another 300 µl cell suspension is given to the test specimens, so that uniform wetting can be ensured. The evaluation of the test results is performed after 72 h.

For optical examination and determination of fluorescence intensity, the agent Resazurin is used. This is a non-toxic, water-soluble, blue redox dye, which is converted by cell activity to pink, fluorescent Resorufin. Therefore with higher cell activity the amount of Resorufin increases, which the amount of Resorufin can be determined by fluorescence intensity measurement. Cell vitality can be determined first optically and in a further step via fluorescence intensity. Optically, the color purple represents a high cell presence and so the local distribution of the cell activity can be determined. The measurement of a high fluorescence intensity indicates that a lot of Resazurin has been converted and thus speaks for a high cell activity.

After optical examination of the specimens under the fluorescence microscope and measurement of the fluorescence intensity, the cells are fixed for further examination using the scanning electron microscope. The cell fixing is performed by subsequently dehydrating the cells through an ethanol series (10%, 30%, 50%, 70%, 85%, 95%, and 100% ethanol, 15 minutes each) in distilled water. Finally, the samples are dried. For this purpose, hexamethyldisilazane (HDMS) functions as a chemical drying agent. The samples are first placed in a mixture of 50% HDMS and distilled water for 60 minutes and then remain in 100% HDMS for 5 minutes.

3. Results

3.1. Surface roughness

To evaluate the different types of surface roughness directly after the SLM process, after milling and after sandblasting, SEM images of the surfaces are shown in figure 3. For the as build samples in figure 3a it can be seen that those samples have a large number of incompletely molten metal powder particles on the surface. Position and size do not follow any pattern. Furthermore, the melt paths and their intersections can also clearly be identified. The samples are characterised by a high surface roughness of about 10 µm and a high standard deviation of the surface roughness for different sections of the surface. The sandblasted surfaces (figure 3b) are characterized by the highest roughness. The large sand particles result in large indentations on the surface with sharp edges. On the milled specimen (figure 3c), the valleys shaped by the path of the ball cutter can clearly be identified. Despite the milling process, individual metal powder particles can still be identified. The distribution of these particles are significantly responsible for the standard deviation of the roughness over the sample.

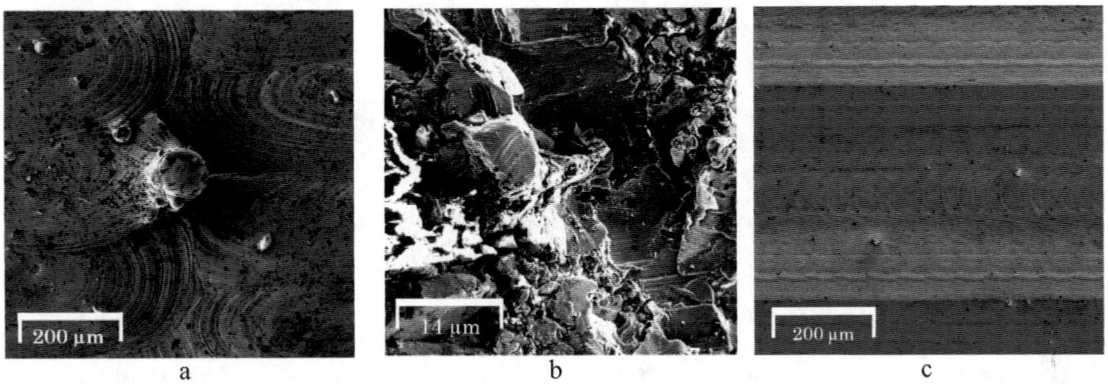

Figure 3: SEM Images of as build surface (a), sandblasted surface (b) and milled surface (c)

The surface roughness R_a measured with a laser scanning microscope for the different post processing before HIP, after HIP and after the etching step the is shown in figure 4. It is obvious to see that the roughness increases continuously with the milling path distance, which is caused by valley structures produced by the ball cutter (cf. figure 2b). With higher milling path distance, the valley gets deeper and therefore the roughness increases. For milling path distances lower than 0.18 mm, the roughness saturates in the range of 1.5 µm, i.e. lower distances only result in longer cycle times but not lower roughness. Generally, the roughness for the milled surfaces is lower as compared to the as build

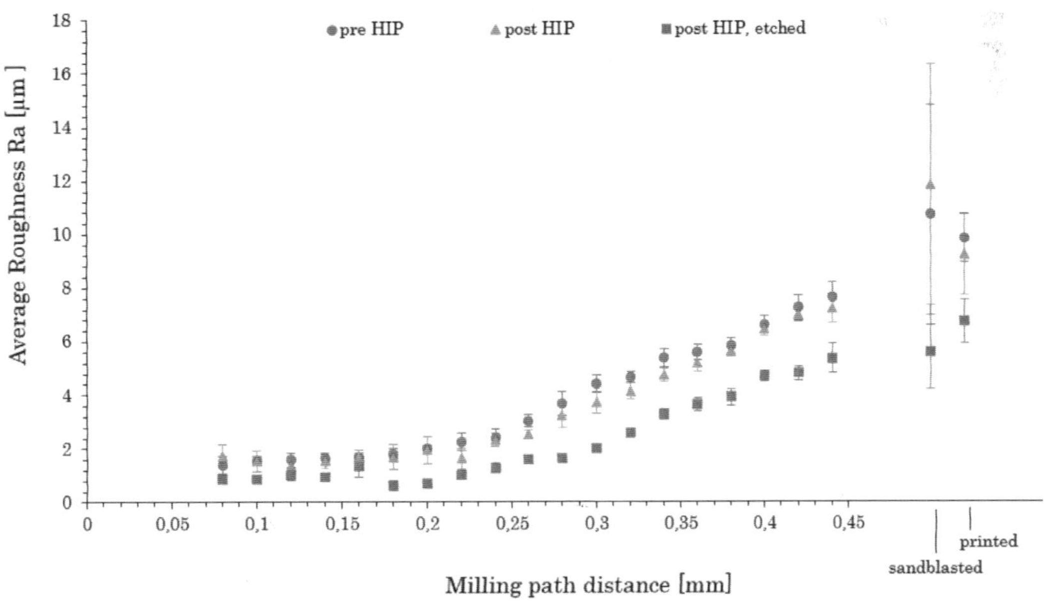

Figure 4: Average roughness as function of the milling path distance

surfaces or sandblasted surfaces which have a similar R_a. The HIP process itself has only minor influence on the roughness but the following etching step reduces the roughness significantly, especially for the samples milled with high milling path distance and the sandblasted samples. The reason for this is that the ridge between the valleys is removed partly together with remaining melt particles.

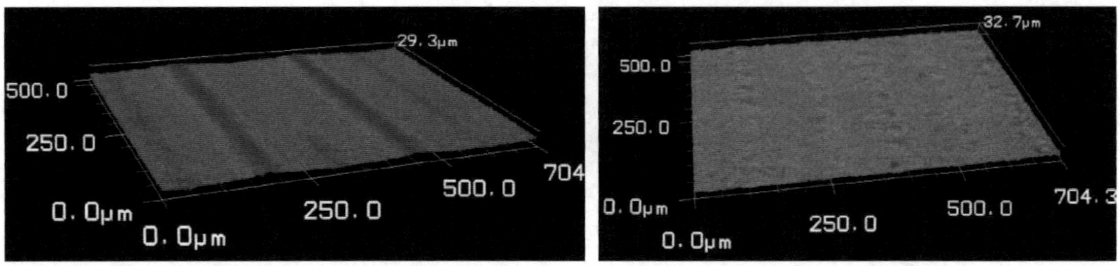

a b

Figure 5: Surface topology with a milling path distance of 0.28 mm (a) and 0.14 mm (b)

To evaluate the reason for the saturating roughness for milling path distances lower than 0.18 mm, scanning electron microscope images are taken to show the topology in figure 5. For a milling path distance of 0.28 mm (figure 5a), the valleys and the ridge between them can be seen clearly. Beside this valley structure, the surface is quite smooth. For the milling path distance of 0.14 mm (figure 5b), the ridge is ruptured, not continuous anymore and the roughness is caused by lined up peaks. These peaks are not removed by lower milling path distance because of the blade rotation so the roughness doesn´t decrease further.

3.2. Cell growth

In order to evaluate the cell growth behaviour, the described cell growth tests are performed with a series of samples with different milling path distances and the as build and sandblasted samples as control group. To handle the typical outliners during cell growth tests, three test specimens each are combined into a category and their characteristic values are averaged. A seventh category represents the comparison group of as build and sandblasted samples. The fluorescence intensity determined here is shown in figure . It can be seen that the fluorescence intensity and therefore the cell growth behaviour decreases with increasing milling path distance.

In our experiments, optimal cell growth conditions, indicating superior biocompatibility, are observed for smooth surfaces as obtained by milling path distances between 0.12mm and 0.18mm. For both, a high median and arithmetic mean fluorescence intensity are measured. In contrast, for milling path distances between 0.24 mm and 0.42 mm, being characterized by a higher surface roughness, the fluorescence intensity is significantly lower as being characterized by a low median. This, in turn, is associated to a lower biocompatibility. Beside the apparent influence of the surface roughness, we observe a pronounced impact of ridges between the milling paths, which are observed for path distances higher 0.18 mm (cf. figure 5 and 6). Though specimen groups with path distances of 0.24 mm and 0.42 mm exhibit an arithmetic mean of the fluorescence intensity of halve the intensity at 0.12 mm, this intensity can be attributed to one significant positive outliner in each group which increases the mean and the standard deviation. Please note, in contrast to the median, the arithmetic mean is no robust statistic because it can be greatly influenced by outliners. In the specimen groups with 0.3 mm and 0.36 mm path distance, these outliners are missing which results in both low median and arithmetic mean. The comparison group with the sandblasted and as build samples reveal a median intensity of less than one third of the best results with milling path distances below 0.18 mm. Also here a positive outliner increases the arithmetic mean significantly but still this is less than half of the intensity as compared to the intensity for milled specimen with less than 0.18 mm distance. In summary, milling path distances of 0.18 mm and below without ridges reveal a good median cell growth behavior while higher milling path distances and the comparison group (as build and sandblasted samples) are characterized with a much lower median cell growth, which is in accordance to Ref. [8].

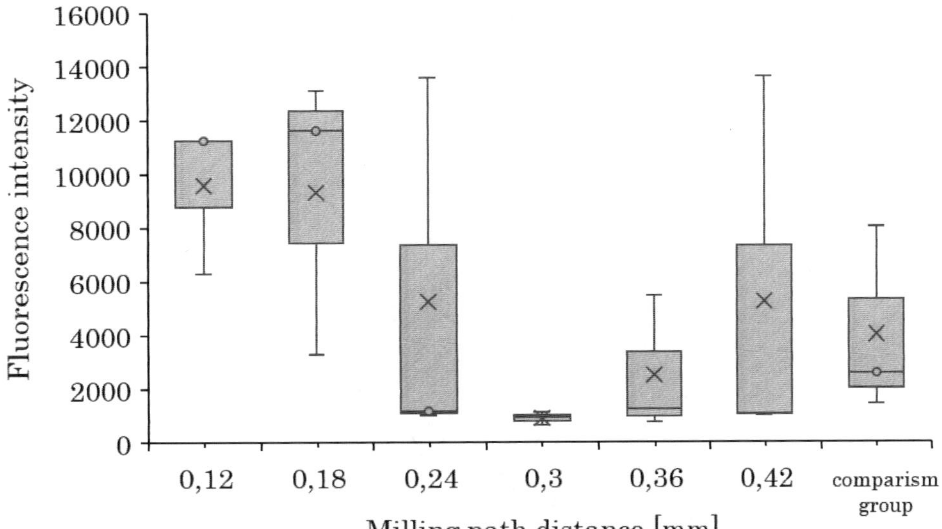

Figure 6: Fluorescence intensity as a function of the milling path distance

To perceive the different cell growth behavior for different milling paths distances, two exemplarily chosen fluorescence images are depicted in figure 7. Fluorescence microscopy reveals vital cells by their blue coloration, indicating vitality of the cells and delivering information in which areas cells grow best. In figure a, a sample with low cell growth is shown which is indicated by a small amount of slightly fluorescent dots which are distributed over the surface. In figure b, small lines without cell activities are visible. These dark lines are the ridges between the milling paths which are obviously not suitable for cell growth. On the other hand, the cells show a good growth in the valleys of the milling path. Apparently, the ridges hamper the cell growth and with milling path distances above 0.18 mm the height

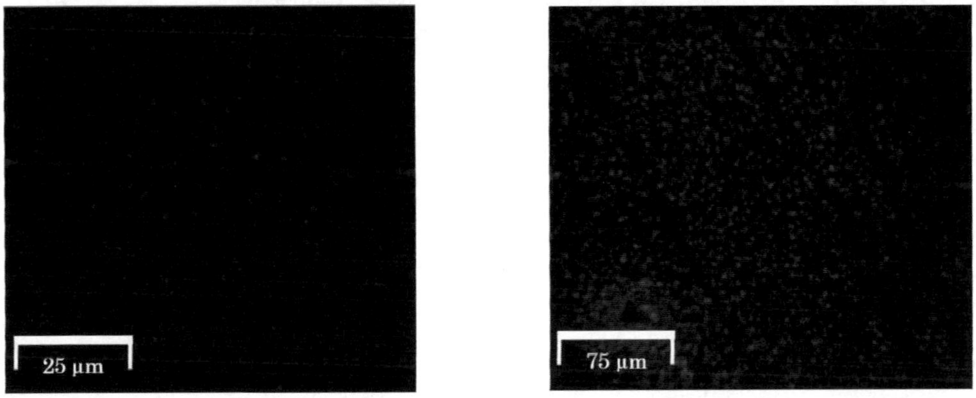

a b

Figure 7: Fluorescence images with low cell growth (a) and high cell growth (b), please note the different scales

and ridges of the ridges increase significantly which is the reason for the low cell growth at high milling path distances.

3.3. Influence of pores

Parts build with the SLM process do always have a certain porosity [20] which depends on the material and laser machine parameter combination. Typical porosities in Ti-6Al-4V with good process parameters are in the range of 0.1 % to 0.3 % [21]. Because open pores are usually not completely closed by the HIP process and while during the milling process, pores can laid open, the surface is characterized by holes. To determine the influence of the pores on the cell growth behavior, SEM images of pores at the surface are taken after the cell growth and fixing process. In figure 7 a pore with a size of about 40 µm and oval shape can be seen and a cell that grew over the hole. The cells are obviously able to grow as a bridge over the hole and can therefore also close holes. As a result it can be stated that the holes produces by the SLM process are no drawback for the cell growth.

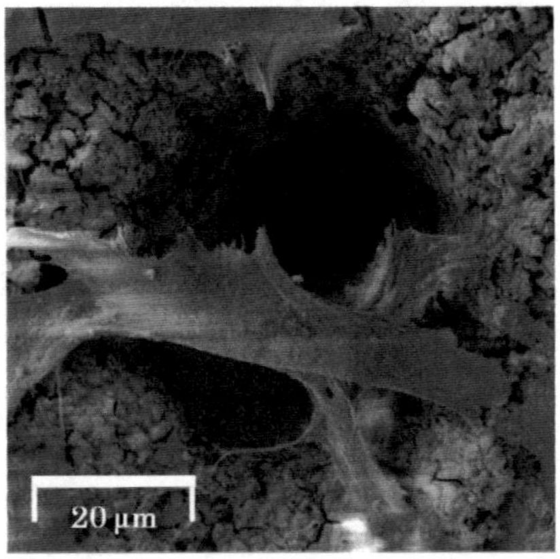

Figure 7: SEM image of cells growing over a SLM caused pore

4. Conclusion

In this contribution, we determine the cell growth behavior on SLM build Ti-6Al-4V with different mechanical post processing for spinal implants applications. The post processing is performed by milling with a ball cutter or sandblasting. For milling, different milling path distances are chosen to achieve a different surface roughness. Before the cell growth test, samples are hot isostatic pressed, removed from the baseplate by EDM, etched and steam sterilized. Roughness measurements show a low surface roughness of 1.5 µm for milling path distances below 0.18 mm and increasing roughness with increasing milling path distance. For the sandblasted and as build samples, the roughness is in the range between 6 µm and 12 µm, respectively. Biocompatibility tests based on fibroblast growth reveal a good cell growth behavior for milling path distances below 0.18 mm with an associated good surface roughness. Samples with higher milling path distances or sandblasted samples are characterized by low median cell growth, however but positive outliners in some groups can increase the arithmetic mean significantly. Fluorescence images show that ridges between the milling paths observed at high milling path distances above 0.18 mm hamper the cell growth and results in a low cell growth. Process failures like pores from the SLM process are yet not hampering the cell growth.

Acknowledgement

This work has been financially support by the German Federal Ministry of Education and Research (BMBF grant 03FH016IN6).

ORCID iDs
B. Adelmann: https://orcid.org/0000-0002-1944-6489
R. Hellmann: https://orcid.org/0000-0001-5592-3388
M. Abb: https://orcid.org/0000-0002-5957-8187

References

[1] A. Edelmann, M. Dubis, and R. Hellmann, "Selective Laser Melting of Patient Individualized Osteosynthesis Plates-Digital to Physical Process Chain," *Materials (Basel, Switzerland)*, vol. 13, no. 24, 2020, doi: 10.3390/ma13245786.

[2] M. Fousová, J. Kubásek, D. Vojtěch, J. Fojt, and J. Čapek, "3D printed porous stainless steel for potential use in medicine," *IOP Conf. Ser.: Mater. Sci. Eng.*, vol. 179, p. 12025, 2017, doi: 10.1088/1757-899X/179/1/012025.

[3] S. Arabnejad, B. Johnston, M. Tanzer, and D. Pasini, "Fully porous 3D printed titanium femoral stem to reduce stress-shielding following total hip arthroplasty," *Journal of orthopaedic research : official publication of the Orthopaedic Research Society*, vol. 35, no. 8, pp. 1774–1783, 2017, doi: 10.1002/jor.23445.

[4] D. Leordean, C. Dudescu, T. Marcu, P. Berce, and N. Balc, "Customized implants with specific properties, made by selective laser melting," *Rapid Prototyping Journal*, vol. 21, no. 1, pp. 98–104, 2015, doi: 10.1108/RPJ-11-2012-0107.

[5] L. Hao, S. Dadbakhsh, O. Seaman, and M. Felstead, "Selective laser melting of a stainless steel and hydroxyapatite composite for load-bearing implant development," *Journal of Materials Processing Technology*, vol. 209, no. 17, pp. 5793–5801, 2009, doi: 10.1016/j.jmatprotec.2009.06.012.

[6] T. Bormann, R. Schumacher, B. Müller, M. Mertmann, and M. de Wild, "Tailoring Selective Laser Melting Process Parameters for NiTi Implants," *J. of Materi Eng and Perform*, vol. 21, no. 12, pp. 2519–2524, 2012, doi: 10.1007/s11665-012-0318-9.

[7] C. A. Engh, J. D. Bobyn, and A. H. Glassman, "Porous-coated hip replacement. The factors governing bone ingrowth, stress shielding, and clinical results," *The Journal of bone and joint surgery. British volume*, vol. 69, no. 1, pp. 45–55, 1987, doi: 10.1302/0301-620X.69B1.3818732.

[8] A. T. Sidambe, "Biocompatibility of Advanced Manufactured Titanium Implants-A Review," *Materials (Basel, Switzerland)*, vol. 7, no. 12, pp. 8168–8188, 2014, doi: 10.3390/ma7128168.

[9] Y. Li, Y. Ding, K. Munir, J. Lin, M. Brandt, A. Atrens, Y. Xiao, J. R. Kanwar, and C. Wen, "Novel β-Ti35Zr28Nb alloy scaffolds manufactured using selective laser melting for bone implant applications," *Acta biomaterialia*, vol. 87, pp. 273–284, 2019, doi: 10.1016/j.actbio.2019.01.051.

[10] M. Niinomi and M. Nakai, "Titanium-Based Biomaterials for Preventing Stress Shielding between Implant Devices and Bone," *International journal of biomaterials*, vol. 2011, p. 836587, 2011, doi: 10.1155/2011/836587.

[11] W. Zhenhuan, D. Yu, L. Junsi, J. Xiaowei, X. Zongyu, L. Li, and X. Xiaoli, "Physiochemical and biological evaluation of SLM-manufactured Ti-10Ta-2Nb-2Zr alloy for biomedical implant applications," *Biomedical materials (Bristol, England)*, vol. 15, no. 4, p. 45017, 2020, doi: 10.1088/1748-605X/ab7ff4.

[12] Y. Iwaya, M. Machigashira, K. Kanbara, M. Miyamoto, K. Noguchi, Y. Izumi, and S. Ban, "Surface properties and biocompatibility of acid-etched titanium," *Dental materials journal*, vol. 27, no. 3, pp. 415–421, 2008, doi: 10.4012/dmj.27.415.

[13] S. Han, F. Salvatore, J. Rech, and J. Bajolet, "Abrasive flow machining (AFM) finishing of conformal cooling channels created by selective laser melting (SLM)," *Precision Engineering*, vol. 64, pp. 20–33, 2020, doi: 10.1016/j.precisioneng.2020.03.006.

[14] J. Matena, S. Petersen, M. Gieseke, A. Kampmann, M. Teske, M. Beyerbach, H. Murua Escobar, H. Haferkamp, N.-C. Gellrich, and I. Nolte, "SLM produced porous titanium implant

improvements for enhanced vascularization and osteoblast seeding," *International journal of molecular sciences*, vol. 16, no. 4, pp. 7478–7492, 2015, doi: 10.3390/ijms16047478.
[15] P. Jamshidi, M. Aristizabal, W. Kong, V. Villapun, S. C. Cox, L. M. Grover, and M. M. Attallah, "Selective Laser Melting of Ti-6Al-4V: The Impact of Post-processing on the Tensile, Fatigue and Biological Properties for Medical Implant Applications," *Materials (Basel, Switzerland)*, vol. 13, no. 12, 2020, doi: 10.3390/ma13122813.
[16] W. Tillmann, C. Schaak, J. Nellesen, M. Schaper, M. E. Aydinöz, and K.-P. Hoyer, "Hot isostatic pressing of IN718 components manufactured by selective laser melting," *Additive Manufacturing*, vol. 13, pp. 93–102, 2017, doi: 10.1016/j.addma.2016.11.006.
[17] M. Ahmadi, Y. Karpat, O. Acar, and Y. E. Kalay, "Microstructure effects on process outputs in micro scale milling of heat treated Ti6Al4V titanium alloys," *Journal of Materials Processing Technology*, vol. 252, pp. 333–347, 2018, doi: 10.1016/j.jmatprotec.2017.09.042.
[18] B. Vrancken, L. Thijs, J.-P. Kruth, and J. van Humbeeck, "Heat treatment of Ti6Al4V produced by Selective Laser Melting: Microstructure and mechanical properties," *Journal of Alloys and Compounds*, vol. 541, pp. 177–185, 2012, doi: 10.1016/j.jallcom.2012.07.022.
[19] E. Eisenbarth, D. Velten, M. Müller, R. Thull, and J. Breme, "Biocompatibility of beta-stabilizing elements of titanium alloys," *Biomaterials*, vol. 25, no. 26, pp. 5705–5713, 2004, doi: 10.1016/j.biomaterials.2004.01.021.
[20] D. Joguet, S. Costil, H. Liao, and Y. Danlos, "Porosity content control of CoCrMo and titanium parts by Taguchi method applied to selective laser melting process parameter," *Rapid Prototyping Journal*, vol. 22, no. 1, pp. 20–30, 2016, doi: 10.1108/RPJ-09-2013-0092.
[21] S. Leuders, M. Vollmer, F. Brenne, T. Tröster, and T. Niendorf, "Fatigue Strength Prediction for Titanium Alloy TiAl6V4 Manufactured by Selective Laser Melting," *Metall and Mat Trans A*, vol. 46, no. 9, pp. 3816–3823, 2015, doi: 10.1007/s11661-015-2864-x.

AUTHOR INDEX

Institute of Physics
Dirac House, Temple Back
Bristol BS1 6BE UK

ISSN: 1757-8981
ISBN 978-1-7138-4614-7

3rd International Symposium on Engineering and Earth Sciences (ISEES 2020)

IOP Conference Series: Materials Science and Engineering
Volume 905

Grozny, Russia
28 - 29 February 2020

IOP
Institute of Physics